AF528248

GOVERNMENT AND NGO INTERVENTIONS IN DRYLAND AGRICULTURE

Government and NGO Interventions in Dryland Agriculture

A STUDY OF TWO PROJECTS IN ANDHRA PRADESH

Editors
M. PUT AND M.P. VAN DIJK

IDPAD

MANOHAR
2000

First published 2000

ISBN 81-7304-336-1

Published by
Ajay Kumar Jain for
Manohar Publishers & Distributors
4753/23 Ansari Road, Daryaganj
New Delhi 110002

Typeset by
A J Software Publishing Co. Pvt. Ltd.
305 Durga Chambers
1333 D.B. Gupta Road
Karol Bagh, New Delhi 110005

Printed at
Rajkamal Electric Press
B 35/9 G T Karnal Road Indl Area
Delhi 110033

Contents

	List of Tables, Figures and Maps	7
	Abbreviations	15
	Glossary	19
	Preface and Acknowledgements	23
1.	Introduction G.A. de Bruijne and M.P. van Dijk	27
2.	An Introduction to Andhra Pradesh D. Adi Lakshmi, A.J. Dietz and M. Put	51
3.	A Profile of Ranga Reddy and Mahabubnagar: A Drought Risk Assessment, Agricultural and Social Change, and the Role of NGOs A.J. Dietz, M. Put and S. Subbiah	75
4.	Project Interventions in Dryland Agriculture: The Examples of the Government of India/World Bank-supported Maheswaram Project and of AWARE's Activities in Amangal *Mandal* M. Put and A.J. Dietz	125
5.	An Introduction to the Study Villages M. Put	185
6	Indigenous and Recommended Technologies for Dryland Agriculture in the Study Villages M. Put	229
7.	Adoption of Recommended (*in-situ*) Soil and Water Conservation Technologies M. Put	269
8.	Adoption of Recommended Crop Management Technologies for Dryland Crops M. Put	299
9.	Summary, Conclusions and Recommendations M. Put, M.P. van Dijk, A.J. Dietz and G.A. de Bruijne	377
	Bibliography	401
	Contributors	411
	Index	413

Tables, Figures and Maps

Tables

Table 1.1 Expenditure per Sector during Various Plans in India (*percentages*) 31

Table 1.2 Selected Agricultural and Rural Development Programmes during Eighth Five Year Plan 33

Table 1.3 Number of Respondents (in Absolute Figures and Related to our own 1991 Census), by Community, by Village 49

Table 1.4 Absolute number of Respondents for the Open Interviews, by Community, by Village 50

Table 2.1 Cultivation data for Telangana, 1913-45 (*in 1,000 ha*) 54

Table 2.2 Area Cultivated with Various Crops in Telangana, 1925-45 and 1969-92 (*averages in 1,000 ha*) 57

Table 3.1 Decadal Averages for Rainfall in mm per month for Hyderabad, 1901-90 85

Table 3.2 Decadal Averages for Rainfall in mm per month for Mahabubnagar town, 1901-80 86

Table 3.3 Hyderabad, Average Monthly Rainfall and Potential Evapotranspiration data 87

Table 3.4 Mahabubnagar town, Average Monthly Rainfall and Potential Evapotranspiration data 87

Table 3.5 Hyderabad 1900-90, Seasonality Assessment 89

Table 3.6 Mahabubnagar town 1900-80, Seasonality Assessment 89

Table 3.7 Rainfall and Drought Risks near the Research Area: Ibrahimpatnam/Maheswaram, 1957-93 96

Table 3.8 Drought Risk Distribution, Comparison between Mahabubnagar and Ranga Reddy/Hyderabad 100

Table 3.9 Rainfall Index Figures for the Rainy Season, Various Stations 102

Table 3.10 Estimated *Kharif* Yields in Mahabubnagar and Ranga Reddy districts, 1981-91 (Index Figures) 103

Table 3.11 Subjective Opinions from Farmers in the Research Villages about the Years 1990/1, 1991/2 and 1992/3 104
Table 3.12 Changes in the Area Under Various Crops, Mahabubnagar and Ranga Reddy districts, 1955-90, Averages per year (*1,000 ha*) 108
Table 3.13 Yields of Food Grains and Oil Seeds in Mahabubnagar and Ranga Reddy districts, 1955-90 (*kg/ha*) 110
Table 3.14 Farmers and Agricultural Labourers in Andhra Pradesh, Telangana, Ranga Reddy and Mahabubnagar districts, 1961-91 (*in millions*) 113
Appendix Mahabubnagar and Hyderabad: Drought Risks 1901-92 121
Table 4.1 Agricultural Recommendations of the Maheswaram Project for Dryland Crops 132
Table 4.2 Budgets (1981 and 1985) and Expenditure (1983/4-1992/3), Maheswaram Project (*Rs million*) 146
Table 4.3 Agricultural Recommendations by AWARE for Dryland Crops in Amangal *Mandal* 163
Table 4.4 AWARE's Sources of Funding (1976-87 and 1988-93) (*Rs million*) 172
Table 4.5 Expenditure by AWARE, by Programme (1988-93) 173
Table 5.1 Households in which the Level of Education of one of its Members is Higher than its Head, by Village, by Community 194
Table 5.2 Pure Dryland Farmers during 1991/2-1993/4, by Village, by Community 200
Table 5.3 Number and Amount of Loans (*in Rs 1,000*) Received for *Kharif* 1992/3, by Source, by Village 211
Table 5.4 Average Area of Dryland and Wetland owned, Laid Fallow and Net Sown Area during *Kharif* 1991/2-1993/4, by Village (*ha*) 212
Table 5.5 Average Area Cultivated with Dryland and Wetland Crops during *Kharif* 1991/2-1993/4, by Village (*ha*) 213
Table 5.6 Average Yield of Selected Dryland Crops (*kg/ha*) during *Kharif* 1989/90-1992/3, by Village 216
Table 6.1 Main Source of Information and Advice for Dryland Agriculture in 1991/2 and 1992/3, by Village (*percentages*) 263
Table 7.1 Rate of Adoption of Recommended Soil and Water Conservation Technologies, by Village 272

Table 7.2 Comparison between Observed Adoption Rates and Hypotheses for Recommended Soil and Water Conservation Technologies 278
Table 7.3 Rate of Adoption of Indigenous and Recommended Soil and Water Conservation Technologies, by Type of Farmer, by Village 281
Table 7.4 Comparison between Observed Adoption Rates and Hypotheses for Summer Tillage for *Jowar* 286
Table 7.5 Rate of Adoption of Summer Tillage for *Jowar*, by Type of Farmer, by Village 287
Table 7.6 Comparison between Observed Adoption Rates and Hypotheses for Contour/Across Sowing 291
Table 7.7 Rate of Adoption of Across Sowing, by Type of Farmer, by Village 292
Table 8.1 Rate of Adoption of Recommended Crop Management Technologies, by Village 304
Table 8.2 Comparison between Observed Adoption Rates and Hypotheses for Intercropping of *Jowar* with Redgram Irrespective of Ratio 308
Table 8.3 Rate of Adoption of Intercropping of *Jowar* with Redgram Irrespective of Ratio, by Type of Farmer, by Village 309
Table 8.4 Comparison between Observed Adoption Rates and Hypotheses for Improved Castor Seeds 316
Table 8.5 Rate of Adoption of Improved Castor Seeds, by Type of Farmer, by Village 317
Table 8.6 Comparison between Observed Adoption Rates and Hypotheses for Seed-rate of Castor 323
Table 8.7 Comparison between Observed Adoption Rates and Hypotheses for Recommended Level of FYM Applied to *Jowar* 330
Table 8.8 Comparison between Observed Adoption Rates and Hypotheses for Recommended Level of FYM Applied to Castor 332
Table 8.9 Rate of Adoption of Recommended Quantity of FYM Applied to *Jowar* and Castor, by Type of Farmer, by Village 333
Table 8.10 Comparison between Observed Adoption Rates and Hypotheses for 50 per cent of the Recommended Quantity of Chemical Fertilizers Applied to *Jowar* 342

Table 8.11 Comparison between Observed Adoption Rates and Hypotheses for 50 per cent of the Recommended Quantity of Chemical Fertilizers Applied to Castor 345
Table 8.12 Rate of Adoption of 50 per cent of the Recommended Quantity of Chemical Fertilizers Applied to *Jowar* and Castor, by Type of Farmer, by Village 346
Table 8.13 Comparison between Observed Adoption Rates and Hypotheses for Early Sowing of *Jowar* 349
Table 8.14 Rate of Adoption of Early Sowing of *Jowar*, by Type of Farmer, by Village 349
Table 8.15 Comparison between Observed Adoption Rates and Hypotheses for Interculturings of *Jowar* 354
Table 8.16 Comparison between Observed Adoption Rates and Hypotheses for Interculturings of Castor 355
Table 8.17 Rate of Adoption of Recommended Interculturings for *Jowar* and Castor, by Type of Farmer, by Village 357
Table 9.1 Average Adoption Rates of Recommended Technologies, by Community, by Village 388
Table 9.2 Average Adoption Rates of Recommended Technologies, by Landownership Class, by Village 388

Figures

Figure 3.1 Drought Risk Assessment for Ranga Reddy (Hyderabad), 1901-92 93-4
Figure 3.2 Drought Risk Assessment for Mahabubnagar (Mahabubnagar town), 1901-80 94-5
Figure 3.3 Ibrahimpatnam/Maheswaram: Long-term Rainfall Trends, 1957-93 95
Figure 3.4 Ibrahimpatnam/Maheswaram: The Variability of Monthly Rainfall, 1957-93 97
Figure 3.5 Ibrahimpatnam/Maheswaram: Drought Risk Assessment, 1957-93 98
Figure 4.1 Organizational Set-up of AWARE and its Target Group Organizations after 1985 166
Figure 4.2 Expected Differences in Adoption Rate of Dryland Agriculture Recommendations, by Village 176
Figure 4.3 Expected Differences in Adoption Rate of Dryland Agriculture Recommendations, by Community, by Village 177
Figure 4.4 Expected Differences in Adoption Rate of Dryland

Agriculture Recommendations, by Landownership Class, by Village 178
Figure 5.1 Holding-size Classes of Dryland (%), by Village (1991/2-1993/4) 197
Figure 5.2 Holding-size Classes of Wetland (%), by Village (1991/2-1993/4) 197
Figure 5.3 Holding-size Classes of Total Holding (%), by Village (1991/2-1993/4) 197
Figure 5.4 Average Dryland Holding (ha), by Community, by Village (1991/2-1993/4) 199
Figure 5.5 Average Wetland Holding (ha), by Community, by Village (1991/2-1993/4) 199
Figure 5.6 Average Total Holding (ha), by Community, by Village (1991/2-1993/4) 199
Figure 5.7 Percentage Farmers owning an Open Well, by Holding-size, by Village (1992/3) 202
Figure 5.8 Percentage Farmers owning a Pump, by Holding-size, by Village (1992/3) 202
Figure 5.9 Percentage Farmers owning a Tubewell, by Holding-size, by Village (1992/3) 202
Figure 5.10 Percentage Farmers owning an Open Well, by Community, by Village (1992/3) 204
Figure 5.11 Percentage Farmers owning a Pump, by Community, by Village (1992/3) 204
Figure 5.12 Percentage Farmers owning a Tubewell, by Community, by Village (1992/3) 204
Figure 5.13 Percentage Farmers owning Draught Animals, by Holding-size, by Village (1991/2-1992/3) 207
Figure 5.14 Percentage Farmers owning Milch Animals, by Holding-size, by Village (1991/2-1992/3) 207
Figure 5.15 Percentage Farmers owning Draught Animals, by Community, by Village (1991/2-1992/3) 208
Figure 5.16 Percentage Farmers owning Milch Animals, by Community, by Village (1991/2-1992/3) 208
Figure 7.1 Observed Adoption Rate (%) of Recommended Soil and Water Conservation Structures on Dryland Plots, by Village (*kharif* 1993/4) 276
Figure 7.2 Observed Adoption Rate (%) of Recommended Soil and Water Conservation Structures on Dryland Plots, by Community, by Village (*kharif* 1993/4) 276

Figure 7.3 Observed Adoption Rate (%) of Recommended Soil and Water Conservation Structures on Dryland Plots, by Landownership Class, by Village (*kharif* 1993/4) 276
Figure 7.4 Observed Adoption Rate (%) of Summer Tillage of *Jowar*, by Village (*kharif* 1992/3-1993/4) 285
Figure 7.5 Observed Adoption Rate (%) of Summer Tillage of *Jowar*, by Community, by Village (*kharif* 1992/3-1993/4) . 285
Figure 7.6 Observed Adoption Rate (%) of Summer Tillage of *Jowar*, by Landownership Class, by Village (*kharif* 1992/3-1993/4) 285
Figure 7.7 Observed Adoption Rate (%) of Sowing Direction on Dryland Plots, by Village (*kharif* 1992/3-1993/4) 290
Figure 7.8 Observed Adoption Rate (%) of Sowing Direction on Dryland Plots, by Community, by Village (*kharif* 1992/3-1993/4) 290
Figure 7.9 Observed Adoption Rate (%) of Sowing Direction on Dryland Plots, by Landownership Class, by Village (*kharif* 1992/3-1993/4) 290
Figure 8.1 Observed Adoption Rate (%) of Intercropping of *Jowar*, by Village (*kharif* 1991/2-1993/4) 307
Figure 8.2 Observed Adoption Rate (%) of Intercropping of *Jowar*, by Community, by Village (*kharif* 1991/2-1993/4) 307
Figure 8.3 Observed Adoption Rate (%) of Intercropping of *Jowar*, by Landownership Class, by Village (*kharif* 1991/2-1993/4) 307
Figure 8.4 Observed Adoption Rate (%) of Improved Castor Seed, by Village (*kharif* 1991/2-1993/4) 315
Figure 8.5 Observed Adoption Rate (%) of Improved Castor Seed, by Community, by Village (*kharif* 1991/2-1993/4) 315
Figure 8.6 Observed Adoption Rate (%) of Improved Castor Seed, by Landownership Class, by Village (*kharif* 1991/2-1993/4) 315
Figure 8.7 Observed Adoption Rate (%) of FYM Applied to *Jowar*, by Village (*kharif* 1991/2-1993/4) 329
Figure 8.8 Observed Adoption Rate (%) of FYM Applied to *Jowar*, by Community, by Village (*kharif* 1991/2-1993/4) 329

Figure 8.9 Observed Adoption Rate (%) of FYM Applied to *Jowar*, by Landownership Class, by Village (*kharif* 1991/2-1993/4) 329

Figure 8.10 Observed Adoption Rate (%) of FYM Applied to Castor, by Village (*kharif* 1991/2-1993/4) 331

Figure 8.11 Observed Adoption Rate (%) of FYM Applied to Castor, by Community, by Village (*kharif* 1991/2-1993/4) 331

Figure 8.12 Observed Adoption Rate (%) of FYM Applied to Castor, by Landownership Class, by Village (*kharif* 1991/2-1993/4) 331

Figure 8.13 Observed Adoption Rate (%) of Chemical Fertilizers Applied to *Jowar*, by Village (*kharif* 1991/2-1993/4) 341

Figure 8.14 Observed Adoption Rate (%) of Chemical Fertilizers Applied to *Jowar*, by Community, by Village (*kharif* 1991/2-1993/4) 341

Figure 8.15 Observed Adoption Rate (%) of Chemical Fertilizers Applied to *Jowar*, by Landownership Class, by Village (*kharif* 1991/2-1993/4) 341

Figure 8.16 Observed Adoption Rate (%) of Chemical Fertilizers Applied to Castor, by Village (*kharif* 1991/2-1993/4) 344

Figure 8.17 Observed Adoption Rate (%) of Chemical Fertilizers Applied to Castor, by Community, by Village (*kharif* 1991/2-1993/4) 344

Figure 8.18 Observed Adoption Rate (%) of Chemical Fertilizers Applied to Castor, by Landownership Class, by Village (*kharif* 1991/2-1993/4) 344

Figure 8.19 Observed Adoption Rate (%) of Recommended Interculturings for *Jowar*, by Village (*kharif* 1992/3-1993/4) 352

Figure 8.20 Observed Adoption Rate (%) of Recommended Interculturings for *Jowar*, by Community, by Village (*kharif* 1992/3-1993/4) 352

Figure 8.21 Observed Adoption Rate (%) of Recommended Interculturings for *Jowar*, by Landownership Class, by Village (*kharif* 1992/3-1993/4) 352

Figure 8.22 Observed Adoption Rate (%) of Recommended Interculturings for Castor, by Village (*kharif* 1992/3-1993/4) 356

Figure 8.23 Observed Adoption Rate (%) of Recommended Interculturings for Castor, by Community, by Village (*kharif* 1992/3-1993/4) 356
Figure 8.24 Observed Adoption Rate (%) of Recommended Interculturings for Castor, by Landownership Class, by Village (*kharif* 1992/3-1993/4) 356

Maps

Map 1.1 India and Andhra Pradesh 29
Map 3.1 Ranga Reddy and Mahabubnagar Districts in Andhra Pradesh 76
Map 3.2 Rainfall Pattern in Ranga Reddy and Mahabubnagar Districts 82
Map 3.3 Rainfall Seasonality in Ranga Reddy and Mahabubnagar districts 89
Map 5.1 Location of the Projects and Study Villages 186

Abbreviations

ADA	Assistant Director of Agriculture
ADATS	Agricultural Development and Training Society
ADSS	Area Development Service Society
AICRP	All India Co-ordinated Research Project
AICRPDA	All-India Coordinated Research Project for Dryland Agriculture
APAU	Andhra Pradesh Agricultural University
ASCI	Administrative Staff College of India
AWARE	Action for Welfare and Awakening in Rural Environment
BC	Backward Class
BDO	Block Development Officer
BINGO	Big NGO
CAA	Community Aid Abroad (Australia)
CAPART	Council for Advancement of People's Action and Rural Technology
CBO	Community-Based Organization
CDO	Cluster Development Officer
CDP	Community Development Programme
CDSS	Cluster Development Service Society
CGIAR	Consultative Group for International Agricultural Research
CPI	Communist Party of India
CPR	Common Property Resource
CRIDA	Central Research Institute for Dryland Agriculture
CROSS	Comprehensive Rural Operations Service Society
DAP	Diammonium Phosphate
DARE	Development Action for Rural Environment
DDS	Deccan Development Society
DDSS	District Development Service Society
DPAP	Drought Prone Area Programme
DRDA	District Rural Development Agency
DWDC	District Watershed Development Committee
EROR	Economic Rate Of Return
FC	Forward Caste
FF	Farmers First
FYM	Farm Yard Manure
GoAP	Government of Andhra Pradesh
GoI	Government of India

GSO	Grassroots Support Organization
HYVP	High Yielding Variety Programme
IAAP	Intensive Agricultural Areas Programme
IADP	Intensive Agricultural District Programme
IDPAD	Indo-Dutch Programme on Alternatives in Development
ICAR	Indian Council of Agricultural Research
ICRISAT	International Crops Research Institute for the Semi-Arid Tropics
IOV	Dutch Operation Review Section, Department of Development Cooperation
IRDP	Integrated Rural Development Programme
ISWDP	Integrated State Watershed Development Programme
JRY	*Jawahar Rozgar Yojana*
KVK	Farm Science Centre
LEIA	Low External-Input Agriculture
MDDS	*Mandal* Development Service Society
MDO	*Mandal* Development Officer
MFAL	Agency for Development of Marginal Farmers and Agricultural Labourers
MLA	Member of the Legislative Assembly
MNP	Minimum Needs Programme
MSO	Membership Support Organization
MYRADA	Mysore Relief And Development Agency
NABARD	National Bank for Agriculture and Rural Development
NGDO	Non-Governmental Development Organization
NGO	Non-Governmental Organization
NREP	National Rural Employment Programme
NWDC	National Watershed Development Council
NWDPC	National Watershed Development Policy Committee
NWDPRA	National Watershed Development Programme for Rainfed Agriculture
NWDPRA	National Watershed Development Programme for Rainfed Areas
PAP	Poverty Alleviation Programme
PDS	Public Distribution System
PRA	Participatory Rural Appraisal
RDT	Rural Development Trust
RHC	Red Hairy Caterpillar
RLEGP	Rural Landless Employment Guarantee Programme
RRA	Rapid Rural Appraisal
SAU	State Agricultural University
SC	Scheduled Caste
SCRDTC	Soil Conservation Research, Demonstration and Training Centre
SFDA	Small Farmers Development Agency
SMS	Subject Matter Specialist

ST	Scheduled Tribe
SWC	Soil and Water Conservation
SWDC	State Watershed Development Cell
SWDPC	State Watershed Development Policy Committee
T&V	Training and Visit system of agricultural extension
TDSS	*Taluk* Development Service Society
TOT	Transfer-Of-Technology
VA	Village Association (local organization or *sangham*)
VANI	Voluntary Action Network India
VASORD	*Vasavya* Society for Rural Development
VDO	Village Development Officer
VEO	Village Extension Officer
ViP	Villages in Partnership
VLW	Village Level Worker
WCADL	Weaker Community's Action for Development and Liberation
WDT	Watershed Development Team
WWFN	World Wide Fund for Nature
YfA	Youth for Action
YIP	Young India Project

Glossary

Adivasi	tribal groups (also called *Girijans*), considered to be the indigenous peoples of the subcontinent
bajra	pearl millet
beedies	country cigarettes
bhagela	agricultural serf often forced to bonded labour
binami names	fake names
Brahmins	most important *varna* in the Hindu caste system
chalka soils	sandy loam (Alfisol)
Chenchu	name of a tribe
chulla	fuel efficient, smokeless stove
dante	indigenous blade harrow, predominantly used in dryland cultivation
deshmukh	feudal landlord, person considered to represent the *Nizam*
despandya	feudal landlord, person considered to represent the *Nizam*
dubba soils	loamy sand (Alfisol)
Goud	the traditional toddy-tapper caste; belongs to the BC community
gramasabha	group meeting in one or a couple of villages
Grama Sevak	Village Development Officer (VDO)
Gram Swaraj	concept of self-sufficient village communities, propagated by Gandhians
guntaka	indigenous blade harrow, predominantly used in dryland cultivation
Harijans	untouchables, Scheduled Castes or *Dalits*; the lowest section of the Hindu caste system
jagirdar	feudal landlord, person considered to represent the *Nizam*
jambu	an agricultural implement used to level plots
jati	(sub)caste

jowar	sorghum
karthe	fortnight. Based on astrology, the agricultural year is divided into 26 fortnights. Farmers are convinced that certain agricultural operations are best carried out in a specific 'auspicious' fortnight.
kharif	south-west monsoon or rainy season; the period between June until October during which dryland is cultivated
khus	Vetiver grass
kirana shop	small provision shop in villages
Krishi Vigyan Kendras	KVKs or Farm Science Centres
Kshatriyas	the second *varna* in the Hindu caste system
Lambada	name of a tribe
maa-bap	when people look for guidance and direction by someone who is considered to be a 'divine' person
Madiga	name of a *Harijan* group
mahasabha	mass meeting during which often (ten) thousands of people join
mahila mandal	women's organization
Mala	name of a *Harijan* group
mali patel	civil servant for revenue collection
mandal	an administrative unit below district level
mandal praja parishad	*mandal* council
nagu loans	loans in kind, mostly grain, for food or as seed
nala	streamlet
neem	tree
Nizam	an independent Muslim sovereign who ruled Hyderabad State
panchayat	elected village council; basic unit of local government
panchayati raj	a three-tier system of local self-government consisting of the village, the *mandal* and the district
panchayat samiti	block council
patta	title deed; a certificate of landownership
pattedars	farmers who own the land they cultivate
patwari	civil servant for keeping and maintaining

	accounts of village landholdings and land revenue
police patel	civil servant for maintaining law and order
poramboke	land owned by the government
rabi	north-east or retreating monsoon; period during October until January with limited agricultural activities on dryland
ragi	finger millet
Reddy	sub-caste of the *Sudra varna*. Under the reign of Muslim sovereigns they formed the rural elite. Even today, Reddies are primarily engaged in agriculture as large farmers, which gives them not only economic and social, but also political leverage.
roti	bread made of all kinds of flour.
sangham	local organization, set up in villages by many NGOs (also AWARE)
sarpanch	village headman and head of the village council (*panchayat*)
seethapal	custard apple
Sudras	the fourth *varna* in the Hindu caste system; by tradition consisting of farming and artisan-service castes.
taluks	the administrative division below the district, renamed *mandal* after an administrative reorganization in 1986
tanda	the habitat of the tribals, mostly situated at some distance of the main village
tempo	a goods carrier carrying people as well, similar to a bus service
Vaisyas	the third *varna* in the Hindu caste system
varna	a distinct class or caste in Hindu religion with its own rules of conduct and behaviour. This system of social ranking is in fact an idealtype and in reality thousands of (sub)castes or *jatis* exist.
vetti	forced labour
zilla (praja) parishad	district council

Preface and Acknowledgements

In this study, two fields of interest meet. The first field covers the interest in semi-arid areas, in particular those located in south India. The second covers the interest in intervention strategies of the government and Non-Governmental Organizations (NGOs) in rural areas. Already in the 1970s the Department of Human Geography of the University of Amsterdam devoted part of its research oń livelihood strategies of people living in these areas and the role of the government (the 'Semi-Arid Lands Project Amsterdam' or SALPA). One of the countries selected for in-depth research was India, in particular the rural hinterland of Tamil Nadu. In 1985, a team of five MA students did fieldwork in some villages in Dharmapuri district, a remote, semi-arid region in that state. One of the findings of this fieldwork was the large differences in technologies and inputs used in dryland (not irrigated agriculture) and wetland agriculture (or irrigated agriculture). Another finding was that the interest of the government was limited to the supply of infrastructural facilities. Farmers said that they had never seen an agricultural extension officer in their village for instance. Starting in 1987, additional research was carried out at the Royal Tropical Institute in Amsterdam. This desk study concluded that agricultural innovations for dryland had been developed in India but that, amongst others, extension was not very successful.

This present study is in fact a continuation of this interest in the problems faced by farming households living in remote, economically backward, semi-arid regions and the policies towards these regions. It also wants to contribute to the discussions which started in the late 1980s on the effect of government and NGO activities. The study focuses on agricultural change in a semi-arid area south of the state capital of Andhra Pradesh, Hyderabad, and analyses in-depth the effect of two intervention strategies: a project financed by the World Bank and implemented by the Andhra Pradesh government and a project implemented by AWARE, one of the largests NGOs in Andhra Pradesh and in India.

The study is part of the 'Livelihood and Environment' programme, a research programme initiated by the Department of Human Geography in 1990. The project on which the present publication is based was developed in cooperation with the Department of Geography, University of Madras. It forms part of the Indo-Dutch Programme on Alternatives in Development (IDPAD) which was set up by the Indian Council of Social Science Research (ICSSR), New Delhi, and currently, the institutional context of the programme has changed at the Dutch side from Nuffic to WOTRO. Additional funding was received by the Netherlands Foundation of the Advancement of Tropical Research (WOTRO) in The Hague. We express our gratitude to both funds for financing this research. IDPAD and WOTRO, however, do not take any responsibility for the facts stated, opinions expressed and conclusions reached in this publication.

This book is the result of a joint research endeavour made by various researchers working in several institutes. On the Indian side, Prof. S. Subbiah (Department of Geography, University of Madras) and Mrs D. Adi Lakshmi carried out part of the research. On the Dutch side, Prof. A.J. Dietz and Prof. G.A. de Bruijne (Department of Human Geography, University of Amsterdam) together with Dr M.P. van Dijk (Department of Regional Economics, Erasmus University Rotterdam) participated in this research. The fieldwork (from November 1991 to March 1994) was coordinated by Mr M. Put (Department of Human Geography, University of Amsterdam), who was also responsible for the major part of the data processing and analysis.

This book could not have been written without the suppport of many people. First of all we want to express our gratitude to the villagers of Debbadaguda, Mansanpally, Ekvaipally and Ghatpally. We feel a deep sympathy for all those petty farmers who till the (dry) soil in such a climatologically, socially and politically, harsh environment. The fieldwork could not have been carried out without the valuable support of our interviewers. First and foremost, we remember Mr M. Shrikant Reddy, a very upright person, who committed suicide together with his Brahmin girlfriend for the sake of their (forbidden) love. We express our gratitude to all other interviewers who managed to live rather uncomfortably (small rooms, no proper beds, unreliable electricity supply, lack of good food and sometimes safe drinking water) for many, many months in the study villages. In alphabetical order: T. Nagesh, B. Siva Krishna Prasad, M. Bhaskara Rao, N. Neelakanta Rao, K. Reddy, M. Sirazuddin, V. Veeraswamy, P. Venkatesham. We are also grateful to S. Diemel who, as a student of the Department of Human Geography, participated in the fieldwork. A special word of thanks goes to Mrs A. van der Wal (Centre for Development Cooperation Services, Free University, Amsterdam) who spent her holidays in February 1994 interviewing many farmers.

A study comparing two projects cannot do without the help of the implementing institutions and the help of the people involved. First of all we want to thank the project officers of the Maheswaram Pilot Project for Watershed Development in Rainfed Area who spent many hours with us discussing the effects of their project: Mr A. Krishna Murti (Project Leader), Mr G.V. Nagaraju Naidu (assistant director) and Mr Hassan (assistant director). In addition, we acknowledge the help given by the other staff and the numerous extension agents of the project for communicating to us the various recommendations made for dryland agriculture and for informing us about their experiences in the field. We also remember with gratitude the details on the various watershed programmes implemented in Andhra Pradesh communicated to us by Mr S.K. Moinuddin, Deputy Director of Soil and Water Conservation, Department of Agriculture, Government of Andhra Pradesh. We have also benefited from the discussions with staff of the Agriculture Unit at the World Bank office in New Delhi, in particular with Mr J. Alexander and Mr M.C. Lodha. We appreciate the assistance rendered to us by the Administrative Staff College of India (ASCI) in Hyderabad, the institute responsible for the monitoring and evaluation of the Maheswaram project. A special word of thanks is directed at Dr M.V.K. Sivamohan, the team leader of the evaluation study, who presented to us all kinds of reports on the project. It enabled us to compare their outcomes and the progress of the project.

Although our relationship had its ups-and-downs, we are indebted to the staff of AWARE: Mr P.K.S. Madhavan (chairman), Mr Satyabhupal Reddy (zonal officer), Mr Ramana (head, research department), Mr Satyanarayana (head, publicity) and all others at the central office in Hyderabad and those in the field. A special word of thanks to the staff at AWARE's Human Resource Development Centre located some 25 km south of Hyderabad, who provided boarding and lodging to us and were an invaluable source of information.

A word of thanks also for the researchers at the Central Research Institute for Dryland Agriculture (CRIDA) in Hyderabad. We owe much to Dr N.K. Sanghi, Zonal Coordinator of the Lab to Land Programme at CRIDA, who aroused our interest in indigenous agricultural practices in the study area and informed us about the various recommendations made for dryland agriculture. He surely belongs to the diminishing group of people enthusiastic about walking in the fields. We acknowledge the help of several researchers at the Centre for Economic and Social Studies (CESS) in Hyderabad. We value the assistance rendered by Dr R. Radhakrishna (then director), Dr S. Sudhakar Reddy, Dr Sheik Galab, Dr G. Reddy and Mr P. Satya Sekhar. A word of thanks to Prof. Narashima Reddy (Department of Economics, Central University, Hyderabad) who commented on an earlier version of this report. We have appreciated the comments on this same version made by two anonymous reviewers as well.

During our research we met quite a number of people from several government departments at various levels, in and outside Hyderabad. With gratitude we remember the discussions we had with and the reports and other data obtained from them. A special word of thanks for Mr Chandramouli, DRDA project director of the District Rural Development Agency (DRDA) in Anantapur district and his colleague, Mr Ram Gopal Reddy, project director DRDA in Ranga Reddy district.

Our study also benefited from the discussions with staff of numerous NGOs, with staff of institutes which support NGOs and with informants familiar with the 'NGO community' in southern India. We thank various persons at Action for World Solidarity, Agricultural Development and Training Society (ADATS), Centre for Environment Concerns (CEC), Deccan Development Society (DDS), Hivos field office Bangalore, MYRADA, National Institute for Rural Development (NIRD), Novib's Bangalore Consultancy Office, Rural Development Trust (RDT), SEARCH, Villages in Partnership (ViP), Weaker Community's Action for Development and Liberation (WCADL) and Youth for Action (YfA).

June 1998

MARCEL PUT
MEINE PIETER VAN DIJK

1. Introduction

G.A. de Bruijne and M.P. van Dijk

Agriculture is of prime concern in India. There is a common feeling that the monsoon not only influences yields but also influences the Indian economy as a whole. A bad monsoon is the nightmare of every government. Hence, food self-sufficiency has been a major goal in Indian politics. That India must be self-sufficient in food is self-evident. Every time newspapers publish reports about the possibility of the import of foodgrains, discussion starts again about the prospects and the state of affairs of Indian agriculture.

Fortunately for the Indian population, India's food production has increased substantially in the last few decades, mainly due to the Green Revolution in irrigated areas. Nevertheless, in many villages located in the semi-arid tracts of India—not the prosperous wetland agriculture with its abundant harvest—dryland agriculture with low yields is a fact of everyday life. Farmers are confronted with a risky environment: rainfall is low and uncertain and this together with soils of low fertility and poor structure impede the intensive growth of crops. Little access to capital, coupled with low income and high debts, also result in low investments in dryland agriculture by those small farmers for whom that type of agriculture is a source of their livelihood. Some of those who, for professional reasons, are engaged in dryland agriculture, believe that because of the innovations which have been developed, dryland agriculture in semi-arid areas may have got over its worst days. Innovations with regard to soil and water conservation and crop management have been introduced from outside these areas. In addition, they refer to the farmers who try to implement positive changes themselves and point to the policies and activities of government agencies and NGOs who intend to promote agricultural production and to combat poverty from outside. The market also witnesses producers and traders introducing new possibilities, that may reach the villages and the farmers.

This study intends to compare government and NGO interventions, and non-intervention with regard to dryland agriculture in four villages in the semi-arid region of Andhra Pradesh, where rivers are not omnipresent, where rainfall is anything but abundant, and where many rural people struggle for their livelihood.

1.1 National Setting

India is a subcontinent, by size and population larger than the European Union, and covers an area of about 3.3 million sq. km. According to the 1991 Census, some 43 per cent of the area is agricultural land while 20 per cent is forest. In 1995, India's population increased beyond 900 million, the second largest in the world after China. Annual population growth during the last decade averaged 2.1 per cent. About 74 per cent of the population lives in rural areas and the relative increase in the urban population is not impressive. It has various climatic zones with marked regional differences: the Himalayan foothills in the north, the well-irrigated areas in the north-west, deserts in the west, the hill tracts in the east, the heavily populated Gangetic Plain, the semi-arid Deccan Plateau, and tropical coastal areas in the south are all different. The country's economic and social conditions, languages, and cultural and religious traditions also show wide regional variations with its political implications. India is not a centralized nation but a federation of 25 states and 7 union territories. The Constitution of India envisages two tiers of government, one at the centre (the union or federal government) and the other at the state level. State governments, headed by a governor appointed by the president, also have different departments, headed by ministers.

The states are administratively divided in accordance with the *panchayati raj* system, which involves a three-tier system of local self-government at the village, block (group of villages) and district levels. In states like Andhra Pradesh, the district population, normally numbering between one to three million people, selects a *zilla parishad* or district council. Being ultimately accountable for planning at the district level, a *zilla parishad* controls the appointed district collector, who is generally in charge of the planning and implementation of policies and programmes of both central and state governments. Below the district is the block level (in Andhra Pradesh the *mandal* level) with *panchayat samities* (block councils), whose compound population normally ranges from 70,000 to 1,00,000, and the village level with village *panchayats* or *gram sabhas* (village councils).

According to World Bank indicators, India should be considered a 'low

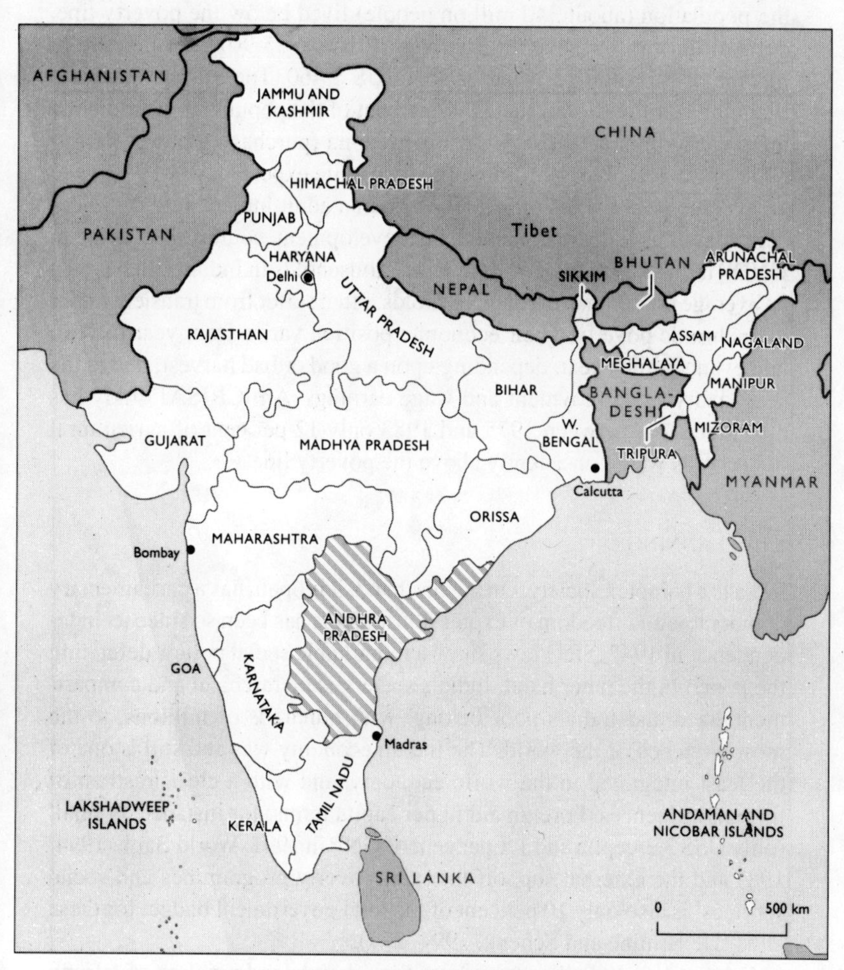

MAP 1.1. INDIA AND ANDHRA PRADESH.

income economy', with a GDP per capita of US $ 300 in 1993. In rural areas as well as in cities many people live in precarious conditions. The World Bank estimated that, at the end of the 1980s, about 40 per cent of the population (about 340 million people) lived below the poverty line, defined during the Seventh Five-year Plan (1985-90) as a household income of Rs 6,400 per year, or about US $ 460. The estimation of the Planning Commission is that 29.9 per cent of the population lived below the poverty line in 1987/8. A real per capita (purchasing power parity) GDP of US $ 1150 in 1991, together with a life expectancy of 59.7 years, an average of 2.4 years of schooling and an adult literacy rate of 50 per cent, classifies India as a 'low human development' country, according to UNDP indicators (UNDP, 1994). Rural households in India, which are on an average poorer than urban households, often suffer from transient rather than chronic poverty. Their economic position varies from year to year, and even within a year, depending upon a good or bad harvest, due to the seasonality of employment and wage earnings. An ICRISAT study has concluded that between 1975 and 1983 only 12 per cent of agricultural households were consistently above the poverty line.

1.1.1 PLANNING

India is a complex society. On the one hand, it is open, has a parliamentary democracy, and freedom of expression and this has been so since its independence in 1947. State laws show a remarkable social policy, defending the poor. On the other hand, India's society is hierarchical and compartmentalized and India's poor belong, with hundreds of millions, to the most wretched of the world. The Indian economy was and still is one of the least integrated in the world economy, and with a clear mistrust of foreign influences. Foreign aid in per capita terms, for instance, is small (only US$ 3.3/capita and 1.1 per cent of GNP in 1991; World Bank, 1994: 198) and the external support for 'anti-poverty programmes and social services' is also only 10 per cent of the total government budget for these aims (De Bruijne and Schenk, 1994: 277).

Independent India has for a long time maintained a policy of 'strong planning', with the state as the driving force of the economy at the central as well as state levels, and an image of 'socialism'. Indian planning has been based on the idea of a benevolent state which, equipped with sufficient information and knowledge, was expected to act solely in society's interest. The Central Planning Commission was set up in 1950 to formulate Five Year Plans which would lay down policy priorities and sectoral

TABLE 1.1. EXPENDITURE PER SECTOR DURING VARIOUS PLANS IN INDIA (*percentages*)

Five-Year Plans	Agriculture and allied activities	Rural develop-ment	Irrigation and flood control	Total agriculture (1+2+3)	Power generation	Village and small industries	Industry and minerals	Transport and com-munication	Social and community services	Total (*Rs million*)
	1	2	3	4	5	6	7	8	9	Abs.
First Plan (1951-6)	10.8	4.0	22.1	36.9	7.6	2.1	2.8	26.4	24.1	19,600
Second Plan (1956-61)	4.7	4.8	11.2	20.7	9.7	3.8	28.3	19.6	18.0	46,000
Third Plan (1961-6)	6.2	3.4	10.9	20.4	14.6	2.8	20.8	24.6	17.4	85,760
Annual Plans (1966-9)	8.3	1.5	12.0	21.7	18.3	1.9	22.8	18.4	16.8	66,250
Fourth Plan (1969-74)	10.5	0.7	12.1	23.3	18.6	1.5	18.2	19.5	19.9	157,790
Fifth Plan (1974-9)	8.2	1.5	12.1	22.2	18.8	1.5	22.8	17.4	17.3	394,260
Sixth Plan (1980-5)	6.7	5.8	10.0	25.5	28.1	1.8	13.7	16.2	16.3	1,092,920
Seventh Plan (1985-90)	5.8	6.8	7.5	21.8	28.4	1.6	13.5	17.3	17.3	2,202,160
Eighth Plan (1992-7)	5.2	9.5	7.5	22.2	26.6	1.5	9.3	18.7	21.7	4,341,000
Average	7.4	4.2	11.7	23.9	20.0	2.1	16.9	19.8	18.8	100.0

Note: Figures for the First to the Seventh Plan relate to expenditure, while the figures of the Eighth Plan relate to the outlay.

Source: GoI (Various Five-Year Plans).

investments for that period. The First Plan covered the first half of the 1950s, and at the moment, the Ninth Plan is in preparation. At present, however, due to the 'liberalization policy', structural reforms aim at reducing government intervention, while at the same time trying to enhance domestic competition and to accelerate the country's integration in the world economy.

The major objectives of the First Plan in the field of agriculture, were to correct imbalances, caused by partition of the country, in the supply of food grains and commercial crops (particularly cotton and jute) and to improve the infrastructure. Agriculture, including irrigation and power, was therefore, accorded the highest priority. Efforts to improve agriculture were spread over the entire country and every villager was to be involved in it through the Community Development Programme.

To achieve its various, and often conflicting, economic and socio-economic objectives like the elimination of rural poverty, agricultural growth, the reduction of the unequal geographical distribution of agricultural production, various governments allocated large sums of money towards agriculture. As Table 1.1 shows, during the different Plan periods the relative amount of those activities has been between 20.4 per cent during the Third Plan (1961-6) and 25.5 per cent during the Sixth Plan (1980-5). Only during the First Plan it was higher, 36.9 per cent. In general, during the first seven Plans, the government invested 26.3 per cent in power, 22.5 per cent in total agriculture, 17.4 per cent in transport and communication, 17.2 per cent in social and community services, and 15 per cent in industry and minerals. So, one cannot say that the main emphasis in the development policy of the government has been on rural development, although more than 70 per cent of the population is living in rural areas. Moreover, within agriculture the larger part of the investments was spent on irrigation and flood control. The investments in improving dryland agriculture have been limited, however.

Table 1.2 gives an overview of the different rural programmes started under the subsequent plans. The date is indicative. Sometimes it refers to the date of announcing the programme, sometimes to the moment the programme really started. Until very recently, the government has held a firm stand in the economic development of the country.

One of the main aims of the Indian government after independence was to be self-sufficient in food at the national level. To reach this goal it implemented amongst others the Intensive Agricultural District Programme (IADP) and the High Yielding Varieties Programme (HYVP)

TABLE 1.2. SELECTED AGRICULTURAL AND RURAL DEVELOPMENT PROGRAMMES DURING EIGHTH FIVE YEAR PLAN

Plan	Programme	Year
FYP 1	Community Development Programme (CDP)	1952
FYP 2	*Panchayati Raj* System	1959
FYP 3	Intensive Agricultural District Programme (IADP)	1961
	Intensive Agricultural Areas Programme (IAAP).	1964
	High Yielding Variety Programme (HYVP)	1966
FYP 4	Small Farmers Development Agency (SFDA)	1969
	Agency for Development of Marginal Farmers and Agricultural Labourers (MFAL)	1969
	Rural Works Programme	1970
	Drought Prone Area Development Programme (DPAP)	1972
	Employment Guarantee Scheme	1972
FYP 5	Minimum Needs Programme (MNP)	1974
	Twenty Points Programme	1975
	Food for Work Programme	1977
	Integrated Rural Development Programme (IRDP), merger of SFDA, MFAL and DPAP	1978
FYP 6	National Rural Employment Programme (NREP)	1980
	New Twenty Points Programme	1982
	Rural Landless Employment Guarantee Programme (RLEGP)	1983
FYP 7	*Jawahar Rozgar Yojana* (JRY), a merger of NREP and RLEGP	1989
	National Watershed Development Programme for Rainfed Agriculture (NWDPRA)	1987
FYP 8	National Watershed Development Programme for Rainfed Areas (NWDPRA)	1992

Note: Years are indicative.
Source: After Prasad (1992: 10).

in the 1960s. The first stage of the strategy pertained to the IADP in 1960/1. In 1964/5 and subsequent years the same approach was extended to several other parts of the country in the form of Intensive Agricultural Area Programme (IAAP). While both the IADP and the IAAP were concerned with the promotion of intensive agriculture, they operated within the limitations set by the existing crop varieties which had at that time a relatively low response to fertilizers.[1] The main planks of the new agricultural strategy were the High Yielding Varieties Programme (HYVP) and the Multiple Cropping Programme, both based mainly on the exotic, high yielding, short duration crop varieties. (The area under high yielding varieties increased from 1.89 million ha. in 1960/1 to 68.11 million ha. in 1990/1.) Other elements of the strategy were the new concepts of irrigation and water management, adoptive research and providing price guarantee to producers.

These programmes laid the basis for the bio-chemical package approach of the Green Revolution. The Green Revolution that started during the late 1960s led to substantial increases in wheat and rice production, particularly in the north-western regions (Punjab, Haryana, western Uttar Pradesh), but also required significant inputs which aggravated the government's budgetary constraints. This Green Revolution strategy aimed at achieving technological modernization. Its main components were: a shift from large-scale irrigation projects to small tube-wells and energized pump sets; adequate provision of credit to those who were considered credit-worthy, i.e. the larger farmers; increased fertilizer consumption; development of commercial sources of energy like electricity and diesel oil; and the development of HYVs that would be sensitive to the application of fertilizer. Land redistribution was no longer emphasized in spite of the fact that less than 30 per cent of the landowners controlled almost 75 per cent of the 182 million ha of arable land. On the contrary, according to Chakravarty (1987: 24), it was openly admitted that it was essential to bet on the strong if agricultural production was to be stimulated. So, it bypassed certain regions and certain groups of people in the rural areas. The higher working capital requirement per unit of output implied that only the relatively better-off farmers in the infrastructurally better endowed regions (i.e. irrigated regions with adequate credit facilities and developed transport and communication network) were able to make full use of the new strategy. Small and landless farmers were not able to make equally profitable use of the new input packages. Since the larger farmers also effectively controlled the rural institutions (including cooperatives, and *panchayati raj* institutions) and the distribution channels for other complementary inputs such as credit, polarization in the rural areas, reflected by skewed land and asset distribution, and growing indebtedness among small and landless farmers increased. This holds, in particular, true for the semi-arid areas. Semi-arid India, defined as that area receiving on an average between 400 and 800 mm of rainfall per year, comprises about 73.5 million ha. or 22.5 per cent of the total geographical area. Some 20 per cent of the Indian population, 160 million people, live in this tract.

In latter Plans, the Indian government emphasized the need for a geographical spread of agricultural development, as well as the need for redistribution and for poverty eradication, without losing its sight on economic growth. During the Fourth Plan, a new orientation was imparted to the agricultural policy. To achieve growth with social justice, a built-in bias was introduced in various agricultural programmes to help the weaker and vulnerable sections and the backward areas. The role of technology as a major input in agriculture was accorded explicit recognition. Besides attainment of self-sufficiency, the Plan envisaged building up of a sizeable buffer stock of food grains and termination of concessional imports of food grains. In that context, various poverty alleviation programmes (PAPs) were implemented during the 1980s by both central and state governments, seeking to improve the income generating capabilities of weaker sections of the population by subsidizing productive assets and to assure an adequate availability of food and other basic needs in dry areas and among poor people. Programmes like the Small Farmers Development Agency (SFDA), Integrated Rural Development Programme (IRDP) and the National Rural Employment Programme (NREP) were set in motion to provide assistance to more vulnerable groups like marginal and small farmers, artisans and landless labourers.

The Drought Prone Area Programme (DPAP) was implemented to stimulate growth in the drier areas, areas hardly touched by the Green Revolution. The major components of the programme are soil and water conservation, and land shaping; afforestation and pasture development; and water resources development.

The DPAP was introduced in 1973. It covers 557 blocks, spread over 74 districts, which account for 12 per cent of the country's population. It is an Integrated Development Programme in the agricultural sector and it aims at optimum utilization of land, water and livestock resources, restoration of the ecological balance and stabilizing the income of the weaker sections of the community. Economic development of these areas would be achieved through activities which, in the long run, contribute actively to creating conditions which mitigate the effects of drought in these areas. Watershed management will receive the highest priority and steps will be taken to promote the cooperative management of the watershed by the people in the area. An important component of the DPAP was the attempt to provide at least a minimum (10 per cent) of the cropped area in each district with irrigation support, even by bringing water from another catchment area. Emphasis had also been given to developing organic sources of manure and higher outlays were provided for setting up biogas plants. Efforts were also made to accelerate the mini-kit seed programme and to strengthen the extension services.

The Integrated Dry Land Agricultural Development scheme was

especially designed to benefit the backward dry farming areas, which covered 68 million ha. of agricultural land. This programme intended to deal with the problems of low productivity and high instability, which characterized this agriculture, both by specially directed research and application of known advances in dry farming techniques.

Dryland areas, or rainfed areas, i.e. those lands which are not irrigated and depend on rainfall only, account for about 70 per cent of the current cultivated area in India. Soil and water conservation work in these dryland areas has been taken up in India already during the British period and was continued afterwards, albeit on a minimal scale. In most cases, the activities that were undertaken, formed part of programmes which were not aimed at the alleviation of the problems in drylands. Until the end of the Fifth Plan (1979), the Indian government spent nearly Rs 5,000 million to treat about 22 million ha. For such a large country this amount and the area treated is quite insignificant.

Until the 1980s, except for the DPAP, dryland agriculture and soil and water conservation activities were not given much attention in planned development. The main activities related to the soil conservation elements of the various programmes, consisted mainly of forming bunds. These bunds often followed ownership patterns rather than topographical patterns. There had been negligible inter-bund treatment or *in-situ* moisture conservation practices. Most activities were carried out not because farmers had asked for it or would benefit, but because they were part and parcel of the programme and targets needed to be reached. Only limited attention was given to improve poor crop management in dryland agriculture. Because improved seeds of major dryland crops were hardly available, production improvements were less than expected.

Recognizing the need for an integrated approach to check environmental degradation and to stimulate agriculture in dryland areas, much greater emphasis was given to the watershed as the basic physical-biological and socio-economic unit for planning. In 1982, the government of India implemented a programme on watershed development in different agro-ecological zones of the country. Nowadays, various programmes are being executed, like those financed by foreign donors and the World Bank, of which the Maheswaram project (one of the subjects of our study) is an example, another programme is the National Watershed Development Programme (NWDP), which is copied at state level and run as the State-level Watershed Development Programmes. In addition there are the Watershed Development Schemes, Catchment Areas of large irrigation projects and Watershed Development Works in DPAP and Desert Area Development

Project Agencies. The number of watershed programmes being implemented at the moment and the number of agencies involved in the execution is such that there is considerable confusion at present. The most expensive schemes are those financed by foreign donors and the World Bank. Latter projects are undertaken with active participation of agricultural universities and the government extension service. Next to these government programmes and projects, various NGOs are executing watershed projects. There are, however, certain major differences between the national and state sponsored programmes and the NGO programmes in respect of size of watershed, cost of development, administration, people's participation and emphasis on treatment of non-arable lands.

The main objectives which have guided the various agricultural programmes of the Indian government are present in the watershed programme as well: to improve and sustain production and productivity of all categories of land and farmers at higher levels and to check environmental degradation. More specifically, the objectives of the various watershed programmes are:

(1) promotion of in-situ soil and water conservation;
(2) optimal use of land to minimize risk in dryland farming, increase in productivity of lands and better returns to all farmers on a sustainable basis through adoption of better technology and cropping patterns and diversification of sources of income;
(3) proper management of non-arable lands to conserve soil and moisture, and store run-off for recharging ground water;
(4) production of fodder, fuel, fibre, timber, and fruits in the watershed and at the same time help to maintain ecological balance.

Only with the watershed approach an integrated strategy was developed in which next to soil and water conservation crop management also had its proper place. It was as if the government had realized that without immediate gains in the form of higher yields, farmers would not be interested in taking up soil and water conservation measures which would yield long-term gains. In a way, these government programmes benefited from the development of more and better improved seed varieties of dryland crops.

Despite all the acclaim for India's success in achieving food grain self-sufficiency, the trend of agricultural growth in value added terms has merely been 2.6 per cent per year in real terms between the beginning of the Green Revolution (late 1960s) and 1990, exceeding population growth by 0.5 per cent only. The fact that GDP growth per year in agriculture between 1966/7 and 1988/9 was limited, compared to 5.2 per cent for the industrial sector and 4.6 per cent for services, helps to explain why poverty is mostly concentrated in the rural areas. Since 1989 the growth figures of food production have varied much: 1990/1 4.2 per cent, 1991/2 1.8 per cent and for 1993/4 3.1 per cent. In absolute terms, the per capita net availability of cereals and pulses per day between 1992 and 1994 was 470 grams, only 5 per cent more than during the period 1970-4.

The main features of the agricultural policy adopted in the various Indian development plans, can be enumerated as:

(1) Land reforms leading to the abolition of intermediary tenures and tenancy reforms.
(2) Increase in the production of food grains, cash and horticulture crops to meet the consumption requirements of the growing population, raw material needs of the expanding industry and for exports.
(3) Spread of high-yielding varieties of inputs.
(4) Extension of irrigation facilities and water management programmes.
(5) Establishment of a system of support prices, procurement and public distribution.
(6) Institutional and organizational arrangements to assist small and marginal farmers, sharecroppers and other so-called weaker sections.
(7) Promotion of agricultural research, education and extension.
(8) Diversion of allied agricultural activities such as animal husbandry and dairy farming, fisheries, forestry.

The uneven agricultural progress in various states can be attributed to the failure of the agricultural policy to extend modern technology and to develop a technology for ecologically handicapped regions. The irrigation potential was not always used fully, nor the potential of rainfed areas, or of the institutional framework to provide the necessary inputs to farmers.

Another weakness of the agricultural policy has been in the non-implementation of land reforms. Security of tenures and ownership rights of benefiting tenants have not been conferred in the majority of states like Andhra Pradesh, Tamil Nadu, Bihar, Haryana, Punjab, etc. Similar laws on ceilings of agricultural land have not been implemented in all the states. The progress of taking over and distributing surplus land has not gone hand in hand with providing assistance to develop the land. Except in Punjab, Haryana and western Uttar Pradesh, no state has completed the work of consolidation of holdings. In the southern and eastern states and Rajasthan not even a beginning has been made.

Institutional arrangement in the form of IRDP, special area development programmes and diversification of allied agriculture activities have failed to increase employment and income of landless labourers, small and marginal farmers and other weaker sections. Huge funds are being spent on agricultural development in every state, but a major part of the physical and financial assets has been concentrated in a few hands, whereas about half of the population in rural areas still lives in dire poverty. The agriculture policy has failed to provide growth with social justice.

This failure has been due to a conflict between the immediate and ultimate objectives in India's agricultural policy. Concentrated efforts are being made to increase agricultural production at a macro level, thereby favouring the productive areas with assured irrigation facilities and superior supply of other inputs. As a result, the geographical imbalance, between rich and poor regions has increased. In the less well-endowed semi-arid areas which lack large-scale irrigation projects, there has been a concentration on the few patches of land which are cultivated with irrigated crops.

1.1.2 RURAL NGOs

From outside, it is not the government alone which intends to combat poverty in dryland areas, but in different forms, Non-Governmental Organizations have focused on the same objectives for many years. The role of such organizations may have become more necessary, as things are changing with the opening up of the economy and a trend of 'withdrawal of the state'. The extensive measures towards a liberalized and market-oriented economy, cause a more savage 'rush for resources', with growing socio-economic disparities as one of the results. There is fear that the state will withdraw from many of its development-oriented and poverty-alleviating activities, further fuelling a socio-economic split between the haves and the many have-nots in Indian society, although for the time being the funds for rural development are increasing instead of decreasing. Society is hardening and lower-middle class based 'communalism' and especially Hindu-fundamentalism is on the increase, to the detriment of, especially, the scheduled castes (SCs or *Harijans*), scheduled tribes (STs or *Girijans*) and poor Muslims of the country. Some regard this as a threat to the basics of Independent India. For instance, Panikkar (1991: 15) states:

> What is at stake is democracy, secularism, rule of law, nationalism and culture; in essence all that Indians fought for and held dear during the last two hundred years. The idea of democracy is replaced by majoritarianism, the concept of secularism is interpreted in religious terms, the rule of law is subverted by public coercion, Indian nationalism is equated with Hindutva, and Indian culture is described as Hindu culture.

It is in this context that a major debate has started about the role of 'civil' or 'civic society' in India: the more or less organized institutions between the individual and the state. Private business can be regarded as part of this 'in-between' level, as well as trade unions and professional organizations. Religious organizations are another part, in its Hindu, Muslim, Sikh, Buddhist, Christian, Jain and other manifestations. Political parties or 'movements' are a third element. Community-based organizations (or CBOs in development jargon) are a fourth element. As partly overlapping and partly parallel agents there are the NGOs, the non-governmental organizations, the non-profit, or 'third sector'. In India they are often called 'voluntary' organizations with either a charitable or development purpose. The latter are more aptly called 'Non-Governmental Development Organizations' or NGDOs.

There have been many attempts to define and categorize NGOs, NGDOs or CBOs (e.g. Shah, 1991). However, organizations legitimizing themselves as NGOs or NGDOs are very diverse. Some have links with the state bureaucracy which are so close that it is often doubtful if they can be regarded as non-governmental at all; others are offshoots of political parties, or 'movements', sometimes closely related to a particular politician; still others have a lot in common with (emergent) businesses, small or even rather big; and many have close links with religious organizations. In south India the activity of various Christian NGOs is a case in point.

Indian NGOs in general are very Indian: with relatively limited sources of finance from abroad and with little involvement from foreigners. This does not mean that the NGO leaders are from the local area and from the same social background as their 'target group'. Many of India's charismatic NGO personalities originate from other states and certainly have a different socio-cultural background compared to the local poor for whom they claim to work. Indian NGOs are also very Indian in another sense: hierarchically structured, run by a few persons, with individualistic and charismatic leadership, often preventing the emergence of seconds in command (De Bruijne and Schenk, 1994: 270-1).

The large majority of Indian NGOs legitimize themselves as 'working for the poor', even if their leadership clearly comes from rich, well-educated and often high-caste or 'outsider' backgrounds. Many refer to a non-violent 'Gandhian' background, with its roots in Indian and partly also Islamic social reform movements of the early nineteenth century onwards (see Alliband, 1983; Sundaram, 1986; Shah, 1991). Given the rather 'social' contents of Indian state law after Independence, many NGOs have concentrated on the 'raising of consciousness' among (sections of) the poor, to increase the access of existing redistributive government programmes or to enforce existing laws. Often this has meant that they were confronted with vested interests. In a number of cases the ideology of 'empowering the poor' became (and in some cases still is) spiced with Marxist terminology and revolutionary actions, straddling the line between coercive pressure and violent action or even armed resistance.

NGOs have become an important force in Indian society, but one should not overestimate their importance; to suggest that NGOs could replace the state seems rather illusory. In India it is most unlikely that the state will be forced into a 'night watch state' only and that its bureaucracy will be easily forced out of positions of power and influence. Moreover, the scale of the NGO sector is rather modest and NGOs are certainly not in a position financially to 'take over major state responsibilities'.

After studying a number of NGOs with rural poverty alleviation programmes, Robinson (1991: 118) concludes that the 'NGOs play a catalytic role in terms of enabling poor communities to define their own development priorities . . . but their activities remain supplementary to those of government, and in some cases they duplicate existing services and programmes'. It is true, but we may add that in other cases problematic gaps are growing because government agencies withdraw from certain services (basic health care and education), the private sector takes over the upper strata (for customers who can afford to pay the real cost or who have insured themselves), and most NGOs and their donors are unwilling to extend their activities in these fields, but instead concentrate their efforts in more productive activities, in employment programmes or in the provision of basic social security. The individualized poverty-alleviation programmes at village level have further weakened the organizational capacity and willingness to take over the responsibility, e.g. basic educational, health and water facilities at the village level. The post-Independence architecture of social and infrastructural services provision is rapidly losing its foundation and NGOs are clearly not in a position to support its renovation.

During the 1980s a growing number of Indian NGOs began to include land improvement strategies in their programmes. Partly it had to do with the growing concern, in India as well as worldwide, for 'the environment' and partly because NGOs began to realize that one-sided attention towards education, awareness, health care and drinking water without paying attention to the productive circumstances of the rural poor would create major problems. Some NGOs which incorporated an 'ecological' approach did this as a sideline to a group approach (based on class, caste or gender). Many adopted an area approach, as a continuation of an 'integrated rural development', or 'integrated village' approach, using administrative areas as the 'organizing principle'. Some NGOs, however, adopted the 'watershed' as the 'organizing principle' and hence everybody living in that watershed area: rich and poor, men and women, SC, ST, BC as well as FC, large farmers and landless labourers are all regarded as the 'target group'. Between these two approaches (group *v.* area) there are harsh controversies, as we experienced while interviewing NGO representatives. NGOs following a group approach are often regarded as 'Marxist' by their opponents; those following an area approach as 'counter-revolutionary', or 'naive'. Therefore one has to emphasize that analysis of intention, policies and activities of a particular NGO is not valid for the whole NGO 'community' as such.

The land improvement approach of NGOs often succeeded in attracting funds from the Indian central or state governments. This, more or less, started with the Seventh Plan (1985-90) when the government decided to experiment with implementation by NGOs; especially in the governmental Wasteland Development and Forestry programmes. For India as a whole during this period out of Rs 20,000 million for the Wasteland Development Programme 20 per cent or Rs 4,000 million was budgeted to be channelled to various NGOs (interview with DRDA Anantapur, July 1993). To reach the goal of complete eradication of poverty, a radical break with the 'top-down' government style was seen as necessary, or as Prasad (1992: 27) formulates it, there is an 'increasing need to evolve appropriate institutions which can be of use in (a) 'bottom-up' approach. It is in this context voluntary organizations (NGOs) are looked upon as possible institutional alternatives which could supplement, if not replace, the present rural administration bureaucracy.

The central government and the respective state governments have been most instrumental in directing policies to accelerate development and alleviate poverty. What are the exact roles of NGOs in this respect? Which roles can be foreseen in the future? These questions are opportune since

NGOs are definitely affected by the current strategic policy changes. The Indian government may design new development policies in which NGOs are mere executive agencies or may rely on NGOs as social 'safety nets' for India's many poor citizens, who will be adversely affected by macro-economic adjustment measures (De Bruijne and Schenk, 1994).

However, at present NGOs—leaving aside the effectiveness of their work—do not cover the whole of poor India at all. In that way also rural development in villages in which neither state organized programmes, nor NGO activities are present, needs attention and analysis.

1.2 Focus of the Study

This study focuses on the ways in which improvements of dryland agriculture in semi-arid areas of India are realized. It aims to analyse the process and the consequences of adoption by farmers of improved technologies for dryland agriculture in semi-arid India. Improved land, water and crop management are the main elements of these improved technologies, which envisage the increase of yields, particularly those of small farmers. The adoption of new technologies at the farm level, and the factors explaining the use of those technologies form a major component of the study.

Innovations not only emanate from within, but are also introduced from outside the village communities. This statement leads us to the other main theme of the study, an assessment of the role of governmental and non-governmental organizations (NGOs) in introducing the improved technologies. The findings of earlier concepts and research results suggest, as the main hypothesis, that governmental organizations tend to reach smaller groups of relatively better-off farmers through a politicized top-down approach introducing the improved technology as a package. NGOs propagate an incremental approach (one improvement at a time), work in a more bottom-up way and reach more and particularly the smaller farmers (Put and Van Dijk, 1989). So, the importance of government and non-government channels at the village level to diffuse these technologies will be examined by analysing the functioning of these 'channels' at village level. A comparison of those two actors, and with a situation where governmental organizations and NGOs are not active, will be made. The study on these themes has taken place in four villages in two districts in Andhra Pradesh, i.e. Ranga Reddy and Mahabubnagar (respectively 2.5 and 3.1 million inhabitants in 1991) (see Map 3.1).

1.2.1 RESEARCH QUESTIONS AND HYPOTHESES

The above formulation of the problem leads to the following main research question:

What effects do government and NGO projects in Andhra Pradesh have on the adoption of new technologies for dryland agriculture by farmers in semi-arid areas, and what causes the different effects?

But this question raises further questions:

- What are the major problems of farmers in Andhra Pradesh with regard to dryland agriculture?
- What is the policy of government agencies and of NGOs towards dryland agriculture in Andhra Pradesh?
- What do the innovations propagated by various government agencies and NGOs consist of?
- What differences exist between the introduction of innovations by government institutions and by an NGO?
- Are innovations also used in a situation where neither the government nor NGOs are, or have been active?
- What factors determine the acceptance or rejections of innovations by farmers — in particular: to what extent do caste and landownership patterns lead to differences in the adoption of new technologies?

Hypotheses to be tested are:

- Compared to government support for irrigated agriculture, less attention is given to dryland agriculture in semi-arid areas. Consequently, information, extension, inputs and special credit schemes are hardly available in these areas.
- In a situation without projects (the 'without case') this policy results in low adoption rates of new technologies. Only larger, wealthier, farmers belonging to higher castes use these technologies, that too in a selective way. They accept only those measures combining little risk with a high probability of larger yields, i.e. improved seeds and fertilizers.
- In government projects there exists a big gap between priorities of the institutions and those of the target group. The diffusion of new technologies progresses faster in government projects than in the 'without

case' because extension, capital and inputs are available. However, the innovations are imposed upon farmers and not really accepted by them (top-down approach). Consequently, farmers are only willing to use the innovations provided inputs are made available to them readily and at acceptable prices and as long as the projects continue. After termination of these projects farmers will partly revert to their indigenous practices.

In NGO projects the gap between priorities of the NGO and the target group is smaller. The projects take the specific circumstances, wishes and needs of the target group into account. As participation of the target group is encouraged (bottom-up approach) and as projects stimulate their awareness about their situation and show possibilities for change, adoption rates will be much higher than in government projects, and particularly so for smaller farmers and for farmers of lower castes.

1.2.2 THE STUDY VILLAGES

Based on the very positive outcomes of most evaluation reports and other documents (e.g. Kumar et al., 1986; Beets et al., 1988) AWARE was asked to represent the NGO approach. These documents claim that AWARE's social and economic programmes have a significant positive impact on the material and general well-being of the target group and praise the participation of the target group from the planning to the implementation phase of projects (the bottom-up approach), the organization of the target group in societies at village and supra-village level, the sustainability of AWARE's activities once it has withdrawn from an area, the dedication of its staff and the presence of a charismatic leader who manages to have a good rapport with the ST and SC community and is able to make the government accessible for these target groups. These claims are validated by a more recent evaluation study (Wils et al., 1994). The impression created by these reports is that, comparatively speaking, AWARE's activities show more positive results than similar endeavours undertaken by the government. AWARE scores high on criteria which are said to explain to a large degree the success of NGOs over government agencies [see, e.g. Robinson, 1991; Farrington et al., 1993(a); Farrington et al., 1993(b)]: proximity to the target groups; possessing location-specific knowledge about their social and economic position and the agro-ecological conditions of their environment; being responsive to their customs, needs and priorities; being more efficient in resource use; promoting the participation of the target groups in project design, implementation and monitoring; being able to develop village and supra-village organizations expressing the interests of the target groups; having a better and more highly motivated staff; having greater flexibility and being innovative. AWARE's scale of activities overcomes one of the comparative disadvantages of NGOs: their often limited area of operations. On the basis of the very positive outcomes of the various evaluation studies, the Dutch organization Novib, one of AWARE's major funding agencies, considered AWARE as an example for other NGOs in India and the rest of the world.

The Maheswaram project was selected because it was considered to be the opposite of everything an NGO stands for, that is, it does not match the criteria like proximity to the target group, described above. Being a World Bank-funded government project it has a top-down approach, with local participation limited to the implementation phase only; its goals are not the improvement of specific target groups but all the people in the area; it concentrates on technologies and not on the human aspects of village life; it is concerned with reaching its targets as described in the Plans and not so much on sustainability and it works with less motivated, less dedicated staff, especially at field level.

Keeping in mind that the Maheswaram project was located not far from Hyderabad, that AWARE had implemented several projects in the Amangal *mandal* and that the agro-ecological characteristics of the study villages had to be similar, we decided to opt for a study in the southern part of Ranga Reddy and the northern part of Mahabubnagar districts. The outcome on variables like average yearly rainfall, importance of irrigation, cropping pattern, transport links with Hyderabad and socio-economic situation is largely the same throughout this region. Within this region, three villages were to be selected for in-depth analysis: one village without any direct project intervention in dryland agriculture or soil and water conservation, one located in AWARE's project area and one in the Maheswaram project area. Before selecting the study villages representatives from AWARE and the Maheswaram project were asked to furnish a list of possible study villages. They were encouraged to include villages where project intervention had been most successful. In order to evaluate the success of project interventions many intervening variables had to be kept more or less constant and were not allowed to differ much. Hence, the villages had to meet the following criteria:

- They should be located at about the same distance from Hyderabad and the *mandal* headquarters and should not be accessible by bitumen road;

- Alfisol (red soils) should dominate;
- The irrigated area (wetland) should be roughly the same in all villages and should be less than 25 per cent of the net cropped area;
- The cropping pattern of the villages should be similar and food crops should dominate;
- The level of economic and social development and the infrastructure of the villages should be almost identical;
- The caste/class structure should be similar between the villages and should reflect the social structure of the state, i.e. FCs, BCs, SCs, and STs should be living in the villages;
- The size of the villages should be more or less similar and should not exceed 500 households or 3,000 inhabitants.

On this basis three villages were selected: Ekvaipally, the AWARE village, Mansanpally, the Maheswaram village, and Ghatpally, the control village. The first two villages had been short-listed by the projects and were accepted by the chairman of AWARE and the team leader of the Maheswaram project as good examples of their respective intervention strategy. The fieldwork started in November 1991. After almost one year, after having communicated the first provisional results, staff in AWARE and the Maheswaram project were not very satisfied with the outcomes. AWARE staff stated that they would protest against the inclusion of negative comments in the report and subtly tried to influence our research staff. This was in vain, however. The staff in the Maheswaram project indicated that they were not happy with the selection of Mansanpally. They gave two reasons: the village territory was only partially situated in the project area and, since it was one of the first villages where project intervention had taken place, the number of activities undertaken was limited and had been implemented half-heartedly as well. Although Mansanpally was short-listed by the project staff and was included in the project evaluations carried out on behalf of the World Bank, we were asked to select another village. In December 1992, while continuing in Mansanpally, the fieldwork started in a second village located in the Maheswaram project area: Debbadaguda.

1.3 Fieldwork Methodology

The fieldwork was carried out by a joint research team consisting of one male researcher of Indian origin selected by the Department of Geography, University of Madras, one female researcher from Hyderabad selected by AWARE and one male researcher from the Department of Human

Geography, University of Amsterdam. They were assisted by a team of Indian interviewers, all having a university degree and of rural background. To circumvent the problem of retrieving unreliable information, the interviewers lived in the study villages during the research. As it proved impossible to select female Indian candidates who were willing to live in the villages for long periods of time, the intention to get one or two female interviewers to participate, did not materialize. Hence, the major part of the fieldwork was undertaken by five male interviewers. Most of them lived between two and three years in the village. Each of them had an 'own' village, where they stayed overnight. To improve the team spirit, the interviewers met each other on various occasions in the villages and in AWARE's Human Resource Development Centre, a large farm and training centre located some 25 km south of Hyderabad. These meetings stimulated the exchange of experience and contributed to a better understanding of the agricultural practices in and the differences and similarities between the villages.

On the basis of the literature (e.g. Walker and Ryan, 1990: 28) and various discussions with informants, it was decided to select our respondents on the basis of community, dividing the inhabitants into FCs, BCs, SCs and STs. In order to get a more reliable picture on project interventions and on the use of innovations it was considered best not to select a representative sample but to select a disproportionate sample of some 25 households from every community in every village. Being interested in the use of innovations in dryland agriculture, respondents had to be farming households, meaning that villagers who did not own and cultivate any land were not eligible for selection. In the event that the number of landowning households belonging to a certain community was less than 25, all households were selected and the number of respondents belonging to the other communities was increased. Unfortunately, since both AWARE and the Maheswaram project were unable to supply details about their beneficiaries, including their names, we were not able to include them in the sample purposely. Our way of sampling means that FC and ST households are overrepresented, to the detriment of BCs and SCs. Since this over-representation relates to all villages, we expect that it will not affect the outcome of our comparison. Since adoption theory (e.g. Rogers, 1995) predicts that FC households will adopt innovations quicker than other households, over-representation of this category of farmers will probably not lead to under reporting of the effects of project intervention. Table 1.3 shows the result of our sampling procedure. The sample

TABLE 1.3. NUMBER OF RESPONDENTS (IN ABSOLUTE FIGURES AND RELATED TO OUR OWN 1991 CENSUS), BY COMMUNITY, BY VILLAGE

Community	Debbadaguda		Mansanpally		Ekvaipally		Ghatpally		Total	
	Abs.	%	Abs.	%	Abs.	%	Abs.	%	Abs.	%
FC	25	Na	23	77	26	67	26	74	100	Na
BC	25	Na	26	24	31	12	31	35	113	Na
SC	25	Na	24	72	29	66	29	43	107	Na
ST	25	Na	25	36	14	45	8	80	72	Na
Total N	100	Na	98	29	100	27	94	45	392	Na

Note: A total listing of all households per community was not available for Debbadaguda.
Source: Own survey.

includes 100 households in Debbadaguda, 98 households in Mansanpally, 100 households in Ekvaipally and 94 households in Ghatpally. Because two households in Ghatpally and one household in Mansanpally were not available during the entire study period, these respondents were removed from the sample, which totals 392 households.

Information on the respondents was collected in various ways. Three rounds of scheduled interviews were conducted, using nine questionnaires, covering topics like household composition, socio-economic status, asset position, financial position, sources of income, sources of information, cropping pattern and experiences with project interventions of AWARE and the Maheswaram project. These three rounds were undertaken in 1991/2, 1992/3 and 1993/4. In addition to the questionnaires, a so-called plotwise schedule was designed to elicit facts related to all activities carried out on each plot the farmer owned and cultivated. It gives detailed information about the agricultural practices on the plot, the inputs used and the yields. These schedules were used in *kharif* (south-west monsoon) and *rabi* (north-east monsoon) 1992/3 and in *kharif* 1993/4 for all kinds of crops on both dryland and wetland. During the study period a considerable number of the plots were visited to check the information and to add new data. All questionnaires and the plotwise schedules were pre-tested in the field in places other than the study villages.

In addition to the questionnaires and the plotwise schedules, open interviews were held with a selected group of 88 farmers, some 20 in every village (see Table 1.4). These farmers were selected on the basics of caste, landownership and the agricultural technology used. These

TABLE 1.4. ABSOLUTE NUMBER OF RESPONDENTS FOR THE OPEN INTERVIEWS, BY COMMUNITY, BY VILLAGE

Community	Debbadaguda	Mansanpally	Ekvaipally	Ghatpally	Total
	Abs.	Abs.	Abs.	Abs.	Abs.
FC	5	4	6	6	21
BC	5	5	6	8	24
SC	6	5	7	6	24
ST	5	5	6	3	19
Total N	21	19	25	23	88

Source: Own survey.

unstructured interviews shed more light on the reasons why a farmer decided to opt for a certain technology and not for another. Topics discussed during these open interviews were the various technologies (indigenous and improved) in dryland agriculture and project intervention by AWARE and the Maheswaram project. These interviews were conducted in February-March 1994 with the assistance of a female geographer of Dutch origin.

Information received from farmers was checked and supplemented by information supplied by various other informants, local and non-local people like elected members and presidents of the *panchayat*, members and presidents of the Village Associations set up by AWARE, influential farmers not represented in the sample, shopkeepers, traders, bankers, extension agents and staff in AWARE and the Maheswaram project. Sensitive information on the activities of both AWARE and the Maheswaram project, was always checked, and more often than not double-checked, on its reliability.

NOTE

1. The Third programme started in 1964/5 with less intensity and less staff. At the moment, about 10 per cent of the total cultivated area is included in the IADP and IAAP.

2. An Introduction to Andhra Pradesh

D. Adi Lakshmi, A.J. Dietz and M. Put

2.1 Andhra Pradesh: Basic Facts

Andhra Pradesh is one of the non-Hindi speaking states in the south of India. It was formed in 1948 on a purely linguistic basis, when all districts having a Telugu-speaking population were merged. Formerly Coastal Andhra and Rayalaseema were part of the so-called Madras Presidency under British rule and Telangana belonged to the semi-autonomous Hyderabad State, that was dissolved in 1948. Andhra Pradesh is the fifth largest state in India, both in terms of area and population. In 1991 some 66 million people inhabited a geographical area of 274,400 sq. km. The population growth in the period 1971-81 and 1981-91 was 23.1 per cent and 23.8 per cent respectively. With a population density of 241 persons per sq. km in 1991, it was less densely populated than India as a whole, having an average of 267. According to the 1991 census 74 per cent of the population lives in rural areas, only 3.6 per cent less compared to 1981. Hyderabad, Andhra Pradesh's capital, had 4.3 million inhabitants in 1991. It is the state's main city in which nearly one quarter of the urban population resides.

Andhra Pradesh can be divided into three regions. The first, coastal Andhra, is the most fertile and the most irrigated zone of Andhra Pradesh, covering about one-third of the state's area. The second region is called Rayalaseema. It is a dry tract in the south, most vulnerable to drought and famine, covering about one-fourth of the state. The area is a plateau with elevations between 150 and 900 m in general, except in most of Chittoor district where elevations range between 600 and 1,200 m. The main soil types in Rayalaseema are red sandy and mixed red and black soils. Rainfall in the region is very erratic and unevenly distributed. Anantapur district receives the lowest average rainfall of this zone and the state: 540 mm/year. Rainfall in most of the districts averages below 750 mm/year.

The third region, the area of our study, is called Telangana. It is also a dry tract, but less arid than Rayalaseema. It is the most backward region of the state, covering about two-fifth of the state's area. Like Rayalaseema, the Telangana region is also an elevated plateau with elevations ranging between 150 m and 650 m above sea-level. Most soils in Telangana are red sandy or red sandy loam, and medium to deep black. Though average annual rainfall in most districts in the Telangana area are above the averages of Rayalaseema, most districts come under the low or medium rainfall regime receiving averages between 700 mm and 900 mm/year. The undulating and rugged surfaces, the poor and unreliable rainfall, the very unevenly distributed forest resources, soil erosion, inadequate transport facilities and the monopoly of Hyderabad metropolitan area in modern developments have rendered major parts of Rayalaseema and Telangana into problem regions. The region is characterized by poor agricultural and economic development, resulting in low standards of living, illiteracy and poor health. All this has resulted in a number of internal troubles, the most noteworthy being the demand for a separate Telangana state and the radical actions of the Naxalites.

Economically, Andhra Pradesh is a backward state, being one of the poorer states in India. The estimated per capita income at current prices in 1990/1 was Rs 4,507 which was below the All-India figure of Rs 4,974 [GoAP, 1993: 557; Statistical Outline of India (1992/3), 1992: 12]. While 29.9 per cent of the total Indian population lived below the poverty line in 1987/8, the figure for Andhra Pradesh was 31.7 [Statistical Outline of India (1992/3), 1992: 28]. In Punjab, the least poor state it was only 7.2 per cent, and in Orissa, the poorest state 44.7 per cent (Economic Survey, 1994/5: 158). In fact, almost 40 per cent of the rural population in the state can be classified as poor or very poor (A social and economic atlas of India, 1987: 218). In fact, except for south Coastal Andhra Pradesh, more than 45 per cent of the rural population was poor to very poor in 1987/8 (Parthasarathy and Durgaparameswari, 1996: 16). Malgavkar (1996) reports that the state ranks low on many non-economic variables as well.

According to estimates made by the Union Ministry of Food and Agriculture, about three-fifths of the area affected by scarcity conditions in India's southern region are concentrated in Andhra Pradesh alone. The affected areas account for 57 per cent of the total area and 41 per cent of the population of the state. The Rayalaseema region, and to a lesser extent parts of Telangana, form the core of the chronic famine area of the state, on account of rocky and unyielding soil and scanty rainfall. Relatively speaking, the highest number of rural poor is found in Telangana (Parthasarathy and Durgaparameswari, 1996: 19). In the relative index provided by the Centre for Monitoring the Indian Economy, Mahabubnagar district ranks lowest within the state (CMIE, 1993: 12).

Although the share of the service sector in the state's Net Domestic Product has surpassed the share of agriculture, in essence Andhra Pradesh has remained an agricultural state. While the share of agriculture in NDP declined from 55 per cent in 1970/1 to 37 per cent in 1990/1, the share of the tertiary sector increased to 42 per cent (GoAP, 1993: 261). On the other hand, 65 per cent of the working population was dependent on agriculture in that year (GoAP, 1993: 18-19). During the different Five Year Plans, percentage-wise, the expenditures on 'total agriculture', has been much more than for India as a whole. In most Plans, the percentage was more than 35 per cent. However, a large part, often more than two thirds, was spent on irrigation.

In 1990/1, the net sown area in Andhra Pradesh was 11.0 million ha or 40 per cent of the total geographical area of 27.4 million ha. Some 16 per cent of the net sown area was irrigated. An additional 2.3 million ha or 8 per cent of the total geographical area, was sown more than once, making the gross cropped area 13.2 ha (GoAP, 1993: 80, 148).

Food crops dominate the agricultural scene in Andhra Pradesh. In 1989/90 the area under total cereals and millets was 6.5 million ha (49 per cent of the gross cropped area) of which paddy, with an area of 4.2 million ha (32 per cent of the gross cropped area) and *jowar* (sorghum), with an area of 1.3 million ha (10 per cent), were the most important crops. Other cereals like *bajra* (pearl millet), small millets, maize, *ragi* (finger millet) and wheat covered less than 5 per cent of the gross cropped area. An additional 1.6 million ha was cultivated with pulses, of which greengram, horsegram, redgram and blackgram were the most important crops. Of the oilseeds, groundnuts and castor were the most significant. Another important cash crop was cotton. Average yields of several crops in 1989/90 were: rice 2,367 kg/ha, maize 2,244 kg/ha, *ragi* 1,212 kg/ha, wheat 794 kg/ha, *bajra* 702 kg/ha, small millets 730 kg/ha, *jowar* 688 kg/ha, and castor 250 kg/ha, showing the low yielding capacity of crops like *jowar*, *bajra* and castor (GoAP, 1993: 88-90, 114).

With an agriculturally rich coastal area, Andhra Pradesh is one of the most important food grain producing states in India. In 1990/1 the state produced 12.6 million tonnes (7 per cent of the all-India food grain production) and ranked fourth among the 25 states and union territories (Observer Statistical Handbook, 1992: 80-1). The Green Revolution has boosted yields in coastal Andhra, where the production of paddy is dominant, in a spectacular way. However, while the average yield of food grains in this region was 1,859 kg/ha in 1989/90, the averages for Rayalaseema and Telangana were only 1,353 kg/ha and 1,371 kg/ha respectively, which shows their relative backwardness in agriculture (GoAP, 1993: 92-3).

Andhra Pradesh is not very well-known for its industries. In fact it lags behind compared to many other Indian states. The share of the secondary sector in the NDP was 17 per cent in 1990/1 (GoAP, 1993: 61). Within the state a clear regional imbalance exists: Hyderabad and surroundings is the motor of industry; other centres are few. In coastal Andhra, agro-industries around Vishakhapatnam dominate the scene. In Telangana, mining of coal, limestone, granites and iron ore are important. Rayalaseema is not well-endowed with mineral resources, and manufacturing is also rather insignificant.

2.2 Agricultural Change in Telangana

The research villages of this study are part of Ranga Reddy and Mahabubnagar Districts, the two southern most districts in the Telangana area and the most drought prone. Telangana is the part of present Andhra Pradesh state, that was part of the Muslim-ruled Nizam state of Hyderabad before 1950 (together with the eastern parts of the present states of Maharashtra and Karnataka). Not much is known about the agricultural history of Telangana during the Nizam's rule, except from an important book written by Ramakrishna Reddy, *Economic History of Hyderabad State*. Before giving some data about the present era, we will summarize his data about the changes in land use during the first part of this century (Ramakrishna Reddy, 1987: 287-93). Table 2.1 gives the details.

Between 1913 and 1945, the area available for cultivation in the whole

TABLE 2.1. CULTIVATION DATA FOR TELANGANA, 1913-45 (*in 1,000 ha*)

	1913	1922	1926/7	1944/5
Total government area	5,664	6,104	6,158	10,749
Available for cultivation (as assessed)	4,095	4,349	4,360	5,381
Used for cultivation (or: 'net area sown')	3,250	3,545	3,584 or 3,646	3,707
Dryland cultivation	2,877	3,147	3,249	3,288
Irrigated cultivation	373	397	397	419

Source: Ramakrishna Reddy (1987: 287-93).

of Telangana increased by 31 per cent to more than 5 million ha. The area actually used for cultivation increased from 3.3 million ha to 3.7 million ha (14 per cent increase), but relatively speaking, it decreased from 79 per cent of the 'available land for cultivation' to 69 per cent. Most of the agricultural land area was used for dryland cultivation (89 per cent in 1913 and still the same percentage in 1945). The irrigated area was, and remained, small in this period (11 per cent).

For the year 1944/5 the figures are a bit more detailed: of the total government-assessed land (1,07,49,000 ha in Telangana), 5,381,000 ha or 50 per cent was regarded as 'available for cultivation', of which 1,674,000 ha or 31 per cent was 'currently under fallow' and 37,07,000 ha were actually used, 89 per cent for dryland agriculture and 11 per cent under some form of irrigation. Of the non-cultivable land 22,11,000 ha were (declared?) forests, 23,50,000 ha were regarded as wastelands not available for cultivation and 8,07,000 ha were regarded as 'cultivable waste, other than current fallows'. In Table 2.2 the available figures about the various crops being cultivated are compared with the crop data of more recent years.

The figures presented in Table 2.2 make it possible to deduce the following:

- The cultivated area increased quite a lot in the early 1930s (probably because farmers wanted to compensate for the loss of income per ha caused by the Depression, by cultivating more land), but the cultivated area contracted in the late 1930s and early 1940s. In the early 1970s the cultivated area had increased again to a level of one-third above the level of the 1920s. However, in the 1970s and 1980s the cultivated land contracted again and nowadays it is not much above the level of the early 1930s.
- Despite the contraction of 'net sown area' in the last thirty years, the 'gross sown area' was rather stable, indicating an increase in double cropping; part of this can be explained by the strong increase in irrigation.
- During this century the irrigated area in Telangana increased from approximately 4,00,000 ha to 14,00,000 ha. In the last thirty years, the absolute and relative importance of irrigation using wells has increased a lot. In the last ten years, the area irrigated by way of using tube-wells increased in particular.
- The area used for dryland agriculture first increased (from 2.9 million ha in 1913, via 3.3 million ha in the mid-1940s to 4.1 million ha in the

early 1970s) but then decreased (to 3.5 million ha in the early 1980s to 2.9 million ha in the early 1990s). The absolute importance of dryland agriculture in terms of acreage is the same now as it was in the beginning of this century.

- The growth of the irrigated area went parallel with the growth of the paddy area: from 2,35,000 ha in 1913 steadily growing to 13,91,000 ha in 1989-92. The productivity shows fluctuations: the average productivity of rice was high in the 1925-30 period (1,115 kg/ha), went down drastically in 1930-5 to 766 kg/ha, and up again in the 1935-45 period (924 kg/ha). Data for later periods suggest that the average yields were between 1,000 and 1,250 kg/ha until the mid-1970s but increased considerably afterwards, to a level of around 1,800 kg/ha in the early 1990s.
- The dryland cereals show a different pattern. *Jowar* first steadily expands (from 8,68,000 ha in 1913 to 15,72,000 ha in the early 1970s) but contracts later, to a level close to the early part of the century (8,76,000 ha in 1989-92). *Bajra* contracts between 1970 and 1990 from 3,27,000 ha to 83,000 ha and seems to follow the same 'fate of extinction' as *ragi* which went down from 70,000 ha to 38,000 ha and the so-called 'small millets'(with their Indian names *Korra* and *Samai*), which virtually disappeared (from 138,000 ha in 1970 to only 14,000 ha in 1990). Probably also the crops traditionally grown as animal feed became much less important. On the other hand maize first contracted and later became more important again, reaching a level of approximately 3,00,000 ha in the 1980s. The productivity of *jowar* shows the same pattern as the productivity of rice, although at a kg/ha level of around 40 per cent of rice. The pulses and grams show a pattern comparable to *jowar*: first an increase and after 1970 a decrease in acreage, although the figures are rather confusing.
- The industrial crops (cotton, groundnuts, tobacco and sugarcane) show a considerable increase during this century: from a level below 2,00,000 ha (mainly cotton) in the 1910s to a level close to 9,00,000 ha in the 1990s (with groundnuts leading, 480,000 ha, followed by cotton, of which the acreage doubled to 350,000 ha and with sugarcane in a third position, 41,000 ha, as the 'irrigated cash crop'). The productivity figures of cotton and groundnuts are very much the opposite of each other during the 1925-45 period. Cotton goes from high (675 kg/ha in 1925-30) to low (359 kg/ha in 1930-45) while groundnuts go from low (only 327 kg/ha in 1925-30) to high (818 kg/ha on average in the 1930-45 period).

TABLE 2.2. AREA CULTIVATED WITH VARIOUS CROPS IN TELANGANA, 1925-45 AND 1969-92 (*averages in 1,000 ha*)

	1925-30	1930-5	1935-40	1940-5	1969-72	1979-82	1989-92
NSA	3,665	4,387	4,076	3,935	4,817	4,507	4,406
GSA					5,151	5,007	5,028
NIA	(397)			(419)	766	979	1,481
GIA					1,031	1,313	1,927
AIW					298	509	862
AITW					0	11	95
Rice	235	368	359	405	894	1,088	1,391
Maize	256	214	153	171	222	299	373
Jowar	868	1,023	1,100	1,213	1,572	1,500	876
Bajra	327	348	346	170	226	190	83
Ragi					70	55	38
Small Millets					138	73	14
Pulses and grams	567	713	664	563	800	785	605
Castor	225	290	274	301			
Cotton	160	190	172	175	103	146	350
Groundnuts	11	63	225	386	404	303	480
Tobacco	11	10	12	11	30	25	23
Sugarcane	0	1	6	8	27	62	41
Food crops	450	428	225	114			
Fruits and							
vegetables/					31	20	
chillies/	72	119	125	84	74	82	
turmeric/					8	15	109
coriander					34	17	
Sesame	166	179	171	180			

Notes: NSA = Net Sown Area; GSA = Gross Sown Area; NIA = Net Irrigated Area; GIA = Gross Irrigated Area; AIW = Area Irrigated under Wells; AITW = Area Irrigated under Tubewells.

Pulses and Grams: this combines the categories 'grams' and 'other cereals and pulses' for 1925-45, 'total pulses' in 1969-82, and 'red and green grams' for 1989-92.

Fruits and Vegetables in 1925-45 includes root crops and maybe chillies, turmeric and coriander, which are presented as separate crops in 1969-82.

Source: 1925-45: Ramakrishna Reddy (1987: 306-24).
1969-91: Agricultural statistics Andhra Pradesh state, primary data.

- About two important traditional crops, castor and sesamum, we do not have current data. Together they still occupied 5,00,000 ha of agricultural land in the mid-1940s.

2.3 Government Dryland Development Activities in Andhra Pradesh

Using the information about Andhra Pradesh in the various government development plans, it is possible to get an idea about shifts in government policies with regard to agriculture and dryland agriculture in particular. Since land reform has not been very important, only 1.4 per cent of the total owned area has been distributed so far (Parthasarathy and Durgaparameswari, 1996: 7), this subject will not be included in our analysis. We will follow the chronological order of the various plans.

The major objectives of the First Plan regarding the area which was to become Andhra Pradesh in the field of agriculture were to correct the acute food shortage and inflation of the late 1940s. The immediate increase in agricultural production became imminent to end inflation, stabilize the economy and prepare the way for a higher rate of development. The objective of increased production was sought to be realized both by intensive and extensive methods of cultivation. The state plan for agricultural development consisted of 34 schemes receiving 9.4 per cent of the outlay.

Soil conservation initiatives were closely linked to improvement of land use in general. Measures for controlling erosion and restoring the productivity of eroded lands can be divided into four classes:

(1) Regulation of land use;
(2) Afforestation and preservation of forests by scientific forest management;
(3) Improvement of land use practices on farm lands like strip cropping, contour bunding, etc.;
(4) Engineering measures, like construction of bunds, terraces, check dams, channels for drainage of surplus water, gully plugging, etc.

The target of Rs 65,900 million fixed for the plan period, had been exceeded by the end of the plan. An amount of Rs 48,80,000 had been spent on this scheme. Overall expenditures under agricultural and rural development also exceeded the plan outlay. The total cropped area of the state increased from 10.2 million ha in 1950/1 to 11.9 ha in 1955/6. The production of cereals had gone up from 3.8 million tonnes in 1950/1 to 5.1 million tonnes in 1955/6.

The Second Plan for Andhra Pradesh, reflected by and large, the priorities adopted in the National Plan except for the outlay on agriculture and irrigation, which had gone up very much compared to the First Plan (to 52.3 per cent of the total outlay). The agricultural programmes intended to provide adequate food to feed the increasing population and to supply the raw materials needed for a growing industrial economy and finally to make available larger exportable surpluses of agricultural commodities. About Rs 5,85,00,000 was originally provided to promote agriculture production in the state. The revised provision was Rs 4,45,00,000. The programmes implemented during this plan were:

(1) Subsidized seed multiplication and distribution of improved seeds;
(2) A rainfed paddy scheme;
(3) Research into castor seeds;
(4) Establishment of a Jowar Research Station.

During this period there was a growing interest in the areas that were not yet used for agricultural production. An area of 1.7 million ha was classified as 'cultivable waste' which was as much as 6.2 per cent of the total area. A sum of Rs 39,88,000 had been set aside for land development.

The area under food grains increased by 2.5 per cent during this plan period. The general index of agricultural production (base 1949-50 = 100) showed an improvement from 126.9 in 1955/6 to 131.3 in 1959/60. However, due to a failure of monsoon rains it declined to 122.5 in 1960/1.

The Third Five Year Plan for Andhra Pradesh had as its main objective self-sufficiency in food grains and an increase in agricultural production to meet the requirements of industry and export. The original plan provision was Rs 10,14,07,000 and the revised figure Rs 9,71,65,000, including the provision for the Agricultural University. The actual expenditure incurred was Rs 10,02,47,000. Various schemes were implemented, during this plan period. Among them the important ones were the distribution of improved seed on subsidy, the intensive agriculture development programme, the intensive agriculture area programme, a pilot scheme for intensive dry farming practices, and a pilot scheme for subsidized distribution of fertilizers for rainfed crops. It is during this Plan period, around the mid-1960s, that attention for dryland agriculture was growing in government circles, but it got much less attention than the 'Green Revolution'. Most attention for the drylands was for soil conservation, not so much for crop improvements.

The Agricultural Department of Andhra Pradesh tentatively presented a programme for soil conservation for the Fourth Plan, which aimed at covering an area of 2,02,000 ha and providing training to 2,300 persons at a cost of Rs 3,67,00,000. It also proposed to cover 4,047 ha of dry and eroded land under an afforestation programme.

It was realized that the execution of soil conservation programmes such as contour bunding would depend on large scale participation of people. Unless adequate incentives were provided, the necessary response from the farmer would not be forthcoming. Until then the entire cost of soil conservation, including the establishment charge of the department together with a 4.5 per cent interest rate, was recovered from the cultivators in 15 annual instalments. A subsidy of 25 per cent was given to the farmers which was adjusted towards the cost of establishment for the extension of soil conservation schemes. It appeared that the cultivators were not generally enthusiastic about taking up soil conservation works under the existing pattern of assistance. The programme evaluation unit also observed that the prevailing method of subsidy did not really help the farmer at all. It was therefore decided that, at least during the Fourth Plan, cultivators should be given greater financial assistance in these schemes. It was regarded as desirable that a 25 per cent subsidy should be provided to the cultivators and no cost of establishment was to be charged to them.

During the Fourth Plan intensive research was to be undertaken in order to increase crop production in the dry areas. An integrated approach was assumed to be needed for the execution of soil conservation works. Each watershed had to be taken as a separate unit and the catchment taken from the higher levels and not from the plains. It was stated that there was an urgent need to bring an integrated legislation for soil conservation throughout the state. In the early 1970s the Andhra and Telangana areas of the state were still governed by two different sets of legislation.

The main objectives for the agricultural sector during the Fifth Five Year Plan for Andhra Pradesh, during the late 1970s, were:

(1) To bring about an increase in production of important agricultural commodities such as food grains, pulses, oilseeds, sugar cane and cotton, for which demand was likely to grow substantially due to a policy of redistribution of incomes that was being adopted;
(2) Major emphasis would be on increasing crop production per ha;
(3) To reduce the impact of the vagaries of monsoon through improved dryland farming techniques and the development of drought resistant varieties, etc.;
(4) To increase irrigation facilities and provide better management in irrigated areas;

(5) Concentration on the small and marginal farmers.

Programmes that received increased attention in this period were the Integrated Dryland Agriculture Development Project, the Drought Prone Area Programme (DPAP) and the Small Farmers Development Agency. The Fifth Plan DPAP strategy revolved around a comprehensive and integrated rural development focus on agricultural and allied sectors. In 7 districts 52 *taluks* were to be covered, on which an amount of Rs 22.060 millions was spent with 50 per cent share from the state sector.

The objectives for the agriculture sector during the Sixth Plan in the early 1980s were to increase the agricultural production in the state. A revised provision of Rs 19,97,00,000 was made for the agricultural sector. There was a decline in food grain production due to unprecedented widespread drought conditions which prevailed in the state and also due to cyclones in one district during the year 1984/5.

Attempts were also made to increase the area under cultivation. Special efforts were made to bring current and other fallows under cultivation to the extent possible. The High Yielding Variety Programme and the Soil and Water Conservation Programme were speeded up, the latter as an integral part of schemes like the DPAP and SFDA.

Agriculture was given higher priority and higher allocations in the Seventh Plan during the late 1980s:

(1) To achieve a growth rate of 6.5 per cent in the state economy;
(2) Concentration on productivity increasing activities to ensure a geographical spread of agricultural development;
(3) The highest priority for dryland farming, horticulture and sericulture;
(4) Full utilization of economic and social infrastructure.

For the first time dryland farming, which used to be a minor part of the government's crop husbandry programme, had been given an exclusive outlay of Rs 400 millions. An innovative programme was contemplated in dryland farming by introducing a package of practices on particular plots, instead of distributing the subsidies and inputs individually and separately. This innovation came in addition to the land development approach in selected micro watersheds and the soil and water conservation programmes. Other proposed items were the further development of cultivable waste lands, popularization of improved agricultural implements and adequate use of fertilizers in dryland farming. It was also planned to strengthen the extension services.

The dryland development programme, the IRDP, the NREP and the RLEGP were continued, just like rainfed farming with World Bank assistance (see Chapter 4). Dryland development initiatives became ever more chaotic, with the National Watershed Development Programme for Rainfed Agriculture (NWDPRA), DPAP, and the Integrated State Watershed Development Programme (ISWDP) all getting considerable funds. The NWDPRA, DPAP and ISWDP will now be discussed briefly.

During the year 1990/1, the NWDPRA was implemented in Andhra Pradesh, covering 94 watersheds in all the districts of the state, except Krishna and Guntur, at a total cost of Rs 68,00,000. It was proposed to cover 32,000 hectares with a crop demonstration and land management programme, and 4,700 ha with dryland horticulture, social forestry, and pasture land development. It was also proposed to distribute 5,880 agricultural implements with 50 per cent subsidy to the farmers. Besides the above, it was also proposed to conduct training courses for farmers on dryland technologies. Farmers of the watersheds that were selected for land improvement measures would also be taken on field trips to agricultural research stations to study the performance of new cropping systems. For all these measures an average sum of Rs 66,00,000 was allotted to each watershed. This was a centrally sponsored scheme. The expenditure would be shared equally by state and Central government. The scheme covered small, and marginal scheduled caste, scheduled tribe and other farmers.

The DPAP wanted to mitigate the effects of drought, through the construction of permanent and semi-permanent works, besides providing employment for rural labourers during the agricultural off-season. During 1990/1 the programme was implemented in 74 watersheds of 8 DPAP Districts of Kurnool, Anantapur, Chittoor, Cuddapah, Prakasam, Mahabubnagar, Nalgonda and Ranga Reddy. The important elements of the programme were:

(1) Contour or graded bunding;
(2) Raising of khus nurseries;
(3) Vegetative bunding with khus;
(4) Stone terracing;
(5) Gully control structures;
(6) Water harvesting structures like check dams, rock fill dams, execution of farm ponds, etc.

In order to expand the dryland development programme to all the districts in the state, the ISWDP was initiated, covering 25 watersheds. The programme would be carried out under different budgetary headings like the land use programme, the soil conservation programme, the dryland farming programme, which were in the end integrated in the ISWDP. The integrated approach to the dryland development on a watershed basis was an investment-oriented and a time-consuming approach. To spread the benefits of dryland farming technology to large areas, some simple dryland techniques were popularized in about 1,500 villages, from 1988/9 onwards.

Notwithstanding the major increase in government funds for dryland development, many observers doubted the effectiveness of these funds since most programmes were very bureaucratic, top-down oriented, often technocratic and without much interest in the 'real life' circumstances of farmers-cum-labourers-cum-non-farm entrepreneurs. Many observers, criticizing the government approach in the state, looked at some of the examples provided by Non-Governmental Organizations (NGOs) for alternatives.

2.4 NGOs in Dryland Andhra Pradesh

In the early 1960s there were very few organizations like the present day NGOs working in Andhra Pradesh, especially in Telangana. There were few local branches of all India organizations in the Telangana area where feudalism and oppression reigned, and those few attracted a lot of suspicion from the upper classes (Rao and Madhavan, 1988: 65-6). According to Mr Ferrer of Rural Development Trust (RDT), an important NGO in Anantapur in the south of the state (personal communication, July 1993) around Independence the village teachers often were the 'one-person NGOs' in the villages and they played important development roles. However, according to him 'today the committed, development assisting and experimenting government teacher is something of the past'. In Mr Ferrer's words: 'almost all are urbanites, salary eaters and heavily unionized against any criticism'. In his view, NGOs have taken their place after about 1970.

A first attempt to get an idea about the 'voluntary sector' in Andhra Pradesh was made in 1988, by Dr Rao, working at the Administrative Staff College of India in Hyderabad and Mr Madhavan, chairman of the largest NGO in the state, AWARE (see Rao and Madhavan, 1988). Their general statements are an interesting combination of 'down to earth' observations and examples of wild reification. On the one hand we read (Rao and Madhavan, 1988: 65, 71, 72 and 79): 'The massive social disorganization . . . can hardly be corrected by isolated and minuscule efforts of voluntary organizations', 'the voluntary agencies are a footnote in Indian development', 'very, very few voluntary organizations in the state do have credibility', 'a very small number of voluntary organizations in Andhra Pradesh are able to retain dedicated workers and mass support on the basis of good work in the field' and 'unfortunately it is said that voluntary agencies became a business'. On the other hand the same study states (Rao and Madhavan, 1988: 54, 56, 58 and 61): 'voluntary organizations (and action) always proved to be the steering wheel—the helm of human communities, guiding the direction of development', 'voluntary action played a significant role in providing turning points in a variety of spheres—social reform, social welfare, education, health, religious reform, and in fact spurred a unique cultural renaissance and a value-based movement for political freedom', 'voluntary action was spurred by a visionary outlook of individuals and their missionary zeal to bring into the mainstream those who were untouched by the process of change, and handicapped by a variety of disadvantages', 'voluntary action is the instrumentality to promote popular participation', 'sky is the limit for voluntary action', 'voluntary action promotes local initiative, (is) more cost-effective, less prone (to) paper transactions and corruption . . . has the added advantage of planning from (the) bottom with due recognition to local needs condition(s)' and 'above all it promotes an ethos of self help and participation'.

Rao and Madhavan (1988: 66) found 26,806 voluntary organizations in Andhra Pradesh in the registers of the Registrar of Societies, but they add: 'many of them are simply paper organizations, youth clubs, *mahila mandalis* (women's clubs) only. Many of these societies which got registered never started functioning. Some met with instant death or infant death.' From the start of registration under the Foreign Contribution (Regulation) Act, in 1976, until 1987 government information showed that 1,369 'voluntary associations' in Andhra Pradesh had registered, 11 per cent of the India-wide figure (Rao and Madhavan, 1988: 141-2). Based on their own survey in 1987, Rao and Madhavan counted 356 of what they call 'development oriented organizations', 92 NGOs in Hyderabad itself, 158 in Rayalaseema (most in Cuddapah District) and 180 in Telangana (of which 18 were in Mahabubnagar and 18 in Ranga Reddy districts) (Rao and Madhavan, 1988: 73). Out of all those, they regarded between 50 and 100 NGOs as trustworthy, genuine NGOs (Rao and Madhavan, 1988: 81) and these included 9 NGOs which could be regarded as rather big in 1987 ['one organization is working in 2,000 villages . . . 3 organizations (are) working in 300 to 500 villages, five more in 100 to 300 villages'] and another 82 in more than 25 villages (Rao and Madhavan, 1988: 78). However, they also noted that the level of funding was increasing. In 1986/7 the officially registered inflow of foreign funds to support Andhra NGOs was Rs 360 million, about 10 per cent of the all-India figure.

To get an idea about the growth and current functions of NGOs in dryland Andhra Pradesh we make use of two recent directories about NGOs, one published by the National Institute of Rural Development in Hyderabad (Charyulu, 1992) and one on *Environmental NGOs in India*, published by the World Wide Fund for Nature, New Delhi, India (WWFN, 1992). These two directories together contain information about 123 active NGOs in the whole of Andhra Pradesh, which are registered, either under the Societies Registration Act of 1860 or under the Andhra Pradesh (Telangana Area) Public Societies Registration Act 1350 Fasli. We will concentrate the analysis on NGOs active in the dryland regions of Andhra Pradesh, which means: the districts in Telangana in the north-western part and Rayalaseema in the south-western part, so excluding the coastal districts. This restricts the analysis to 96 active NGOs. It is certainly not a complete overview, and since 1992 there will have been changes as well. However, the data does give an impression of the most important NGOs active in dryland Andhra Pradesh in the early 1990s.

The development of NGO activities in dryland Andhra can be described as four waves (resembling what happened in India as a whole). The first wave of NGO establishment started around 1966 and lasted until the beginning of the 1970s. Most of the new NGOs had a rural orientation and a Christian background. This is certainly true for the Rural Development Trust, which was established in Anantapur district by a Catholic priest in 1969, for the Village Reconstruction Organization, which was started by Jesuits in Guntur in coastal Andhra in 1969, but soon worked in the whole of Andhra Pradesh and beyond, and for the Methodist Church in Hyderabad who started a separate Water Resources Unit in 1968, later becoming one of the early 'professional NGOs' (under the name of WAMERT from 1977 onwards). Another professional NGO was started in Hyderabad in 1973 as a cooperative of unemployed university graduates, the Natural Resources Development Cooperative Society. There was also the start of a Gandhian NGO, Sevashram, in Warangal district in 1972 and of a farmers' pressure group, the Anantapur District Farmers Forum in 1966. It is interesting to note that none of the NGOs in dryland Andhra Pradesh ever used 'community development' as their mission.

The second wave started around 1975. It was probably not unrelated to the start of the Emergency period in India which lasted until 1977 and created a change of attitude towards the 'government'. There were much more 'left-wing' NGOs started during this period, oriented to the situation of the tribal population, the scheduled castes, the landless and the small and marginal farmers. NGOs which started with a tribal orientation were the Girijana Seema Welfare Association, in 1975, in Khammam district and the Action for Welfare and Awakening in Rural Environment, AWARE, also starting in 1975, in Mahabubnagar district. AWARE soon developed into the biggest NGO of the state and later also included the scheduled castes among its 'target groups'. Examples of NGOs directed towards the landless and marginal farmers were the 'Comprehensive Rural Operations Service Society, CROSS' (1975, Nalgonda and Warangal districts), the Weaker Community's Action for Development and Liberation WCADL (1975, Mahabubnagar district) and the Cuddapah District Yuvajana Harijana Christava Girijana Sarvodaya Sangham (1976). The cyclone calamity of 1977 along the coast resulted in the establishment of relief NGOs which later also covered some of the dryland parts of the state (e.g. the Gandhian-based Arthik Samata Mandal and the Good Samaritan Evangelical Social Welfare Association). AWARE was also involved in this major relief operation.

The third wave started in 1980 and lasted until 1984. Suddenly a lot of new NGOs came into being. We counted at least thirty. A few new NGOs still started in the spirit of the 'uplifting of the oppressed' (e.g. the Mass Education and Organization Society in Anantapur, for the landless, the Girijana Prajala Sangham in Khammam for tribals and the Young India Project and its offshoot, the women-oriented Shri Karmika Aivyamata Sangha for agricultural labourers in Anantapur). The large majority of the new NGOs were (integrated) rural development project groups—a main issue during those days—often having a core of university graduates, and often oriented to a small number of *mandals* of one district. Many of them mention their region of operation in their NGO name (e.g. the Cuddapah District Rural Welfare Development Society, or the Chittoor Rural Development Society). Although most aimed at improving the lot of the poor, there was no inclination to use a class- or caste- or gender-based exclusivity and one also fails to see the *sangham*, or local organization of the oppressed, as an approach to mobilization. NGOs increasingly became avenues of professional employment for those who could handle donors and the local poor at the same time. It were these NGOs who gradually changed their focus from health care, education, drinking water and sanitation to environmental concerns on the one hand,

and to income generation on the other, also because the government, through IRDP was providing (some) funds. We also see the establishment of the first purely 'environmental NGOs' (e.g. the Development Action for Rural Environment) and the increase of attention for women's issues (and the establishment of a women's NGO, the Hyderabad Zilla Mahila Mandalula Samakhya). In the urban areas a new phenomenon started: the 'civil action groups', like the Consumer Education Centre, in Hyderabad and the Warangal Consumer Council.

The fourth wave is the most recent, and started in 1985. There were less newcomers and most of them were one of three types: purely women's NGOs (e.g. the Women's Association for Development Action, in Chittoor), the purely environmental NGOs (like the Centre for Environmental Concerns, the Society for Nature Conservation, both in Hyderabad, or the Environmental Society of Tirupati, covering Rayalaseema as a whole) and the research NGOs (like the International Association of Extension for Rural Development in Tirupati, the Study Action Group for Participative Integrated Development in Khammam or the Action for Community Service Society in Chittoor). Here we see targets formulated like 'to provide a forum for rural development scholars, students and businessmen for undertaking research on rural reconstruction projects' (WWFN, 1992: AP-2) or 'to act as a rural development consultancy unit for application of science and technology' (WWFN, 1992: AP-14). Many of the environmentally-oriented NGOs also developed research and consultancy wings. Quite a number of NGOs became more and more employment (and income) shelters for academics. The link with the state, partly as funding agency for NGO activities, has also grown after 1985. Since the Seventh Plan (1985-90) considerable government sums (Rs 2.5 billion in India as a whole) have been set aside for NGO-implemented activities. It was preceded by a request from Prime Minister Mrs Gandhi in 1982 to all the chief ministers to enlist the co-operation of voluntary agencies in the field of rural development. Unlike other states there was no attempt by the state government of Andhra Pradesh to constitute consultative committees of voluntary agencies (Rao and Madhavan, 1988: 60, 67-8). Also the relationships between the various NGOs were and are not very good ('they meet but hate each other. They work and criticize one another. They speak strongly and compromise quickly', Rao and Madhavan, 1988: 74-6).

Concentrating on the major activities undertaken by NGOs in 1992 as mentioned in the directories of Charyulu and WWFN, the following overview can be made:

(1) *research/study/organizing seminars and workshops*: 17 NGOs;
(2) *integrated rural development*, and using it as a 'catch word': 10 NGOs;
(3) *rural social development*: 48 NGOs, of whom 5 (partly) dealing with relief work, 6 explicitly dealing with improving the social status of tribals—AWARE is among them; 29 dealing with children's education, adult education or literacy programmes; 35 with health care—6 explicitly with herbal medicine and 2 with family planning—3 with the handicapped, 11 with drinking water, 9 with rural sanitation, 5 with rural housing and 4 explicitly with 'women's welfare'. Unlike the situation in the Andhra coast we do not find many NGOs built on explicit Gandhian principles 'to help mould citizens of selfless disposition and of unbending courage and character' (as the charter says of one of the oldest Gandhian NGOs, the Pinakini Satyagraha Asramamu Trust, in Nellore district, which started in 1921; Charyulu, 1992: 72).
(4) *agricultural development and land management*: 31 NGOs, of whom 6 explicitly with a mission to 'promote modern or scientific agricultural technology' and 4 to promote the contrary: permaculture, organic farming or criticizing the use of chemical fertilizers. We see an explicit attention paid to the problems of droughts and dryland agriculture in at least 4 NGOs. Some of AWARE's more important activities can also be located under this heading of 'agricultural development and land management'.
(5) *rural environment*: 42 NGOs, of whom 27 deal with afforestation (often as 'social forestry'), 19 with renewable energy (biogas, solar or wind energy) and often with the introduction of 'smokeless chullas'. 8 NGOs can be regarded as 'environmental research consultants'. 7 NGOs are of the 'nature conservation type'.
(6) *rural non-agricultural economic development*: 27 NGOs, of whom many try to develop 'cottage industries' and 'rural crafts' and 10 are explicitly dealing with women's productive activities.
(7) *urban problems*: only 6 NGOs, 4 of them dealing with the urban environment and only one with urban housing.
(8) *empowerment*: 2 NGOs try to form unions of agricultural labourers. A few NGOs deal with debates about 'democratization'. 6 NGOs are providing legal aid, 10 try to stimulate scheduled caste or scheduled tribe members or landless agricultural labourers to unite and fight for their rights within the legal boundaries (this includes AWARE).

Looking at the location of NGO headquarters and regions of intervention we can conclude that geographically the distribution of NGOs is rather skewed (if the two inventories give a good picture). There is an obvious concentration in Hyderabad, partly of organizations covering major parts of the state and partly the urban-oriented or consultancy-oriented types of NGOs, close to 'where the business is'. The second concentration of NGOs can be found in the south, with Chittoor a leader and within Chittoor a concentration in and around the university town of Tirupati. There is a remarkable NGO emptiness in the north-western part of Andhra Pradesh, in the districts of Adilabad, Nizamabad and Karimnagar. Quite a number of key persons state that this is because of the presence of guerrilla fighters in the area ('if there are Naxalites, there are no NGOs').

For 26 NGOs in dryland Andhra Pradesh the inventory of Charyulu (1992) also gives information about the sources of money. It is not at all complete, but it gives an indication. The Central Government of India and the State Government of Andhra Pradesh are very important as donors. Only 2 NGOs do not get any funds from Indian government sources. Out of the 26 NGOs studied, 19 get money from CAPART in New Delhi, the Council for the Advancement of People's Action in Rural Technology, an organization created as an intermediary between Delhi and Indian NGOs. But other Central Government institutions are important as well for 14 NGOs: like the Ministry of Human Resources Development, the University Grants Commission, the Ministry of Health and Family Welfare, the Ministry of Environment and Forestry, the National Wasteland Development Board and others. And 8 NGOs also get funds from or through the AP government, e.g. from the District Rural Development Agency, or the SC Corporation, or the ST Corporation. Finally 2 NGOs, including AWARE, succeeded in getting funds from private banks. Some do get funds occasionally from Indian-based trusts, but this information is unclear.

For many NGOs foreign funding is very crucial. Out of the 26 NGOs for which we have financial information only 4 do not get any foreign funds. For these, mainly Gandhian NGOs it is a matter of principle not to become dependent on foreign funds.

A minority of the NGOs (about 15) can be regarded as 'membership NGOs', where part of the funds come from members' contributions. AWARE is the largest among them, with between 40,000 and 50,000 members (but not all in Andhra Pradesh). But RASS in Rayalaseema (3,600 members), FSCS in Nalgonda (2,000 members), RDO in Cuddapah (1,300 members) are also considerable organizations with membership contributions which should not be overlooked as a source of NGO income. Some NGOs cleverly make financial use of their assets too: as bank savings to get interest, property to get rent, training courses for sale, consultancy services for honoraria, and by marketing agricultural products and other things produced in NGO premises. We feel that the 'marketing of NGO assets as a source of institutional income' is on the increase and some NGOs can be regarded as 'proto-businesses' now. Looking at the NGO sector in dryland Andhra as a whole, this is clearly still in its infancy.

Information about the financial turnover of NGOs is lacking. Without any doubt AWARE can be regarded as the financial giant among the NGOs active in dryland Andhra Pradesh. Government data about 1985 show that there were three NGOs with foreign contributions of more than Rs 10 million per year: AWARE (Rs 35 million), the World Missionary Evangelism of India in Hyderabad (Rs 19 million) and Sri Satya Sai Central Trust in Anantapur (Rs 16 million). For comparison, the large Bangalore-based MYRADA received Rs 15 million from abroad (Rao and Madhavan, 1988: 157). AWARE's annual turnover can be estimated to have grown to Rs 300 million in 1992 (interviews with ICCO, NOVIB and RDT representatives). The inventories give a 'permanent AWARE staff' of 700-800 full-time employees in 1992, more than 1,000 part-time employees and 40,000-50,000 members. The other big NGOs in Andhra Pradesh, the RDT in Anantapur, CROSS in Nalgonda and Warangal, Andhra Mahila Sabha in Hyderabad, the WCADL in Mahabubnagar, YIP in Anantapur and Rayalaseema Seva Samithi in Chittoor also have more than 100 employees each, but probably an annual turnover which is less than 10 per cent of AWARE's, between Rs 20 million and 30 million (we do not have information of the other two 'BINGOs' WMEI and SSSCT).

In terms of development expenditure, the government is still much more important than NGOs. According to Mr Chandramouli, DRDA in Anantapur District, who has vast knowledge about the state as a whole, 'even AWARE's money is a pittance compared to the government outlays in the areas where AWARE is working'. According to him, NGOs can never replace the government. He warns against making too positive an assessment of NGOs: 'many have a bad track record; suddenly somebody declares himself a director of an NGO, gets foreign funds, which means an air-conditioned car, a posh office and computer/fax'. 'In the first years, often 90 per cent of the money goes to establishment.' To avoid donors raising eyebrows tall claims are made about the effect. In Cuddapah district he was told 6,000 voluntary organizations are registered; 'not a single one can really be trusted now'. 'Before 1990 some were good, but the large majority are "dormant" and the others were in the grip of factional struggle and corruption in 1993.'

Like elsewhere in India, NGOs in Andhra Pradesh started adopting 'environmental' and 'eco-agricultural' issues from mid-1980s onwards, including the improvement of dryland agriculture and 'watershed development'. Most clearly this could be seen in the driest part of the state, in Anantapur district, where two NGOs can be regarded as the pioneers, RDT and MYRADA. Elsewhere in the state a number of other NGOs developed a variety of dryland development approaches, the most important being GRAM in Adilabad and Nizamabad in the north; DDS in Medak and Ranga Reddy, north-west of Hyderabad; and WCADL, ViP and YFA in Mahabubnagar, besides AWARE, which will be presented in chapter four.

Our analysis of dryland activities by NGOs in Andhra Pradesh leads us to the following observations. The first obvious conclusion is that the NGO involvement in dryland agriculture is relatively recent. For the older NGOs like RDT, MYRADA and WCADL it followed an emphasis on other goals and it is still a part of a wider involvement: in awareness building, adult education, health and nutrition projects. Only the NGOs which were established more recently (DDS, YFA, ViP in its own way and GRAM a bit in between), show a clear focus on 'the environment' as their central issue. The leadership of the older NGOs are the brand of dedicated social workers who gradually became convinced of the importance of 'sustainable dryland agriculture', while the leadership of the recent NGOs consists of academic professionals, who have discovered a 'niche'. Even then the 'narratives' are very different as the examples of the 'radical intellectuals' of DDS versus the 'businessmen' of ViP and YFA illustrate. Both groups get support from a variety of foreign environmentally-minded donor organizations and in recent times from the central and state governments in India. The government-directed 'dryland development programme' tries to involve some NGOs (RDT, YFA) on an experimental basis to use a small part of the relatively huge sums of government money for this programme. The experience of MYRADA, in Anantapur and in neighbouring Karnataka, seems to provide important guidance, although not always acknowledged, partly because of its good reporting of experiences. The government-NGO connection also works the other way around: former civil servants join the staff of NGOs and bring in expertise gained in the state-directed programmes. They also bring in their criticism of the top-down and technical approaches (e.g. RDT).

A comparison between the government approach to dryland

development and the approach of most of the NGOs in the late 1980s reveals the difference between an 'area' (e.g. watershed) approach and a group approach. GRAM deliberately concentrated its work on SC and some ST farmers; DDS on small and marginal farmers who mostly happened to be SC and ST; RDT and WCADL (for an exposition on WCADL, see next chapter) started with a focus on SC farmers as well. All four gave a lot of emphasis on the creation, or the revival (DDS) of *sanghams*, either community (=SC-)-specific ones (as in the case of GRAM mostly) or with the attempt to unite different 'oppressed' communities. In the case of DDS these combined *sanghams* should be women's organizations, while WCADL built up both gender-separate and integrated grassroots organizations within the villages. Mostly these *sanghams* also have a savings and credit function and this phase generally precedes the dryland management activities. In the case of RDT, strong *sanghams* can then play a role as building blocks of watershed committees. RDT tried, but did not always succeeded, in developing rules to avoid unequal access to benefits for the larger and FC farmers. In other cases, like DDS and WCADL all larger-scale organizations, either at the village level or at the micro-watershed level, are deeply mistrusted (as corrupt, FC- and large-farmer dominated) and sometimes avoided. MYRADA presents an interesting deviation. They do not deliberately concentrate their group formation activities on specific target groups, but on affinity groups within each of the villages in a watershed, forming credit management groups, before becoming land management groups. They try to involve large farmers (and hence in most cases forward community representatives) as well, forming village-broad watershed-implementation committees for a clear 'area approach'. Their approach led to fierce battles with the more radical NGOs in the region like YIP and ADATS. However, MYRADA is very aware of the cleavages and tensions at the village level and in practice they combine the group and area approach and certainly in the early stages the attention on the poor groups is evident and also towards the benefits to landless people in the villages. This is completely different in the recently-established NGOs, of which we will give two examples in the next chapter: ViP and YFA. In their approach 'the village' and 'the people' are the categories, and there is no mention of any intra-village caste or wealth differences. YFA, however, concentrates on the formation of women's groups, called thrift groups and *sanghams* as well, and they do it purely for efficiency reasons: they claim that it is more easy and more effective to work with organized women, they are less inclined to corrupt practices and they are closer to the land and crop management work that should be the key to village improvements. It is a 'discovery' which they share with their ideological counter-NGO, DDS.

Some land development activities can partly be regarded as an attempt at re-peasantization of people belonging to the landless poor through land acquisition and organized land management. A number of NGOs were involved in the acquisition of non-utilized wastelands for former landless agricultural labourers. These lands were often steep and stony and to make them productive a major labour and capital investment was first needed in stone and bush clearing and for bunding. It depended very much on local landscapes and positions of the reclaimed lands vis-a-vis the lands of other, often large farmers, and how the investments in wasteland recovery had an impact on others. The same is true for the investment in tanks. Sometimes it was possible to restrict the NGO activities to a particular group of (very) poor (new) farmers, but mostly a larger-scale, and often area-wide organizational involvement was necessary to avoid conflicts and improve the agro-ecological efficiency of the measures.

A second large controversy in the work of NGOs concerns the type of technology to be selected. NGOs with a 'mission' to make government funds available for its clientele, either by being intermediaries (like WCADL or ViP) or by mobilizing group action (like GRAM), are confronted with the strong attempts of small and large farmers alike, to first and foremost, develop irrigation and secondly to get 'modern' inputs (fertilizers in particular). It took most of them time to discover the negative ecological consequences of the one-sided support for irrigation and fertilizers. GRAM put sustainable dryland farming centre stage from 1991 onwards, trying to minimize ground-water consumption, fuel and fodder shortages and external crop inputs. RDT also discovered that investments in tanks only meant that farmers changed their cropping pattern from dryland crops to the cultivation of rice and even sugarcane and irrigated vegetables, weakening instead of strengthening the dryland focus in its programme. WCADL, until 1993, even went as far as to try and get bore wells from the government for farmers belonging to their member *sanghams*, resulting in lowering ground-water tables jeopardizing the water conservation elements in their dryland farming programme. Even DDS, so much involved in 'permaculture' and other forms of 'ecological farming' had to comply with the wishes of their groups to, at least, start micro-irrigation projects.

In dryland development the NGOs show a very diverse combination of land and crop improvement programmes, either in favour or very much against the watershed concept (MYRADA and RDT *vs* DDS and

WCADL). The NGOs fighting the watershed approach or technology criticize it because, according to them it will mean that now the last remaining wastelands will be grabbed by the already rich farmers; it only helps the rural rich and it provides more cheap food for the urban rich. They are convinced that this form of eco-development will mean that poverty eradication no longer is the central focus of development programmes.

If we look at the technology actually propagated, most NGOs did include farm boundary bunding in their approach, but they were hesitant to introduce on-farm contour bunds and contour ploughing, because many farmers resisted the labour and risk implications for themselves and especially for their animals. RDT tried to develop an half-way house solution, the 'adjusted contour' and MYRADA also tried very much to find solutions between the 'textbook approaches' and farmers' reluctance. YFA always tried to involve farmers in their 'agro-eco-security' approach by organizing 'participatory resource assessments', with attention for the agro-technical, labour and risk reasons farmers have to avoid some of the so-called 'obvious' improvements. It is interesting to see that most of the NGOs are involved in some forms of tree planting (social forestry, on-farm forestry, agro-forestry), though with some hesitation for the same reasons. It is also interesting that despite the attention on better manure and intercropping practices, an integrated approach is often missing at the level of the organization and certainly at the level of the *sangham*. For instance, the animal husbandry aspect of the approach is often missing and if they are included it takes the form of loans for milch animals only. One may fear that the recent sophistication in thinking about dryland development in the top echelons of organizations like MYRADA, RDT and especially DDS does not reach the field-level workers who will be most pressed to serve the farmers' demands for more irrigation and more and cheaper fertilizers. It may well be true that the circuit of donor - NGO top level conversation and exchange of opinions is rather far from the circuit of farmer-NGO field worker contacts. For the same reason one may fear that the 'narratives of success', so necessary to get additional funding and for NGO survival, are not truly related to the real activities and changes on the ground. As we have seen, the DDS leadership is quite aware of this knot and they, in fact, doubt that any 'objective' monitoring and evaluation of agro-ecological experimentation by NGOs themselves could be possible. Even MYRADA, with its strong wish to document what is happening, does not go much deeper than 'output' measurement, avoiding the effect and impact questions.

3. A Profile of Ranga Reddy and Mahabubnagar: A Drought Risk Assessment, Agricultural and Social Change, and the Role of NGOs

A.J. Dietz, M. Put and S. Subbiah

3.1 An Introduction to Ranga Reddy and Mahabubnagar Districts

The study villages are located in Ranga Reddy and Mahabubnagar districts of Andhra Pradesh (see Map 3.1). Ghatpally and Mansanpally are located in Maheswaram *mandal* and Debbadaguda in Kandukur *mandal* of Ranga Reddy district and Ekvaipally is situated in Amangal *mandal* of Mahabubnagar district. Both districts are part of the Telangana region. Ranga Reddy was formed in 1978 when the urban areas of the twin cities of Hyderabad and Secunderabad, and a few other urban settlements were separated from their rural hinterland. The geographical area of Ranga Reddy district was 7,500 sq. km in 1991, with a population of 2.6 million, of which 53 per cent lived in rural areas. The population density was 340 per sq km. The geographical area of Mahabubnagar district was 18,400 sq. km, with a population of 3.1 million in 1991, of which 89 per cent was rural. The population density was 167 per sq. km. Literacy levels in 1991 were 49 per cent in Ranga Reddy and 30 per cent in Mahabubnagar, showing the influence of Hyderabad in Ranga Reddy. The average for the entire state was 44 per cent (GoAP, 1993: 3-4, 22).

The two districts, being a part of the Deccan plateau, are rather hilly with elevations ranging mostly between 200 and 700 m above sea-level, having isolated granite hills and two ranges: the Rajkonda and the Anantagiri. Soils are mainly red sandy Alfisols, though in some *mandals* mixed red and black soils (Vertisols) are present. The National Commission on Agriculture defined 28 rainfall zones in India, and they included Maheswaram and Kandukur *mandals* of Ranga Reddy district and

MAP 3.1. RANGA REDDY AND MAHABUBNAGAR DISTRICTS IN ANDHRA PRADESH.

Amangal *mandal* of Mahabubnagar district in the same zone. The annual rainfall in this zone varies between 600 to 800 mm. In general, August and September account for about 50 per cent of annual rainfall (National Commission on Agriculture, 1976: 15).

Together the total area of the two districts is 26,00,488 ha. In 1990/1 a little less than half of it was cultivated (46 per cent; a 'net sown area' of 11,84,414 ha of which 8 per cent was 'double cropped', giving a 'gross sown area' of 12,78,609 ha). Another 26 per cent of the total area was statistically described as 'fallow land', while only 2 per cent was described as 'cultivable waste'. Other types of land use were: forest land (15 per cent), grazing land (3 per cent), non-agricultural uses (6 per cent), barren and uncultivable waste land (3 per cent) and 'miscellaneous tree crops and groves' (1 per cent) (GoAP, 1993: 83-4). On the whole, this is a zone with relatively little irrigation: the irrigated area in Ranga Reddy district is only 21 per cent of the net sown area, while it is 18 per cent for Mahabubnagar (GoAP, 1993: 150). This means that around 80 per cent of the cultivated land can be regarded as 'drylands'. As we will see later in this chapter this certainly means: 'risky drylands'.

Food crop production dominates in the two districts, with 68 per cent of the 'gross cropped area'. In 1989/90, the main crops in the districts were: sorghum or *jowar* (around 31 per cent of the gross cropped area in the two districts combined), paddy (around 16 per cent), pulses (around 13 per cent), castor (around 9 per cent), pearl millet or *bajra* (around 2 per cent) and groundnuts (around 18 per cent, more important in Mahabubnagar). In that same year the yields of these crops in Ranga Reddy and Mahabubnagar were respectively: for rice 2,259 and 1,865 kg/ha, for *jowar* 876 and 631 kg/ha, for *bajra* 340 and 427 kg/ha, for castor 468 and 230 kg/ha and for groundnuts 1,024 and 807 kg/ha. It shows that Ranga Reddy district, with its relatively higher rainfall, has a somewhat higher yield for most crops. Compared with the average yields in Andhra Pradesh, except for *jowar*, groundnut and castor cultivated in Ranga Reddy district, all crops yielded less than the average (GoAP, 1993: 122-3). More details will be given later.

In the two districts combined, 2,80,000 ha were irrigated in 1989/90. Almost 2,00,000 ha of this irrigated area was under paddy, another 30,000 ha under 'other food crops' and 50,000 ha under non-food crops. Sorghum, millets, castor and groundnuts were almost completely dryland crops. Although (mostly irrigated) paddy was cultivated on half the area of (non-irrigated) *jowar* and *bajra*, total rice production in 1989/90 was 4,00,000 tons as against 2,90,000 tons of these two other crops. In addition, the two districts produced almost 100 million tons of other food grains and 3,30,000 tons of oilseeds (mostly groundnuts), most of it as dryland crops (GoAP, 1993: 92ff.).

There exists a rather unequal distribution of landholdings in the two districts. While 39 per cent of the landholdings in Ranga Reddy district were below one ha in 1986/7, they accounted for 8 per cent of the area cultivated. The figures for Mahabubnagar were 34 per cent of these marginal holdings, which accounted for 6 per cent of the area. Small farmers, having a holding between one and two ha, totalled 23 per cent and 22 per cent respectively. The area cultivated was 12 per cent in Ranga Reddy and 11 per cent in Mahabubnagar. On the other hand, large farmers, owning holdings of ten ha and more, formed 4 per cent of all farmers in Ranga Reddy and 5 per cent in Mahabubnagar. The area cultivated was 29 and 27 per cent respectively (GoAP, 1993: 142-3).

Compared to Mahabubnagar and other districts in Telangana, Ranga Reddy is rather well endowed with roads and railways. This is clearly a result of its proximity to Hyderabad. The same applies to various supporting institutions. City and district buses ply in this area, connecting various villages with the capital. Pylons and transformers of the power grid can be seen everywhere. Electricity has paved the way for the distribution of electrical pumps for irrigation. The development of animal husbandry is supported by artificial insemination centres, veterinary hospitals and dispensaries. Nationalized banks, branches of the Hyderabad District Cooperative Central Bank, multipurpose cooperative societies, and Farmers Service Cooperative Societies are located in various bigger centres in the districts, often the *mandal* headquarters. The relative nearness to Hyderabad is apparent in certain ways in all the villages. Nevertheless, the psychological distance is still rather long for certain groups of villagers, particularly for the economically poor and socially backward farmers.

The legacy of the very oppressive Muslim-ruled Hyderabad State can still be felt in the two districts, as in the whole of Telangana. Although during the late 1940s and early 1950s the big landowners and tax collectors in Mahabubnagar and Ranga Reddy lost some of their economic and political power, the social inequality is still considerable. It took a long time before the government of Andhra Pradesh started to really assist the small-scale and marginal SC and ST farmers and a lot of it was done half-heartedly and against a lot of rather effective opposition from the rich and powerful. It also took a long time before NGOs started to mobilize the poor against the existing poverty and oppression.

This is not to say that government development activities were non-existent in the research area. Before the start of the Maheswaram project in 1984, various other government programmes, for instance, in the fields of agriculture, infrastructure, education, etc., had been implemented in and outside the project area. The agricultural programmes did not follow an integrated strategy. Soil and water conservation activities, for instance, were implemented in isolation, not even together with improved crop management practices. These soil and water conservation activities (implemented by the Agricultural Department as part of its normal programme) consisted mainly of the construction of graded bunds with waterways. Most of these works were undertaken around 1970-5. Fairly large-sized bunds were formed, employing manual labour (Sitapathi Rao et al., 1990: 16). Farmers in the area informed us that most of these bunds had almost washed away before the commencement of the Maheswaram project in 1984. These bunds were not built in three of our study villages Ghatpally, Mansanpally and Debbadaguda. A few bunds were built in Ekvaipally, though. In contrast, most of these bunds are still in good shape.

3.2 Drought Risks in Dryland Agriculture of Mahabubnagar and Ranga Reddy Districts

3.2.1 A BRIEF REVIEW OF RAINFALL AND DROUGHT STUDIES ABOUT ANDHRA PRADESH

Quite a few agroclimatological studies have been undertaken in Andhra Pradesh. Subrahmanyam and Sastry (1969) made a study about aridity and droughts in Vishakhapatnam district along the northern Andhra coast. Singh (1979) studied the annual potential water deficit in Andhra Pradesh as a whole. Hema Malini did a Ph.D thesis on 'Studies in Eco-climatology of Andhra Pradesh' (1979) and later wrote an article (1981) about the patterns of aridity and the spread and severity of drought in the state in the years 1966-9. Application of the cumulative deviation technique from which severity indices were analysed gave the duration and severity aspects of droughts. Agrometeorological studies for the optional and planned use of surface irrigation and rainfall sources were carried out by Venkataraman (1981). Moisture adequacy indices were computed for 26 stations in the state by Subramanian and Vidyanath (1984). The spatial patterns of moisture adequacy and water deficit were compared with the spatial patterns of intensity of cropping and of irrigation and crop diversification. The relationship between climate and crops in Rayalaseema has been analysed by Venkateswara Rao (1984). An analysis of the rainfall patterns, droughts and water resources of Cuddapah district, part of Rayalaseema, was studied by Krishna Reddy (1990). Kumar (1991) studied the correlation between agricultural potential and development in the drought-prone area of Anantapur district, also in Rayalaseema. He included the relationship between soil type and utilization potential with regard to surface and subsurface water resources. In their important book *Village and Household Economies in India's Semi-Arid Tropics* (1990: 33-9), Walker and Ryan included an analysis of rainfall pattern and the consequences of variability of Aurepalle village in Mahabubnagar district. With the exception of the last study, no other study placed the consequences of the seasonality and variability of rainfall for agriculture in the centre of its approach. In our study we try to do this, focusing on the concept of drought risks. We concentrate on the two districts of our case studies, Ranga Reddy and Mahabubnagar, in the southern part of Telangana.

3.2.2 RAINFALL VARIABILITY IN MAHABUBNAGAR AND RANGA REDDY

Rain-dependent crop farmers in relatively dry areas have to cope with a variety of climate-related risks. The average low rainfall situation only allows a narrow range of crops, and these are mostly crops with relatively low yields. Lack of crop diversity shrinks the ability to diversify the agricultural economy. Average low yields of these crops, especially if combined with land or labour constraints, result in low food security margins. However, for rainfall-dependent farmers in dry areas the most problematic element of farming is not the average low rainfall and the resultant average low yields, but the large fluctuations in rainfall, from place to place, from year to year and within the rainy seasons.

The rainfall stations closest to the four study villages are, Ibrahimpatnam (data for 1957-86), Maheswaram (data for 1981-3 and 1985-93) and Kandukur (data for 1985-92) in Ranga Reddy and Kalwakurthi (1982-92) and Amangal (1988-92) in Mahabubnagar. However, there are no temperature data or evaporation measurements or evapotranspiration assessments. Long-term data for rainfall, evapotranspiration and temperature are available for the district headquarters, Hyderabad airport (1901-73) and Hyderabad ICRISAT Station (1974-92) and Mahabubnagar town (1901-83). For the period 1960-91 we also have data for the average rainfall situation of Mahabubnagar district, based on approximately twenty rainfall stations. For these stations and some additional ones in Ranga Reddy district we also have average monthly rainfall data, based on a variety of measurement years. If we use data for rainfall and potential evapotranspiration for the district headquarters and use the typology suggested by UNESCO (1979), based on the P/ETP classification, Ranga Reddy district (with headquarters in Hyderabad) and Mahabubnagar district are in the 'semi-arid to semi-humid zone', with an aridity index of around 0.45 for Hyderabad town (P 796 mm; ETP 1758 mm) and below 0.50 for Mahabubnagar town (P 834 mm; ETP 1676 mm). Both Hyderabad and Mahabubnagar are located in relatively wet parts of the district. Map 3.2 shows that the area in which the study villages are located can be found in the driest parts of both districts, with an aridity index around 0.40 (P 675 mm; ETP estimate: 1700 mm). The research area is not the driest part of Andhra Pradesh, though. Anantapur district, located in the southern part of Andhra Pradesh, has an aridity index of 0.30 which is clearly semi-arid. Towards the east, coastal Andhra Pradesh is sub-humid, with aridity indexes of 0.60 to 0.70.[1]

The average rainfall figures for this century are around 800 mm for Hyderabad Station and 830 mm for Mahabubnagar Station, not much different. However, the data for decadal averages show considerable differences. In Hyderabad the period between 1920 and 1950, and the 1920s in particular were relatively dry. After 1950 the figures show a gradually improving average rainfall situation, with the best decadal average during the 1980s. The rainfall data for Mahabubnagar show that the 1920s were exceptionally dry and the 1950s exceptionally wet. For Mahabubnagar town the data are missing for the 1980s; there are indications, though that the 1980s were rather a bad period there (see Tables 3.1 and 3.2).

In Hyderabad, the most extreme years have been 1915 and 1916, with 1431 and 1425 mm on the positive extreme, and 1972 with 416 mm on the negative extreme. In Mahabubnagar the most extreme years have been 1916 on the positive extreme, with 1399 mm and 1918 on the negative extreme, with 400 mm of rainfall.[2] If we take all the annual rainfall figures and make five groups of 20 per cent each the 'good' and the 'bad' periods stand out clearly and show some difference between the northern part (Hyderabad) and the southern part (Mahabubnagar). In Mahabubnagar, the first 30 years of the century are relatively bad, with seven very good and good years on the one side against eighteen very bad and bad years on the other side. During the 1930s, 1940s and 1950s the rainfall situation was rather positive: 15 plus; 7 minus years. In the period between 1961 and 1983 the balance was still positive: there were 11 plus years against 8 minus years. In Hyderabad, the first three decades show a negative

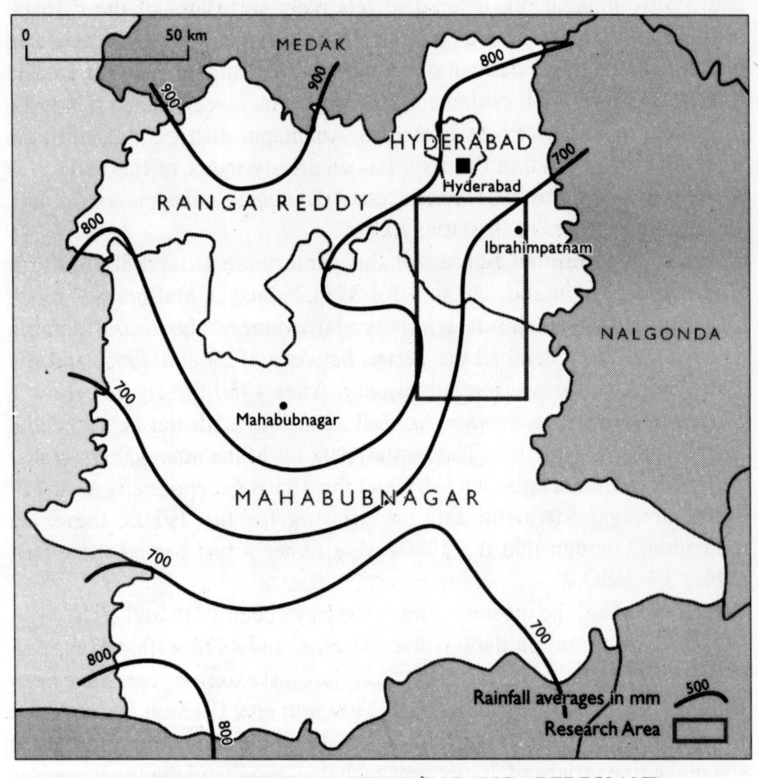

MAP 3.2. RAINFALL PATTERN IN RANGA REDDY AND MAHABUBNAGAR DISTRICTS.

situation as well with 11 very good and good years, against 16 very bad and bad years. The period between 1930 and 1960 taken as a whole shows a balanced picture: 10 good/very good and also 10 bad/very bad years. The recent thirty years show a positive picture with 15 good/very good and 10 bad/very bad years.[3] Before we go on to analyse these figures and trends in detail, we first want to present the seasonality picture.

3.2.2 RAINFALL SEASONALITY IN MAHABUBNAGAR AND RANGA REDDY

The rainfall pattern of Ranga Reddy and Mahabubnagar is clearly seasonal. The greater part of the rainfall is received during the south-west monsoon, from mid-June to mid-October. Average monthly rainfall and evapo-transpiration figures are given in Tables 3.3 and 3.4 for Hyderabad and Mahabubnagar respectively. They show that there are only three months in which ETPm>Pm: July, August and September. The rainy season in Ranga Reddy and Mahabubnagar generally starts during June and ends in September or October. This period is called *kharif*. In Mahabubnagar generally the rains start slightly earlier than in Ranga Reddy. In 30 per cent of the years during this century the period July-October had a higher rainfall than the period June-September. Prior to the real start of the rainy season pre-monsoon showers are often received during May, 'which helps in seeding the sorghum crop in the light textured soils of the region' (AP-DoA, 1987: 3). The period between October and December is called *rabi*; it is a period with much less rainfall although some farmers try to get a second harvest. The period of January and February is called the 'winter' and it is relatively cold and very dry. Some harvesting of castor takes place during this period, but generally it is the agricultural slack season. The period from March to May is called 'summer', the hot and dry season. Most of it is an agricultural slack season too. At the end of this period all farmers start field preparations.

In the literature we find a number of 'seasonality indexes'. The most simple method (for tropical areas) (Walsh, 1981), regards all months with less than 100 mm of average rainfall as 'dry months'. In that case, the area of Ranga Reddy/Mahabubnagar has eight to nine dry months per year on an average. Again, the southern part of Andhra Pradesh is more extreme, with Anantapur district having ten dry months. And coastal Andhra Pradesh towards the east is less extreme, with six to seven dry months. Foeken (1994: 29-30) gives three other seasonality assessments, two based on Walsh (1981) as well. The first one is called the (relative) *seasonality ration*, defined as the difference between mean rainfall of the wettest month and the mean rainfall of the driest month, divided by mean annual rainfall. Values range from 0 for equal distribution, to 1 for complete concentration of rainfall in one month only. If we use this approach for Ranga Reddy/Mahabubnagar it gives a seasonality ration of 0.21 for Hyderabad and 0.23 for Mahabubnagar. This suggests a rather weak seasonality. The second one, developed by Foeken himself, looks at the three months with most rainfall and the three months with least rainfall. The difference between M_3max and M_3min is then divided by the mean annual rainfall. Foeken calls it the '*Degree of Rainfall Seasonality*', DRS, with values from 0 per cent (equal distribution) to 100 per cent (all rainfall in three months). For Hyderabad the DRS would be 60 per cent and for Mahabubnagar 63 per cent, which suggests a rather clear seasonality. Finally Foeken gives Walsh's '*seasonality index*', defined as the sum of the absolute deviations of mean monthly rainfall from the overall mean, divided by mean annual rainfall. Its value ranges between 0 (for rainfall equality) and 1.83; a theoretical maximum indicating all rain to be in one month only. For Hyderabad the seasonality index would be 0.88; for Mahabubnagar a bit more outspoken, 0.93.

We go one step further. We have based our own '*seasonality assessment*' on the statistical 'index of diversity' as well, but adjusted to arrive at values between 0, for maximum distribution and 1, for maximum concentration.

$$D = \frac{{}^{12}\Sigma_1 |x_i - x_a|}{2\,(n-1)\,x_a} \rightarrow (n = 12) \rightarrow S = \frac{{}^{12}\Sigma_1 |x_i - x_a|}{22\,x_a}$$

Where x_i = value for month i (1 for January, 12 for December).
Where x_a = average monthly value for the particular year.

Tables 3.5 and 3.6 present the data for Hyderabad and Mahabubnagar Town, adding a decadal presentation as well. In addition to the 'seasonality assessment', a more simple measure is also presented: the percentage of rain falling in the 'rainy season', here defined as 'June-September or July-October, whichever is highest'.

On a decadal basis we can conclude that for Hyderabad and Mahabubnagar the seasonality assessment figure has always been around 0.50. In Hyderabad there was an exceptional period of less concentration during the 1930s, making the already dry decade even more problematic for agriculture. In Mahabubnagar there was an exceptional period of more marked concentration of rainfall during the 1960s, resulting in the highest-

TABLE 3.1. DECADAL AVERAGES FOR RAINFALL IN MM PER MONTH FOR HYDERABAD,* 1901-90

Year	J	F	M	A	M	J	J	A	S	O	N	D	TOT	KS**
1901-10	11	6	5	32	30	111	144	159	177	74	14	8	771	591
1911-20	7	11	22	27	35	122	193	136	185	81	39	6	864	636
1921-30	18	10	1	20	18	85	147	127	164	79	36	6	711	523
1931-40	3	22	12	31	38	99	166	113	152	70	34	10	750	530
1941-50	3	9	13	18	22	119	157	140	176	54	32	4	747	592
1951-60	0	1	15	20	31	101	179	190	160	97	17	2	813	630
1961-70	3	4	12	27	33	90	197	147	205	71	12	14	815	639
1971-80	5	7	5	22	36	111	139	215	165	94	21	1	821	630
1981-90	9	5	21	25	32	123	195	177	168	83	31	4	873	663
Average	7	8	12	25	31	107	169	156	172	78	26	6	796	604

Note: * Hyderabad Airport (Begumpet) until 1972; Hyderabad ICRISAT from 1972 onwards (there are indications that the ICRISAT station receives slightly more rain than the Begumpet Station).

** *Kharif* season (June-September).

Source: ICRISAT files.

TABLE 3.2. DECADAL AVERAGES FOR RAINFALL IN MM PER MONTH FOR MAHABUBNAGAR TOWN, 1901-80

Years	J	F	M	A	M	J	J	A	S	O	N	D	TOT	KS**
1901-10	5	11	1	22	21	129	161	186	188	73	12	9	816	664
1911-20	5	4	9	13	44	122	192	129	170	83	35	3	799	600
1921-30	8	8	3	16	28	96	166	107	175	66	23	1	695	544
1931-40	7	22	11	22	35	119	192	161	183	73	28	1	854	655
1941-50	6	3	13	12	18	106	170	149	210	78	30	5	800	635
1952-60*	0	1	8	25	62	120	257	208	210	98	12	2	1001	693
1961-70	2	3	4	17	38	113	245	152	221	66	10	12	883	730
1971-80	1	4	5	13	62	102	165	213	136	106	20	0	827	616
Average	4	7	7	18	39	113	194	163	187	80	21	4	834	642

Notes: *Data for 1951 are missing.
**KS = *Kharif* season, June-September, generally starting earlier than in Ranga Reddy.

Source: ICRISAT files.

TABLE 3.3. HYDERABAD, AVERAGE MONTHLY RAINFALL AND POTENTIAL EVAPOTRANSPIRATION DATA

	J	F	M	A	M	J	J	A	S	O	N	D	Ta	Tr
P	7	8	12	25	31	109	168	157	170	77	26	6	796	625
ETP	110	130	182	198	220	198	140	136	120	124	104	100	1758	592
P/ETP	0.06	0.06	0.07	0.13	0.14	0.55	1.20	1.15	1.42	0.62	0.25	0.06	0.45	1.06

Note: Ta = Annual Total; Tr = Total in the rainy season (P based on annual figures for June-September or July-October, whichever is highest; ETP based on June-September).

Source: ICRISAT files. Rainfall based on 1901-92; potential evapotranspiration (most probably) on 1930-60.

TABLE 3.4. MAHABUBNAGAR TOWN, AVERAGE MONTHLY RAINFALL AND POTENTIAL EVAPOTRANSPIRATION DATA

	J	F	M	A	M	J	J	A	S	O	N	D	Ta	Tr
P	4	7	7	18	39	113	194	163	187	80	21	4	834	671
ETP	120	133	177	183	198	158	124	119	117	121	117	111	1676	518
P/ETP	0.03	0.05	0.04	0.10	0.20	0.72	1.56	1.37	1.60	0.66	0.18	0.04	0.50	1.30

Note: Ta = Annual Total; Tr = Total in the rainy season (P based on annual figures for June-September or July-October, whichever is highest); ETP based on June-September.

Source: ICRISAT files. Rainfall based on 1901-50 and 1952-80; potential evapotranspiration (most probably) based on 1930-60.

ever rainfall average during the *kharif* season and an extended period of relatively favourable conditions for agriculture. This is confirmed by the percentage of rain in the rainy season: this has always been around 80 per cent, with indeed a lower figure for Hyderabad during the 1930s (73 per cent) and a higher figure for the 1960s in Mahabubnagar (83 per cent). If we look at the decadal rainfall averages for the rainy season we can again see that the 1950s and 1960s have been exceptionally good and the 1920s (and the 1930s in Mahabubnagar) very bad. The 1970s and 1980s were relatively good around Hyderabad but not around Mahabubnagar. Map 3.3. presents a seasonality assessment for the whole of Mahabubnagar and Ranga Reddy, based on average figures.

3.2.4 THE ASSESSMENT OF DROUGHT RISKS: A METHODOLOGY

Although the average conditions might suggest a climate that easily enables a sorghum and millet cropping pattern, the huge variability of the rainfall totals, its distribution over the months, and the insecure start of the rainy season all present big risks for the farmers who cultivate dryland crops. This is a well-documented fact for all dry climates (cf. Reitsma et al., 1992). The standard deviation of rainfall in the driest parts of the Deccan, with rainfall around 600 mm is around 200 mm (e.g. see Bakker, 1992: 75). According to the Department of Agriculture of the state government of Andhra Pradesh 'it has been observed that the potential of dry land crop(ping) is not significantly reduced even when only two-thirds of the normal rainfall is received provided it is well distributed' (AP-DoA, 1987: 3). Any drought risk assessment should therefore look at the actual rainfall distribution over the months.

We have developed a drought risk index, for a uni-modal climate, based on two earlier indexes, developed by FAO (1980) and by Bailey (1979). The FAO approach, developed for its Agro-Ecological Zones Project used monthly rainfall figures and monthly potential evapotranspiration (ETP) figures and allowed soil moisture storage of a wet month for a following dry month. Bailey's method used monthly rainfall figures and mean monthly temperature figures to arrive at a 'moisture index'.

For the FAO a month was regarded as sufficiently wet for a crop that is adapted to dry conditions when $P_m > 0.5\ ETP_m$. Also when P>ETP, the surplus rainfall can be accumulated for the next month as soil moisture storage (with a maximum storage period of one month). For Bailey the moisture index was constructed as $S_i = 0.18P/(1.045)^t$, where P is in cm and t is in °C. He called months 'wet' when

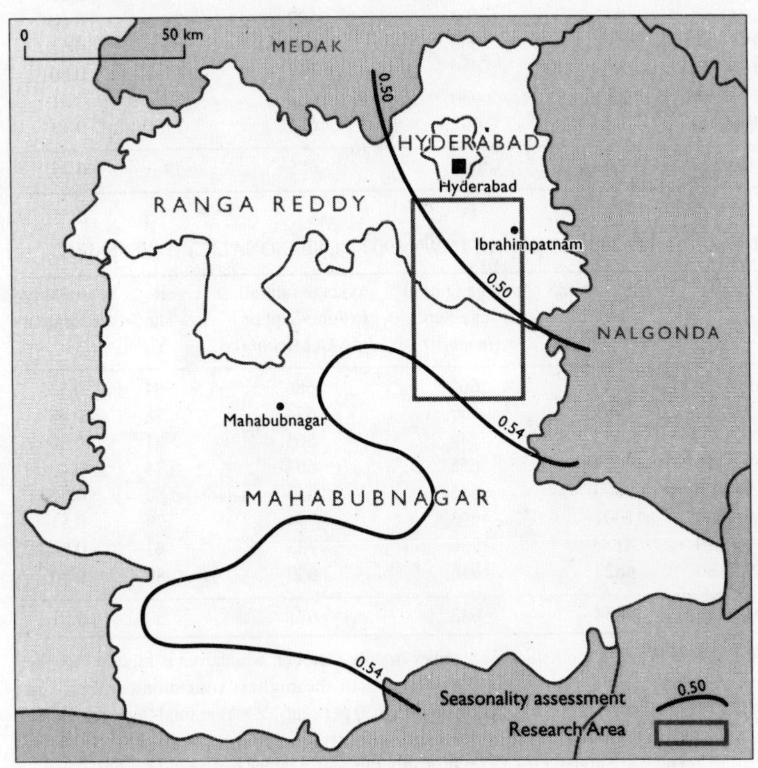

MAP 3.3. RAINFALL SEASONALITY IN RANGA REDDY AND MAHABUBNAGAR DISTRICTS.

TABLE 3.5. HYDERABAD 1900-90, SEASONALITY ASSESSMENT

Decade	Average annual rainfall (*in mm*)	Average rainfall in June-Sept (*in mm*)	Average rainfall in June-Sept or July-Oct (*in mm*)	R ... in % Y	Seasonality assessment
1901-10	771	591	620	80	0.49
1911-20	864	636	662	77	0.45
1921-30	711	523	551	77	0.47
1931-40	750	530	547	73	0.42
1941-50	747	592	592	79	0.50
1951-60	813	630	663	82	0.52
1961-70	815	639	662	81	0.50
1971-80	821	630	658	80	0.51
1981-90	873	663	673	77	0.48
Average	796	604	625	79	0.48

TABLE 3.6. MAHABUBNAGAR TOWN 1900-80, SEASONALITY ASSESSMENT

Decade	Average annual rainfall (*in mm*)	Average rainfall in June-Sept (*in mm*)	Average rainfall in June-Sept or July-Oct (*in mm*)	R ... in % Y	Seasonality assessment
1901-10	816	664	686	84	0.53
1911-20	799	600	624	78	0.48
1921-30	695	544	563	81	0.50
1931-40	854	655	667	78	0.48
1941-50	800	635	643	80	0.52
1952-60*	1001	693	794	79	0.51
1961-70	883	730	733	83	0.54
1971-80	827	616	659	80	0.50
Average	834	642	671	80	0.51

Notes: R = Rainy period: June-September or July-October, whichever is highest (for the century as a whole June-September had the highest four-month rainfall in 70 per cent of the years, July-October in 30 per cent); Y = Year total (average for the decade); S = Seasonality assessment (see text).

* There are no data for 1951; probably this was a rather bad year, bringing down the decadal figures somewhat.

$S_i > 0.81$ and 'dry' when $S_i < 0.53$. In between 0.53 and 0.81 months are called 'neutral' (see Dietz, 1987 for earlier applications).

Our drought risk index works with a scale from zero to ten, where 'ten' indicates the most severe risk of crop failure due to drought. It is adapted to the situation of the study area:

0 At least four consecutive wet months, according to Bailey's

method; normal start of the rainy season (Mahabubnagar and Ranga Reddy: June);

1 Same; late start of the rainy season: first wet month only in July;

2 At least four consecutive rain months according to Bailey's method (that is: neutral or wet); normal start of the rainy season;

3 Same; late start;

4 At least three consecutive rain months (Bailey); normal start;

5 Same; late start;

6 According to Bailey's method there are no three consecutive rain months; but according to the less strict FAO method there are; normal start;

7 Same; late start;

8 According to the FAO method there are no three consecutive rain months; however a dry month is sufficiently compensated by storage of rain of the previous month (storage: P-ETP; available rain = $P_m + P_{m-1} - ETP_{m-1}$); normal start of the rainy season;

9 Same; late start;

10 Not enough rainfall or storage for crop growth (or only a start of a rainy season in August).

In addition to a drought risk, farmers sometimes experience the opposite: extremely high rainfall. This can damage the crop. When excess rainfall occurs during the fourth (and fifth) month of a rainy season and especially if it comes in rain storms, with very high rainfall concentrations in a day, we can speak about a 'harvest spoilage risk'. Walker and Ryan (1990: 49) rightly observed in their Aurepalle case study in Mahabubnagar, to the south-west of our research area 'too much rainfall received at the wrong time can have a devastating effect on yield.... Abundant and sustained rain in late September and early October precipitated a sharp rise in insect and disease infestation.' However, one needs more detailed rainfall data for any useful assessment. We will not include any analysis about 'harvest spoilage risks' here. Farmers generally try to avoid late harvests, knowing that the rainfall concentration in the last part of the rainy season can be high and can be potentially threatening for crops. This is one of the reasons why all farmers try to sow in the last week of May or the first week of June; before the start of the rainy season; they try to harvest in September and not in October. If the situation in September/October after the *kharif* period is still good some try to get a second *rabi* harvest, somewhere in December. For castor most farmers are more relaxed: they start in July, and harvest in the December-February period.

3.2.5 ASSESSING THE DROUGHT RISKS IN MAHABUBNAGAR AND RANGA REDDY: THE RESULTS OF THE ANALYSIS

The table in the annex to this chapter and Figures 3.1 and 3.2 give the details and the results of the analysis of drought risks in Mahabubnagar and Ranga Reddy, based on the rainfall situation in the district headquarters. Rainfall stations in Hyderabad and Mahabubnagar town are rather far from the research area and they are both located in the better parts of Ranga Reddy and Mahabubnagar districts respectively. Long-term rainfall data for rainfall stations near the research villages are lacking, with the exception of Ibrahimpatnam (1957-85), which is near Maheswaram for which we have detailed data for 1986-93. However, there are no evaporation and temperature data for the Ibrahimpatnam/Maheswaram station. Precise drought risk assessment is therefore not possible. We will apply the Hyderabad data to the situation in Ibrahimpatnam/Maheswaram for an estimate (see Table 3.7). Additional information is given about the long-term trends, the variability of monthly rainfall and the drought risk assessment in Figures 3.3, 3.4 and 3.5.

It is obvious from these data that the rainfall situation in the research area is much worse compared to Hyderabad or Mahabubnagar town stations. The average annual rainfall is not around 800 mm, but 638 mm. The average rainfall during the rainy season is only around 500 mm. The average drought risk is moderate, with an average assessment index of 7.2, which is considerably higher (and hence more problematic) than in the area around the district headquarters. Looking at the worst years (with a drought risk higher than 5 and a - or -- position) 1963 was even worse than in Mahabubnagar, but in Ibrahimpatnam it was followed by two additional very bad years. The period between 1963 and 1965 must have been terrible in the research area. As everywhere else the years 1971 and 1972 were also bad and in the period between 1977 and 1986 there were seven bad years, and especially the period between 1984 and 1986 must have been problematic. Also 1992 and 1993 were bad.

DROUGHT RISKS IN MAHABUBNAGAR

The distribution of the drought risks in Mahabubnagar (based on the town data) shows that there were eight years with virtually no drought risk (cat. 0 and 1), 22 years with very light drought risks (cat. 2 and 3), 17 years with light drought risks (cat. 4 and 5), 19 years with moderate drought risks (cat. 6 and 7), 9 years with severe drought risks (cat. 8

Hyderabad, 1901-20

YR	1	2	3	4	5	6	7	8	9	0	1	2	3	4	5	6	7	8	9	0
++			x					x						x	x	x				
+									x										x	
0						x				x							x			
-		x					x					x								
- -	x			x	x						x		x					x		x
DR	6	5	1	8	10	2	6	2	2	6	9	5	6	0	0	6	3	7	6	7

Hyderabad, 1921-40

YR	1	2	3	4	5	6	7	8	9	0	1	2	3	4	5	6	7	8	9	0
++											x									
+					x		x						x							
0	x			x						x					x			x		
-		x	x					x	x					x		x				x
- -						x						x					x		x	
DR	4	5	7	3	7	7	2	3	9	2	6	9	4	5	9	3	7	3	7	7

Hyderabad, 1941-60

YR	1	2	3	4	5	6	7	8	9	0	1	2	3	4	5	6	7	8	9	0
++							x			x					x					
+			x										x	x		x	x	x	x	
0		x		x	x				x											x
-								x			x									
- -	x					x						x								
DR	9	4	6	6	2	7	5	5	6	5	7	7	3	3	0	2	4	5	2	10

Hyderabad, 1961-80

YR	1	2	3	4	5	6	7	8	9	0	1	2	3	4	5	6	7	8	9	0
++		x					x			x					x			x		
+			x		x								x	x		x				x
0	x					x													x	
-				x				x	x								x			
- -											x	x								
DR	7	5	2	8	7	3	2	9	7	2	7	10	4	6	3	7	9	4	5	2

(contd.)

Hyderabad, 1981-92

YR	1	2	3	4	5	6	7	8	9	0	1	2
++	x		x					x	x			
+											x	
0		x								x		x
-				x		x						
- -					x		x					
DR	0	6	3	5	9	4	4	2	5	5	2	4

FIGURE 3.1. DROUGHT RISK ASSESSMENT FOR RANGA REDDY (HYDERABAD) 1901-92

Mahabubnagar, 1901-20

YR	1	2	3	4	5	6	7	8	9	0	1	2	3	4	5	6	7	8	9	0
++			x			x								x		x	x			
+								x	x	x										
0															x					
-		x		x	x						x	x							x	
- -	x						x						x					x		x
DR	9	10	1	6	6	0	6	7	2	2	5	7	10	2	3	2	2	10	6	10

Mahabubnagar, 1921-40

YR	1	2	3	4	5	6	7	8	9	0	1	2	3	4	5	6	7	8	9	0
++																		x		
+												x			x					x
0					x	x	x	x		x			x	x						
-			x	x							x						x		x	
- -	x	x							x							x				
DR	8	9	7	5	5	5	2	6	7	2	9	5	0	6	5	4	7	0	3	2

Mahabubnagar, 1941-60

YR	1	2	3	4	5	6	7	8	9	0	1	2	3	4	5	6	7	8	9	0
++							x						x	x	x	x			x	
+				x						x							x			x
0						x			x									x		
-			x		x			x				x								
- -	x	x																		
R	10	6	9	8	2	4	0	5	6	2	nd	7	1	2	0	2	2	5	2	6

(contd.)

Mahabubnagar, 1961-80

YR	1	2	3	4	5	6	7	8	9	0	1	2	3	4	5	6	7	8	9	0
++	x	x		x											x					
+						x	x			x				x		x		x		
0					x			x	x				x						x	x
-																	x			
- -			x								x	x								
DR	4	2	10	2	6	5	2	9	2	2	6	8	6	5	1	8	4	5	4	5

FIGURE 3.2. DROUGHT RISK ASSESSMENT FOR MAHABUBNAGAR (MAHABUBNAGAR TOWN)

and 9) and 7 years with extreme drought risks (cat. 10).

There is an obvious negative correlation between the rainfall typology for the rainy season, and the drought risk. For the ++ years the drought risk is only 1.4 on the scale of 0-10, for the + years 4.1; for the o years

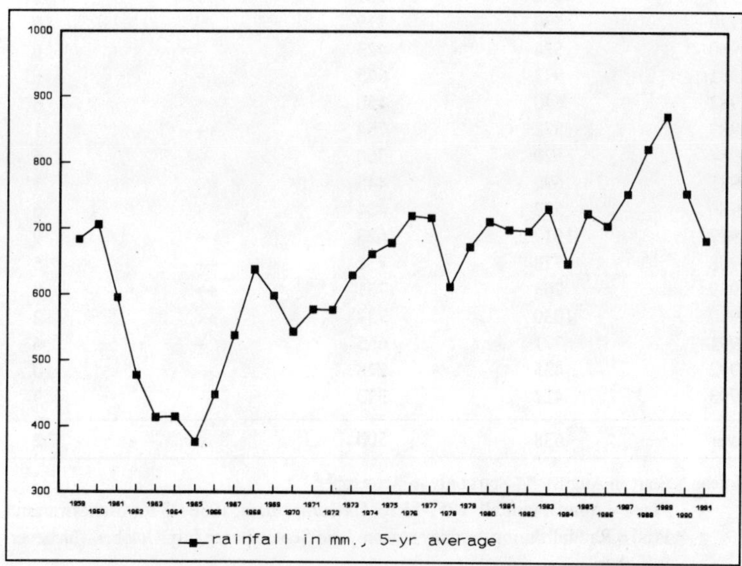

FIGURE 3.3. INRAHIMPATNAM/MAHESWARAM: LONG-TERM RAINFALL TRENDS, 1957-93.

TABLE 3.7. RAINFALL AND DROUGHT RISKS NEAR THE RESEARCH AREA: IBRAHIMPATNAM/MAHESWARAM, 1957-93

Year	P(an)	P(rs)	P(rs)typ.	DR
1957	711	629	+	6
1958	754	458	o	9
1959	682	597	+	5
1960	602	451	o	8
1961	667	512	+	7
1962	822	688	++	3
1963	202	143	- -	10
1964	95	88	- -	10
1965	283	283	- -	10
1966	671	496	o	3
1967	633	597	+	9
1968	560	452	o	10
1969	545	390	-	6
1970	782	611	+	6
1971	471	387	-	10
1972	363	239	- -	10
1973	728	491	o	7*
1974	544	478	o	9**
1975	1044	891	++	3
1976	630	504	o	4
1977	449	384	-	9
1978	932	707	++	2
1979	530	379	-	10
1980	524	423	-	6
1981	932	645	+	6
1982	640	450	-	8
1983	872	764	++	1
1984	520	360	- -	6
1985	690	445	-	9
1986	522	424	-	6
1987	1019	623	+	9
1988	778	731	++	5
1989	763	731	++	9
1990	1030	737	++	2
1991	771	655	+	6
1992	435	278	- -	10
1993	422	373	- -	9
Aver.	638	500	o	7.2

Notes: * Start in August. ** Start only in September.

P(an) = Annual rainfall; 1957-85 for Ibrahimpatnam; 1986-93 for Maheswaram.

P(rs) = Rainfall during the rainy season: June-September or July-October whichever is higher.

P(rs)typ = See P(rs). Typology: - - = < 379 mm.; - = 379 to 450 mm.; o = 451 to 511 mm. + = 512 to 688 mm.; ++ = > 688 mm.

DR = Drought Risk Assessment, based on methodology explained in the text.

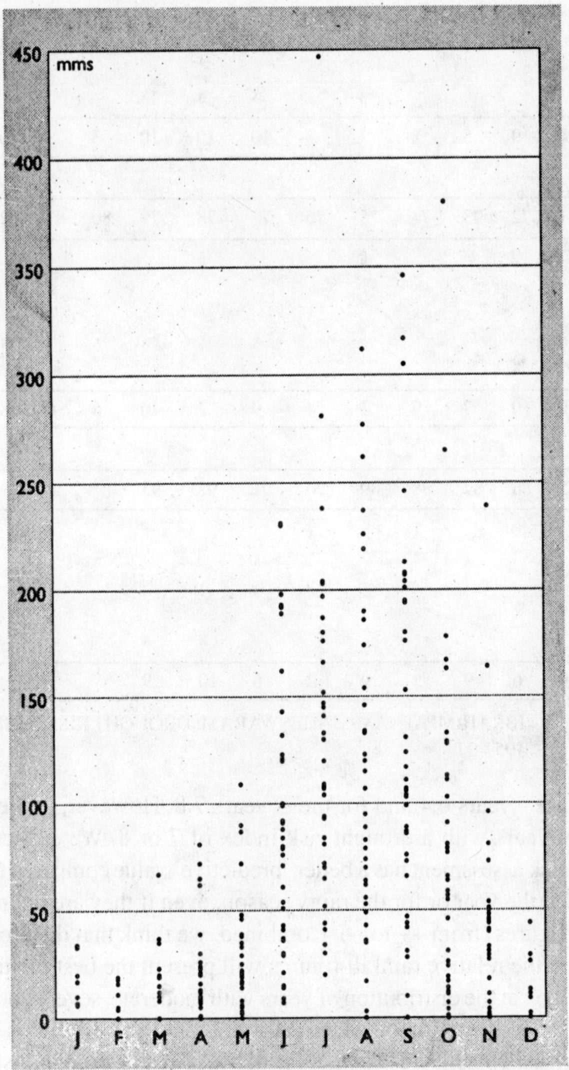

FIGURE 3.4. IBRAHIMPATNAM/MAHESWARAM: THE VARIABILITY OF MONTHLY RAINFALL, 1957-93.

YR	57	58	59	60	61	62	63	64	65	66	67	68	69	70
++						x								
+	x		x		x						x			x
0		x		x						x		x		
-													x	
--							x	x	x					
DR	6	9	5	8	7	3	10	10	10	3	9	10	6	6

YR	71	72	73	74	75	76	77	78	79	80	81	82	83	84
++					x			x					x	
+											x			
0			x	x		x								
-	x						x		x	x		x		
--		x												x
DR	10	10	7	9	3	4	9	2	10	6	6	8	1	6

YR	85	86	87	88	89	90	91	92	93
++				x	x	x			
+			x				x		
0									
-	x	x							
--								x	x
DR	9	6	9	5	9	2	6	10	9

FIGURE 3.5. IBRAHIMPATNAM/MAHESWARAM: DROUGHT RISK ASSESSMENT, 1957-93.

4.5; for the - years 6.4 and for the -- years 7.8. However, in the + group there are years with a drought risk index of 7 or 8. We expect that the drought risk assessment has a better 'predictive' value compared to rainfall figures for the year or for the rainy season, even if they are interpreted as relative figures (from ++ to --). Combined, we think that the drought risk index and the relative rainfall figures will present the best picture.

If we look at the distribution of years with moderate, severe and extreme drought risks the period between 1901 and 1930 and the 1940s show the highest risks. There were less years with considerable drought risks during the 1930s, and from 1950 onwards.[4] Farmers have adapted themselves, albeit with enormous difficulties, to these years with drought risks. If there is a drought year following a relatively good year and again followed by a year with adequate rainfall most farmers will manage. A major problem exists though if two or more drought years follow each other. Figure 7.1 shows that rows of at least two good or very good years (++ or +), with low drought risks (less than 3) only existed in the early part of the century (in 1909-10 and 1916-17) and a long period in the 1950s (1953-57); none afterwards. Rows of at least two bad or very bad years (- - or -), with drought risk scores higher than 5, took place in 1901/2, 1904/5, a disastrous period between 1918 and 1923, 1941-3, 1971/2 and 1982/3, probably continuing until 1987. Farmers who experience a number of bad years in a row will find it difficult to feed their families and they might be forced to do more work elsewhere or to get food on credit, resulting in a worsening debt trap. They will probably lose animals if they have any, which will be detrimental for their future ploughing and manuring capacity. There are indications that farmers in Ekvaipally, the southern most of the research villages, indeed perceived a worsening rainfall situation during the 1980s and that they—unlike farmers in the more northern villages—have adapted their farming system accordingly. They intercrop sorghum or millet with redgram, like most farmers in the area, but they have added cowpeas to the intercrop mix, a more drought resistant crop with fast returns.

3.2.7 DROUGHT RISKS IN RANGA REDDY

Based on the evidence for Hyderabad, Ranga Reddy farmers during this century experienced 5 years with virtually no drought risks (DR = 0 or 1), 24 years with a very light drought risk (DR = 2 or 3), 22 years with light drought risks (DR = 4 or 5), 28 years with moderate drought risks (DR = 6 or 7), 10 years with severe drought risks (DR = 8 and 9) and 3 years with extreme drought risks (DR = 10). Table 3.8 shows a relative comparison between Mahabubnagar, Ranga Reddy and Ibrahimpatnam/ Maheswaram.

It is clear from the comparison that the situation in the southern area (Mahabubnagar) was more extreme: both, the chances of no drought risks and the chances of extreme drought risks were higher than in the northern area. It is also clear that the drought risks in Ibrahimpatnam/Maheswaram are much higher than in the district headquarters.

In Ranga Reddy (Hyderabad) there is again a high but not complete correlation between the drought risk scores and the rainfall typology. For the very good years (++) the average drought risk appeared to be 2.8 with a range from 0-6; for the good years (+) 4.0 with a range from 2 to 7; for the average years (o) 4.7 with a range from 2 to 7 and two extreme cases of a drought risk of 9 and 10; for the bad years (-) 6.1 with a range from 3 to 9 and for the very bad years (--) 7.6 with a range from 4 to 10. Again it is clear that a relatively low rainfall in the rainy season holds a higher chance of a drought risk, as assessed by our drought risk assessment methodology.

Based on the evidence for Hyderabad we can say that between 1901 and 1920, 1931 and 1950 and again between 1961 and 1980 there was a more than 50 per cent chance of a moderate, severe or extreme drought risk. The 1920s with 40 per cent chance, the 1950s with 30 per cent chance and especially the 1980s with only 20 per cent chance of Drought Risk scores upwards from 5 were positive exceptions. However, in the south-eastern, driest parts of Ranga Reddy district, around Ibrahimpatnam and Maheswaram the chances of a moderate, severe or extreme drought risk were much higher: 80 per cent in the 1960s and 70 per cent in the 1970s and 1980s.

Periods with two or more bad/very bad years in a row (- or --) occurred in 1901/2, 1904/5, 1911-3, 1922/3, 1928/9, 1936/7, 1939-41, 1951/2, 1963-5 in Ibrahimpatnam but not in Hyderabad, 1968/9 in Hyderabad but not in Ibrahimpatnam, 1971/2, 1979/80 in Ibrahimpatnam but not in Hyderabad, and 1984-6 in Ibrahimpatnam-Maheswaram and Hyderabad respectively. Also 1992-3 should be included in the list of two bad years in a row. If a period of two or more bad or very bad rainfall years coincides with a moderate, severe or extreme drought risk score, we can safely conclude that these were extremely difficult periods for the farmers. As figure 3.1 shows, for Hyderabad these extremely difficult periods existed in 1904/5, 1911-3, 1939-41, 1951/2, 1968/9 and 1971/2 and not afterwards. In Ibrahimpatnam/Maheswaram 1963-5, 1979/80, 1984-6 and 1992/3

TABLE 3.8. DROUGHT RISK DISTRIBUTION, COMPARISON BETWEEN MAHABUBNAGAR AND RANGA REDDY/HYDERABAD

Drought Risk Assessment		Mahabubnagar (%)	Ranga Reddy/ Hyderabad (%)	Ranga Reddy/ Ibrahimpatnam/ Maheswaram (%)
0 & 1	none	10	5	3
2 & 3	very light	27	26	14
4 & 5	light	21	24	8
6 & 7	moderate	23	30	27
8 & 9	severe	11	11	27
10	extreme	9	3	21
Total		100	100	100

should be added and 1968/9 subtracted (Figure 3.5). These are the data for the northern area.

Problems would be worse if the southern area was also to experience such a lengthy problematic period at the same time. As we have seen, in Mahabubnagar the periods of two or more years with both a bad or very bad rainfall situation in the rainy season and a relatively high drought risk score as well were in 1901/2, 1904/5, 1918-23, 1941-3, 1971/2 and 1982/3, probably continuing up to 1987. During this century, overlapping periods of extremely bad circumstances for two or more years in a row only occurred in 1904/5 and 1971/2 for the whole area. This should warn us to base far-going conclusions on only one rainfall station. Elsewhere (Dietz, Put & Subbiah, 1998) we have presented a further elaboration of our drought risk assessment approach for Ranga Reddy district.

3.2.8 A MORE DETAILED LOOK AT THE MOST RECENT PERIOD

Most situations of dry years with a high drought risk occur in some places and not in others, and even within small areas, farmers in some places get more rainfall at adequate times than others. So, it is useful to look at the more recent period with more attention towards regional detail. Since the mid-1980s more rainfall stations have come into being which could throw more light on the geographical scale of 'bad' or 'good' years. Finally we can also look at farmers' opinions for the most recent years for our research villages. It will also provide the context for the assessment of the detailed study results in the years 1990-3, which will be presented in Chapters 5-7. For the recent period we have monthly rainfall data from rainfall stations closer to the research villages; we also have assessments made by officials working at the district level giving average yields for various crops; and we have subjective judgements from the farmers in the research villages about the cropping seasons 1990/1 to 1992/3. These can guide us to judge the agricultural position in the most recent years. The evidence is given in Tables 3.9, 3.10, and 3.11.

Assessment of the Individual Years

1981: Very good rainfall in the north, a bit above average in the south; with a late start and a light drought risk there. However, the harvest in the south was a little above average and in the north below average; maybe an excess of rainfall during the month of August (306 mm)?

TABLE 3.9. RAINFALL INDEX FIGURES FOR THE RAINY SEASON, VARIOUS STATIONS

YR	Mahab	Kalwa	Amang	Hyder	Ibrah	Kandu	Mahes
1981	106	nd	nd	138	125	nd	nd
1982	90	89	nd	90	87	nd	nd
1983	128	105	nd	141	148	nd	nd
1984	77	100	nd	77	70	nd	nd
1985	63	nd	nd	59	86	61	70
1986	63	nd	nd	81	82	74	77
1987	88	74	nd	68	nd	65	113
1988	140	103	nd	135	nd	102	133
1989	123	99	119	137	nd	140	133
1990	95	93	109	87	nd	144	134
1991	127	136	116	100	nd	115	119
1992	nd	nd	79	86	nd	nd	51
1993	nd	nd	77	nd	nd	nd	68
100 =	554 mm	565 mm	499 mm	663 mm	515 mm	548 mm	549 mm
base years	81-91	82-4 & 87-91	89-93	81-92	81-86	85-91	85-93

Notes: Mahab = Mahabubnagar; average figures for the district, compiled by the District headquarters, probably based on more than 20 rainfall stations; Kalwa = Kalwakurthy; Amang = Amangal; Hyder = Hyderabad; Ibrah = Ibrahimpatnam; Kandu = Kandukur; Mahes = Maheswaram.
nd = no data.
Rainy season figures: June-September or July-October whichever is higher.

Source: Mahabubnagar and Hyderabad: ICRISAT files; Kalwakurthy and Amangal: Mahabubnagar District Handbook of Statistics, various years; Kandukur and Maheswaram: Ranga Reddy District Handbook of Statistics, various years; and raw data; Ibrahimpatnam: AP-DoA (n.d.: 5, ii and iii).

1982: Below average to average rainfall in the north, but with a dry spell in August, causing a moderate to severe drought risk and below average harvests. In the south, below average rainfall but well distributed, resulting in a very light drought risk and above average harvests.

1983: Very good rainfall in the north and good in the south, although rather short, resulting in a light drought risk. August was a very wet month (305 and 306 mm in the district headquarters and 186 mm in Ibrahimpatnam near the research villages), followed by another exceptionally wet September in the north (Hyderabad: 287 mm; Ibrahimpatnam: 207 mm), resulting in below average harvests.

1984: Bad rainfall in the north (however, with a light to moderate drought risk) and very bad in the south (with a severe drought risk). Harvests were far below average, especially in the south.

1985: A second very bad year, now both in the north and in the south, with severe drought risks. Yields were estimated to be below average in the south, but strangely enough not in the north.

1986: A third bad year in a row, although with lower drought risks compared to 1984 and 1985 (moderate in the south; light to moderate in the north). The harvest of *jowar* in the north and of *bajra* in the south were estimated to be above average, however. Only the *jowar* harvest in the south was assessed as bad, but not as bad as in 1984.

TABLE 3.10. ESTIMATED *KHARIF* YIELDS IN MAHABUBNAGAR AND RANGA REDDY DISTRICTS, 1981-91 (INDEX FIGURES)

Year	Sorghum (*jowar*) in Mahabubnagar district	Sorghum (*jowar*) in Ranga Reddy district	Pearl Millet (*bajra*) in Mahabubnagar district
1981	115	82	103
1982	110	96	99
1983	93	83	112
1984	61	75	61
1985	90	110	85
1986	89	120	112
1987	118	113	117
1988	78	72	63
1989	132	109	126
1990	148	132	148
1991	67	104	74
100 =Index	495 kg/ha	725 kg/ha	338 kg/ha

Source: Mahabubnagar District and Ranga Reddy District Handbooks of Statistics, various years, for Ranga Reddy data for *bajra* are only given from 1985 onwards and not very relevant. For both districts also data for *jowar* during the *rabi* or second season are given as well as data for fingermillet (*ragi*). These figures are probably partly referring to irrigated fields. Most probably the *kharif jowar* and *bajra* are not irrigated and can give a useful indication of rainfall-dependent yields. Some perspective can be added for Mahabubnagar district. In the period 1987-92 there were 2,46,000 ha under *jowar* during the *kharif* season and 45,000 ha also during the *rabi* season. In addition there were 25,000 ha under *bajra* in the kharif season and virtually nothing during the *rabi* season and 29,000 ha under *ragi* during the *kharif* season out of which 1,000 ha also during the *rabi* season [Mahabubnagar district, 1993 (Handbook for 1991-2): 49-50].

TABLE 3.11. SUBJECTIVE OPINIONS FROM FARMERS IN THE RESEARCH VILLAGES ABOUT THE YEARS 1990/1, 1991/2 AND 1992/3

Village	Best Rainfall Year for Dryland Crops (% of farmers)			Worst Rainfall Year for Dryland Crops (% of farmers)			
	1990/1	1991/2	same	1990/1	1991/2	1992/3	same
Deb	88	7	5	1	0	94	5
Mans	96	4	0	0	3	97	0
Ekv	82	19	0	1	0	99	0
Ghat	96	0	4	0	0	96	4
Total	90	8	2	1	1	96	2

Notes: Deb = Debbadaguda (n = 88); Mans = Mansanpally (n = 94); Ekv = Ekvaipally (n = 81); Ghat = Ghatpally (n = 90); Total n = 353.
Source: IDPAD surveys, coordinated by M. Put.

1987: And a fourth bad rainfall year (although not in Maheswaram with regard to the rainfall during the rainy season; however, a dry spell in September caused severe drought risks there). A moderate drought risk in the southern area was mainly caused by a late start and by the dry spell in September. However, this year the district statisticians assessed the yields to be above average in both districts.

1988: The period of four years of bad rainfall had come to an end, with exceptionally high rainfall in 1988, in the south (and in the research area as well) a bit short. The local rainfall stations near the research area showed a less extreme rainfall peak, compared to the area of the district headquarters. Yields, however, were low for *jowar* and even lower in the case of *bajra*, probably because of excess rainfall.

1989: This was a year with good rainfall, although with light and not negligible drought risks, especially in the south, due to a dry spell in August. In Maheswaram the drought risk was even severe. For the districts as a whole yields were very good in the south and good in the north.

1990: A year with average to high rainfall in the north (although starting very early, in May, around Hyderabad followed by lower rainfall in June and July, but not so in the research area) and a bit below average in the south, but with low drought risks. Yields were very good and also almost all farmers in the research area regarded this year as good.

1991: Rainfall was above average and drought risks only very light, although in Maheswaram it was moderate. In the north yields were above average, but, according to farmers in the research area, not as good as in the previous year. In the south yields were bad, although some farmers in the southern research village Ekvaipally regarded this year as better than 1990.

1992: Rainfall data are less complete here, but the evidence for Hyderabad, Maheswaram and Amangal shows a rather bad year, in Maheswaram even with extreme drought risks. Virtually all farmers in the research villages regarded this year as the worst of three years.

1993: Probably this year was also very bad. In Maheswaram there were severe drought risks, especially because of a dangerous dry spell in August.

3.2.9 CONCLUDING REMARKS

The success of attempts to improve dryland agriculture by external agencies are, of course, influenced by the rainfall position in the period preceding, during and after the project activities. AWARE started in Ekvaipally and other villages in the Amangal area in the mid to late 1970s, during a period with relatively good agricultural prospects. It was followed by a few worrying years, in which harvests were low, and measures to introduce more 'drought awareness' probably well received. This was also a time when the World Bank-supported Maheswaram project had its first activities in some villages in the north. After 1987 the rainfall situation improved a lot and farmers even experienced excess rainfall, sometimes resulting in low yields of sorghum and millet, because of too much rain. When this research project started collecting the first data, for the year 1990, farmers were not in a bad shape and the year 1990 probably was one of the best years possible. After 1990 the situation must have deteriorated considerably, with 1992 and 1993 being rather bad years. With this background information we will describe the research villages in Chapter 5. After that we come to the core of the study: the impact of agro-technological interventions on farmers' behaviour with regard to soil, water and crop management and the changes in the ways they cope with drought risks. But before doing so, we will describe the dryland improvement activities of the Maheswaram project and of AWARE and we will give details about the project agencies.

3.3. Recent Trends in the Cropping Pattern of Ranga Reddy and Mahabubanagar District

Sorghum (*jowar*) was and still is the most important crop in the cropping pattern of both districts, as can be seen in Table 3.12. Until the early 1970s the area under *jowar*, *ragi* and *bajra* combined increased by 21-5 per cent in a span of twenty years. After 1975 the area under these crops decreased considerably in both districts, and the total area in 1990 in Mahabubnagar was even below the level of the early 1950s. The decrease of the millets, in particular, is even more clear if we add the (scant) data for the so-called 'small millets', which have virtually disappeared. The relative importance of sorghum and millets in the cropping pattern in Mahabubnagar was shrinking from 54 per cent of the total recorded cropping area in 1965-70 to 45 per cent in 1985-90 and in Ranga Reddy from 50 per cent to 46 per cent.

The area under grams fluctuated during the whole period to around 110,000 ha in Ranga Reddy and around 50,000 ha in Mahabubnagar, mostly inter-cultivated with sorghum and millets. In Mahabubnagar the more recent period saw a rather drastic decrease of the total area under grams. However, if we look at the crop details we see a major shift from a dominance of horse gram and black gram to a more important position for red gram and at a lower level from bengal gram to green gram.

The area under (irrigated) paddy shows a strange fluctuation: from 1,30,000 and 55,000 ha in the early 1960s back to 1,00,000 and 40,000 ha in the early 1970s, up to 1,50,000 and 50,000 ha in the late 1970s but down again to a level of 1,00,000 and 40,000 ha in the late 1980s (and up again later to approximately 1,50,000 and 60,000 ha in the mid-1990s).

The total 'net irrigated area' went up from 94,000 and 49,000 ha around 1970 to 1,68,000 and 65,000 ha around 1990. Most of it was under paddy. According to the (raw) sources for these data the 'net sown area' went down, from 10,25,000 and 3,71,000 ha around 1970 to 9,04,000 and 3,09,000 ha around 1990 and the decrease for the area under dryland crops must have been even more, because of the increase of irrigated land from 9 to 19 per cent of the net sown area in Mahabubnagar and from 13 to 21 per cent in Ranga Reddy.

Two crops were and still are important as non-food grains crops, mostly as 'cash crops': castor and groundnuts. Four other 'cash crops' have always been considerably less important: cotton, tobacco, sesame and chillies. Together these six dryland 'cash' crops improved their position in the cropping pattern from 1,60,000 ha or 21 per cent of the recorded crops (gross cultivated area) in the early 1960s to 2,70,000 ha or 34 per cent in the late 1980s in Mahabubnagar and from 41,000 ha or 15 per cent to 47,000 ha or 17 per cent in Ranga Reddy.

The total production of food grains (rice, sorghum, the millets and the grams) increased from a level of 3,30,000 and 1,50,000 tons in the early 1960s and a stagnating 3,15,000 and 1,40,000 tons in the period between 1965 and 1975 to a much higher level of 4,50,000 and 2,10,000 tons in the late 1970s and 4,30,000 and 2,05,000 tons in the early 1980s but down again to 3,85,000 and 2,00,000 tons in the late 1980s. Per capita the production of food grains was 200 kg around 1960 in Mahabubnagar and 165 kg in Ranga Reddy (excluding the urban population of Hyderabad-Secunderabad town). It went down to 160 kg in Mahabubnagar and 120 kg in Ranga Reddy in the early 1970s and up again to 190 kg in Mahabubnagar and 140 kg in Ranga Reddy around 1980. The late 1980s saw a dangerous decrease in the level of food grains production per capita within the two districts, to only 140 kg in Mahabubnagar and 90 kg in Ranga Reddy. Generally, a level of 180 kg/capita is regarded as a necessary consumption level for food grains if they cover 80 per cent of the food needs. In the early 1960s both districts were close to food grain self-sufficiency. In the late 1980s they only covered an average of 50 (Ranga Reddy) to 75 (Mahabubnagar) per cent of the necessary food grains. In

TABLE 3.12. CHANGES IN THE AREA UNDER VARIOUS CROPS, MAHABUBNAGAR AND RANGA REDDY DISTRICT, 1955-90, AVERAGES PER YEAR (*1,000 ha*)

1 = Mah 2 = RR		1955-60	1960-5	1965-70	1970-5	1975-80	1980-5	1985-90
jowar	1	285	296	307	360	317	310	286
	2	98	99	105	126	118	124	112
ragi	1	40	42	49	32	51	44	34
	2	10	11	13	10	15	13	9
bajra	1	49	43	51	59	56	42	44
	2	9	8	8	11	9	7	3
sm. mi.	1			(92)		(50)		(11)
	2			(20)		(11)		(1)
grams	1	97	118	105	102	117	104	79
	2	49	48	52	52	54	54	53
maize	1	0	0	0	0	0	0	0
	2	5	6	7	8	9	1	5
paddy	1	...	130	106	98	151	133	98
	2	...	154	45	40	53	48	41
castor	1	70	76	73	94	68	74	105
	2	31	29	73	34	23	27	25
cotton	1	4	9	10	9	6	6	8
	2	1	0	0	0	0	0	8
g. nuts	1	133	64	124	152	104	169	145
	2	6	4	7	9	5	7	7
sesame	1	6	5	3	3	2	2	2
	2	6	5	4	6	4	5	3
sugarc.	1	0	0	0	0	0	0	0
	2	1	1	1	1	2	2	1
chilli	1	12	6	7	8	6	8	5
	2	7	3	3	4	3	5	4
tobac.	1	2	3	4	5	9	6	8
	2	1	0	0	0	0	0	0
Total*	1	...	793	840	921	886	898	812
	2	...	268	273	301	294	302	272

Notes: * Excluding the small millets.
sm. mi. = small millets; g. nuts = groundnuts; sugarc. = sugarcane; tobac. = tobacco.

Source: Primary data provided by CESS, Hyderabad.

bad years it approached even the 50 per cent level in Mahabubnagar.

Table 3.13 gives the changes in the productivity per ha for the various food grains. Most crops experienced a considerable intensification. Only the figures for *bajra* show a stagnating yield level per ha, and the figures for grams until the 1980s. The yield/ha level in Ranga Reddy is generally higher than in Mahabubnagar and improvements also seem to happen earlier in this district surrounding the Hyderabad metropolis, than in the more remote Mahabubnagar district. However, in both cases, the intensification of virtually all crops does not compensate for the decrease in cropping area in the last twenty years. Intensification would have been caused by land and crop management improvements, but we should also be aware of the improved (average) climatic conditions in the 1980s compared to the 1960s, which were described in section 3.2.5. If these climatic conditions were to take a downturn again, we fear that the production of food grains per capita will further fall drastically below the food needs. If we add the expected market problems for the most important 'cash crop', groundnuts, the situation warrants much more reason for concern than the rather rosy picture which is currently given by most of the government and NGO representatives we interviewed.

Finally we will give a few comments on the productivity changes of the two most important 'cash crops' of the area: castor and groundnuts. The yield per ha of castor showed a gradual increase from the mid-1950s to the mid-1970s in both districts, with Ranga Reddy at a lower level compared to Mahabubnagar. After 1975 the crop experienced a drastic fall in productivity, after which there was some improvement, but never reaching the level of the early 1970s. Groundnuts showed a rather stagnating yield level in Mahabubnagar, with quite large fluctuations. In Ranga Reddy, though, the yield level did increase gradually from 1965 onwards until it reached an all-time high of more than 850 kg/ha, much higher than in Mahabubnagar and 75 per cent more than in the late 1950s in Ranga Reddy itself. However, in absolute terms the production in Ranga Reddy and the acreage under groundnuts was only a fraction of both in Mahabubnagar.

3.4 The Caste and Labour Situation of Ranga Reddy and Mahabubnagara Districts

In 1981 the total population of Andhra Pradesh was 53.5 million of which 20.2 million were living in the Telangana area. In 1991 the state's population had increased to 66.5 million of which 26.1 million were living

TABLE 3.13. YIELDS OF FOOD GRAINS AND OIL SEEDS IN MAHABUBNAGAR AND RANGA REDDY DISTRICTS, 1955-90 (*kg/ha*)

1 = Mah 2 = RR		1955-60	1960-5	1965-70	1970-5	1975-80	1980-5	1985-90
jowar	1	383	429	428	384	480	503	529
	2	578	611	501	454	677	669	784
ragi	1	433	615	430	517	604	630	663
	2	455	605	539	509	624	748	778
bajra	1	339	371	327	293	294	351	341
	2	334	666	304	378	387	417	409
grams	1	211	218	233	194	200	217	270
	2	181	220	214	266	244	320	380
paddy	1	...	1032	1128	1254	1511	1589	1794
	2	...	1174	1243	1489	1827	1720	1962
castor	1	162	175	270	280	122	188	195
	2	121	150	223	223	180	196	191
g. nuts	1	580	659	500	544	671	637	572
	2	492	729	566	655	787	848	861

Source: Primary data provided by CESS, Hyderabad.

in Telangana. Mahabubnagar had 2.4 million inhabitants in 1981; Ranga Reddy 1.6 million. In 1991 this had increased to 3.1 million and 2.6 million respectively, while the urban district of Hyderabad had grown to have 3.1 million inhabitants. Mahabubnagar's growth had been much slower than that of Ranga Reddy's, part of which was due to an overflow of the rapid expansion of Hyderabad (Mahabubnagar plus 27 per cent, and Ranga Reddy plus 65 per cent in the 1981-91 period).

Almost one-fifth of the population of the two districts consisted of 'scheduled caste' (SC community) people in 1991 (18 per cent in Mahabubnagar and 17 per cent in Ranga Reddy) which is a bit more compared to Telangana or Andhra Pradesh as a whole (16 per cent). The number of 'scheduled tribes' (ST community) people was much less: 2,30,000 in Mahabubnagar in 1991 (7 per cent of the total in the district) and 1,10,000 in Ranga Reddy (4 per cent). This is below the average for Telangana (9 per cent are 'tribals') and for Andhra Pradesh as a whole (6 per cent are 'tribals'). Until 1981 there was a gradual increase in the percentage of both SCs and STs because of a differential (higher) natural increase compared to 'forward caste' people and people belonging to the 'backward classes' (BC community). If we look at the population increase between 1981 and 1991 for Mahabubnagar (plus 27 per cent) this is still the same although much more clear for the ST community (plus 53 per cent) than for the SC community (plus 30 per cent). In Ranga Reddy the situation was completely different, though: there the ST and particularly the SC community grew less fast than the average population: plus 57 per cent and plus 39 per cent as against the overall growth of plus 65 per cent (Rao and Madhavan, 1988; GoAP, 1993: 42-3).

Out of 2.4 million people counted in 1981 in Mahabubnagar, about 43 per cent were regarded as 'workers', or 10,49,000 people. Of all these workers, 5,10,000 were counted as 'cultivators' (people whose main occupation was farming, either as owner or as tenant of the land), almost 49 per cent. Another 4,00,000 people were 'agricultural labourers' (it does not necessarily mean that they did not own or rent land as farmers but that their main income was from working as agricultural wage labour for other farmers). This is 38 per cent of all workers. Of the remaining group of 1,39,000 workers in Mahabubnagar probably half worked in the 'household industry' (mostly in the rural and urban informal sector) and the other half in other occupations, partly as civil servants for the government. In 1991 the number of 'workers' had increased to 1.5 million, of whom 5,90,000 were 'cultivators' 6,30,000 were 'agricultural labourers' and 2,80,000 were 'other workers'. The differential growth of these three categories is striking: against an overall increase of 43 per cent, cultivators only increased by 16 per cent, agricultural wage labourers by 58 per cent and other workers by more than 100 per cent in those ten years.

In Ranga Reddy 6,80,000 people were regarded as 'workers' in 1981, of whom 2,30,000 or 34 per cent as farmers ('cultivators') and 2,00,000 or 29 per cent as agricultural labourers. The remaining group is bigger than in Mahabubnagar, and close to Hyderabad town, the number of people in non-agricultural wage labour jobs would have been considerable. In 1991 the total number of workers in Ranga Reddy had increased to 10,10,000 people (plus 49 per cent), of whom 2,60,000 were cultivators (plus 13 per cent only), 2,90,000 were agricultural wage labourers (plus 45 per cent) and 4,70,000 were 'other workers' (plus 88 per cent) (Rao and Madhavan, 1988: 87-8; GoAP, 1993: 16-17).

Not all 'cultivators' in both districts belong to the forward castes (FC) or to Backward Classes and not all SC and ST people are 'agricultural labourers' mainly, but there is a high correlation. However, there are far more 'agricultural labourers' than there are SCs and STs in Mahabubnagar and Ranga Reddy. This means that for some FCs and a considerable number of BCs, agricultural wage labour is a main source of income.

The number of agricultural wage labourers in both districts has grown considerably and consistently from 1961 to 1991 (see Table 3.14). The number of cultivators dropped between 1961 and 1971 and went up again between 1971 and 1991, which could, maybe, partly be explained by different statistical definitions. Comparing 1991 with 1961, the growth in Mahabubnagar was considerable (+ 44 per cent), and in Ranga Reddy a little higher than in Telangana as a whole and in AP (Ranga Reddy + 24 per cent; Telangana + 21 per cent and AP + 5 per cent). However, the growth of the number of cultivators was much below the population growth in both districts and in Telangana and the state as a whole. In absolute figures the number of cultivators increased by 1,80,000 in Mahabubnagar and the number of agricultural labourers by 3,60,000 in those thirty years. In Ranga Reddy there were 50,000 additional cultivators and 1,50,000 additional agricultural labourers.

3.5 The Experiences of Land-Use Oriented NGOs in Mahabubnagar and Ranga Reddy Districts

3.5.1 NGOS IN MAHABUBNAGAR AND RANGA REDDY DISTRICTS

Compared to some other districts in Andhra Pradesh (e.g. Anantapur and Chittoor), NGO activities in Ranga Reddy and Mahabubnagar districts, are much more modest and also more recent, with the exception of WCADL in the north-western part of Mahabubnagar and AWARE's early activities in the Amangal *mandal* of north-eastern Mahabubnagar. Currently, in Mahabubnagar besides AWARE and WCADL, NGOs like Youth for Action (YfA), VASORD and ARADS are the most important ones in financial terms, followed by NGOs like Villages in Partnership (ViP), AIRDS, BGMS, and CRYD. In Ranga Reddy the situation is not very clear. AWARE, VASORD and the Christian Concern Mission are active in some areas, but there is also a likelihood of 'contact overflow' of the NGOs with their bases in urban Hyderabad or its twin city Secunderabad. The spatial coverage of NGOs is not complete at all. In both districts there are quite some *mandals* without any NGO interference.

In 1990 the district collector of Mahabubnagar started organizing the various NGOs in the district in a 'platform'; 19 NGOs were involved. One of the results was a gentlemen's agreement among the NGOs that in a village where an NGO was working no other NGO would start activities. That is the reason why no other NGO started activities in Amangal *mandal*: because it was 'claimed' by AWARE, even if the actual AWARE involvement was very minimal after 1990.

TABLE 3.14. FARMERS AND AGRICULTURAL LABOURERS IN ANDHRA PRADESH, TELANGANA, RANGA REDDY AND MAHABUBNAGAR DISTRICTS, 1961-91 (*millions*)

	1961		1971		1981		1991	
	Cult.	Agric. Lab.	Cult.	Agric. Lab.	Cult.	Agric. Lab.	Cult.	Agric. Lab.
AP state	7.49	5.34	5.80	6.83	7.41	8.33	7.89	11.63
Telangana	2.85	1.72	2.36	2.18	3.06	2.79	3.44	4.00
- Mahab.	0.41	0.27	0.36	0.35	0.51	0.40	0.59	0.63
- RR	0.21	0.14	0.18	0.17	0.23	0.20	0.26	0.29

Source: Rao and Madhavan (1988: 87-8); GoAP (1993: 16-17).

There are relatively good contacts between NGOs and the Mahabubnagar authorities. The government channels part of its current funds into the land improvement programme through a few NGOs. UNICEF also tries to use local NGOs for its programmes on girls' education, and CAPART (an organization linked to the Central Government of India, see Chapter 2) recently gave Rs 1 million to NGOs to construct village link roads (interview with Dr Swamy ViP and Mr Wesley, of the Fellowship for Professional Workers, July 1993).

We will present three case studies of NGOs in Mahabubnagar which in different ways deal with the environment and agriculture. In Chapter 4 we will then continue with the analysis of AWARE, for a long time the most important NGO in Mahabubnagar, if not in Andhra Pradesh.

3.5.2 THE EXAMPLE OF THE WEAKER COMMUNITY'S ACTION FOR DEVELOPMENT AND LIBERATION (WCADL)

The founding members of WCADL are Mr and Mrs Francis, active in social work since 1962. In 1976 they selected three villages in north-western Mahabubnagar, which was regarded as one of the most backward areas in Andhra Pradesh. WCADL claims to be the first NGO with a comprehensive socio-economic programme in Mahabubnagar district.

The basic principle is to form *sanghams* of SC (and ST and some BC; not exclusive, but strongly focused on the really poor: small and marginal farmers plus casual and bonded labourers). Women are strongly encouraged to participate in these *sanghams*. Sometimes separate women's *sanghams* are formed, and sometimes there are specific women's activities in mixed *sanghams*. It is not their aim to organize women against men. WCADL wants 'gender unity which is needed to fight oppression'. In almost twenty years of dedicated work there are WCADL *sanghams* now in 225 villages and WCADL is well known all over Andhra because they have a Telugu journal about social change with quite a large circulation range. WCADL has 35 permanent and 150 part-time employees. Many are recruited from among the unemployed youth in the area.

The German donor agency Brot für die Welt and a special Sangham Support Group in Cologne, Germany, supports WCADL with a 'core fund' for administrative purposes and with project funds; no longer with additional money for a revolving fund: that has to finance itself. Members from the Cologne group visit the NGO area almost every year and there are close personal ties.

Adult education has always been the core activity of WCADL. Between 1976 and 1992 the adult education programme covered 550 villages in 18 *mandals*; 51,600 adults participated. WCADL has two training centres: in Kothur and in Rajapur; at Rajapur there is also a farm. Adult education training covers small-scale manufacturing (e.g. in 1987-90, 61 women completed a course in tailoring and they all now earn a living by stitching clothes, with earnings between Rs 10 and 15 per day. It also covers political awareness training and training on legal affairs like equal pay for equal work; land ceilings; child marriage; atrocities against women. One of the major impacts is that wages have gone up in all villages with *sanghams* and more than in villages without *sanghams* (this is a general claim of NGOs working with the *sangham* structure). WCADL also has a health care centre at Rajapur since 1986 and an active involvement in community health care.

Literacy classes also deal with agricultural improvements: dry farming, water conservation, organic farming, compost, moderate use of chemical fertilizers, and the importance of tree plantations. To stimulate afforestation, saplings are distributed and farmers are encouraged to go to the Governmental Horticultural Department to get more saplings as well as advice.

Recently, income generation through productive improvements in agriculture has become more important. Apart from training, WCADL started revolving loan funds: until 1993, 3,223 families benefited in 114 villages. Funds were made available for a.o. wells, land levelling, bunding, supply of seeds, fertilizers, milch animals and sheep. The total sum received from the German donor agency Brot für die Welt, for the revolving fund was Rs 3.3 million. WCADL claims that because of the revolving character—although gradually slowing down due to farmers repaying later than planned—the money distributed among *sangham* members was estimated to have been at least 10 times as large. That would mean that Rs 33 million were put in the land improvement programme, that is: Rs 10,000 per benefiting family. The impact of the revolving fund should not only be measured by looking at the direct benefits but also by acknowledging that it becomes less necessary to borrow money from the village moneylenders and become trapped in bondage.

Sangham farmers are also supplied with seeds from WCADL's own farm or WCADL acts as an intermediary (e.g. in buying Aruna castor seed; a castor seed production programme was also undertaken by *sangham* farmers in collaboration with the seed corporation, ASW and CRIDA).

When soil and water conservation measures were included in the education programme, it was noticed that farmers predominantly practise bunding and levelling in order to create wetland. Despite discussions at *sangham* meetings everybody neglected their dryland. WCADL first offered interest-free loans (to be used for anything) if people agreed to level their dryland: for every acre a family received a loan of Rs 400. Together 100 acres were levelled by 81 families until mid-1993. Monitoring of the yield in these fields showed an increase of 50 per cent after the introduction of the land improvement measures. After the harvest 80 per cent of the 81 families paid back the loan. Many others asked for this facility later. Now WCADL uses another strategy: to be entitled for the *sangham*'s revolving credit fund, people have to have their fields levelled and make bunds, but many people have now started levelling and bunding their fields on their own. WCADL also encourages farmers to plant trees, but this is not a condition. Generally farmers only want to practice 'square bunding' or field/property boundary bunding. WCADL only encourages farmers to practise contour bunding if the slope of the field is steep, but they leave it to the farmers to decide.

Asked about the problem what would happen with the 'areas in-between' if *sangham* farmers with non-contiguous land were to construct bunds in their land, the answer was that most of this land belong to Reddies and they often follow suit (so there is no need to offer special subsidies or assistance to them). If not, their land will deteriorate or at least its value will become less. The WCADL *sanghams* do not regard that as their problem.

Until 1993 more than 300 wells and bores and in-well bores were financed, using the revolving loan fund. However, it was gradually discovered that with the ground-water level going down emphasis on micro-level water harvesting was not so useful. There was a shift to water conservation by bunding and levelling. Also, they discovered that water harvesting schemes were dominated by the not-so marginal farmers; but they did finance tank and well desiltation programmes proposed by *sanghams*.

WCADL does not believe in the 'poverty-oriented' approach of the government watershed programme; according to them the location of the percolation tank is crucial. Only farmers with a little more land will agree to a part of it becoming a tank (and they gain because they get silt and fish water). So the richer farmers have to be involved. They are close to the village elite and they will always try to influence the selection procedure regarding the location so that the lands/wells which will be favoured most will be the ones that belong to the rich farmers.

A problem in the area is the pollution of tanks and wells by industrial activities, some of which are extremely dirty. The water can no longer be used for drinking purposes or for watering animals. In some cases the water was so polluted that the land could not be irrigated (discussion with Mr C. Francis and Mrs L. Francis, July 1993 and WCADL, 1992).

3.5.3 THE EXAMPLE OF VILLAGES IN PARTNERSHIP (ViP)

ViP is a typical example of a group of academically educated professionals who have founded an NGO, meanwhile undertaking consultancy work alongside the NGO activities. Not ideology but 'business in disguise' seems to be their creed, not very different from the top brass of the district administration. They started in 1987 and became fully operational in 1990.

ViP's approach is to train 'villages' to take development in their own hands and to get access to government funds. A village can apply to become associated with ViP; if accepted they get assistance from a full-time manager hailing from the area, trained by ViP. After two years, the person's salary has to be paid fully by the village. ViP has 20 trained village managers now, working in 6 *mandals*. There are one to ten projects per village. Until 1993 ViP enabled the 20 participating villages to get access to Rs 10 million of government funds.

ViP gets Rs 4.5 million from foreign funds per year, mainly from the German Ministry for Development Cooperation, channelled through the German funding agency Caritas. ViP got a three-year open budget; originally personnel costs were also included (core funding). ViP now tries to get its own finances, e.g. by organizing rural tourism for which they have made a brochure.

A lot of ViP village projects deal with land improvements. Many of these projects focus on the construction of irrigation wells. In 1993 there were also six tank desiltation projects in which ViP provided transport and the villages paid for diesel and labour. Other agricultural projects deal with the organization of inputs; agricultural training at *mandal* level; the organization of field trips (e.g. to ICRISAT, dairy farms, horticultural projects in Anantapur and 'wormiculture' projects in Chittoor); checkdams and feeder canals (interview with Dr Swamy, July 1993).

3.5.4 THE EXAMPLE OF YOUTH FOR ACTION (YFA)

Youth for Action is another relatively young 'area development' NGO. It started in 1986 with a mission to 'promote collective community action in the development of drought affected, resource scarce, environmentally degraded lands of Mehoobnagar [*sic*] district in Andhra Pradesh'(YfA brochure 1993) and working primarily through women's associations. Activities include construction and maintenance of water harvesting structures, soil and water conservation, some irrigation, seed distribution, sericulture, fodder cultivation, tree plantation and social forestry, solar energy, biogas, protection of common lands and promotion of dryland agriculture with an emphasis on sustainability and renewal of the village ecological system. But the women *sanghams* are also meant to earn their own funds, so income-generating activities outside dryland agriculture (e.g. brick making) and thrifts are stimulated as well (YfA, 1992/3). The approach is based on the Participatory Resource Appraisal and YfA claims a pioneering role in this respect. It also adopted a 'problem-oriented in-farm research' approach with the 'lab-to-land programme'.

Main donors are NORAD, OXFAM, SIDA, the Rockefeller Foundation, the German Friedrich Ebert Stiftung and the Indian government (e.g. through CAPART) and there are close links with research institutes like CRIDA, ICRISAT, The Indian Council of Agricultural Research and other 'researchers in the scientific temples' (YfA, 1993: 46). The annual donor funding comes to Rs 4 million (YfA, 1992/3: 43).

The NGO has a self image of a 'professional centre for the improvement of people's lives through various self managed institutions, local groups and local governments' (YfA Mission Statement Brochure, 1993). They try to work through village *sanghams*, each and every one of them to be registered. Women *sanghams* are encouraged with 'women members belonging to all castes' (YfA, 1992/3: 9). From very well presented workshop publications and other public relations material it becomes clear that the focus is not on particular groups of relatively poor people (STs,

SCs or agricultural labourers, etc.). Continuously 'the people', or 'the community' are presented as the social entities, with a focus on women as the implementers of programmes. The marginalization of 'the poor' is seen as caused by the segmented approaches to forestry, agriculture and water systems and the solution is an 'emphasis on close links between forestry, food production and natural resource management systems', a 'holistic approach' (YfA, 1993:1-2). Agro-eco security, as it is called, should be the central concern. It is advocated that local governments, the *panchayati raj* institutions, should function 'like NGOs', and local, village-level institutions should be revived, breaking with the election system ('the very election process divides the community': 4). Old systems of land management are cited as examples of good land management: 'local zamindars' (the tax collectors under the oppressive feudal system) 'played an important role in management of natural resources' (p. 2). These 'community' systems of local resource management under the responsibility of the 'aristocratic elites' 'have been weakened to be replaced by State authority for regulation, revenue collection and maintenance which made people depend on the outside agency for upkeep of the natural resources and the welfare of the people' (p. 5). Local bodies should be directly funded and NGOs could then play a major role at the village level, and at the level of the 'watershed communities', assisting the 'micro planning and the participatory management process'. Part of the funds should come from 'the industrial corporate sector who enjoyed much of the share of our natural resources. They have to contribute a lot' (p. 48). The goal is the 'greening of rainfed areas by maximizing rainwater conservation and eco-restoration in fragile eco-system(s), where local production systems have ceased to be sustainable' (p. 49). The local NGO-type of management structures would be responsible for the training of people to look after the common property resources, to develop sustainable fodder production, protecting enclosures for village woodlots and grass reserves for stall feeding ('social fencing') and vegetative conservation measures. 'Land shaping, terracing and levelling should be avoided as they are costly measures and also cause soil erosion in the first few years' (p. 52). The involvement of the 'government machinery' should be as minimal as possible. The community itself should be the 'donor' of programmes, by organizing hamlet-wise thrift groups who generate resources. Any government or foreign contribution should be through a 'memorandum of understanding' with the watershed committee and the NGO (p. 53). Is this an NGO building its power base as a modern (or better 'postmodern') zamindari?

3.6 Conclusion

Although not the driest parts of Andhra Pradesh, Ranga Reddy and Mahabubnagar districts are clearly very risky areas for dryland agriculture, and this is most obvious in the study area, the driest and most risky parts within the two districts. Social and economic inequality is rather extreme, and the livelihood security of small and marginal farmers of the SC and ST caste groups, in particular, is continuously threatened, and more so if they do not have irrigated fields. Government attention towards these vulnerable groups and for dryland agriculture has been meagre and late. Also, NGO involvement has been rather small and rather late. However, since the early 1980s both on the part of government agencies and NGOs attention for dryland improvements has grown. Within the NGO community two approaches can clearly be discerned: a target group approach (as with AWARE, see Chapter 4, as also with WCADL, described in this chapter) and an area approach (with Villages in Partnership and Youth for Action as examples).

NOTES

1. UNESCO suggests the following categories for P/ETP: > 0.75 humid; 0.60-0.75 sub-humid; 0.50-0.60 semi-humid; 0.40-0.50 semi-arid to semi-humid; 0.25-0.40 semi-arid; < 0.25 arid.
2. The data for the year 1963 show only 178 mm, but here there might have been a mistake in the report.
3. For Hyderabad the categories are based on the annual rainfall data for 92 years (1901-92); for Mahabubnagar on the annual rainfall data for 90 years (1901-50 and 1952-91). The categories are:

	Mah	Hyd
++	>945	>925
+	814-944	824-925
o	744-813	708-823
-	593-743	639-707
- -	<593	<639

4. In the 1950s there were only two years with drought risk indexes of 6 or higher. But probably the year 1951 should be added; it is the only year for which there are no data for Mahabubnagar. However, the 1951 data for Hyderabad show a year with low rainfall and a moderate drought risk.

APPENDIX

TABLE 1. MAHABUBNAGAR AND HYDERABAD: DROUGHT RISKS 1901-92

	Mahabubnagar			Hyderabad		
YR	DR	P_{rs}	P_{rs}typ.	DR	P_{rs}	P_{rs}typ.
1901	9	470	- -	6	440	- -
1902	10	499	-	5*	533	-
1903	1	1020	++	1	1117	++
1904	6	604	-	8	449	- -
1905	6	570	-	10	426	- -
1906	0	954	++	2	619	o
1907	6	486	- -	6	455	-
1908	7	824	+	2	933	++
1909	2	727	+	2	689	+
1910	2	703	+	6	539	o
Av.	4.9	686	o	4.8	620	o

Note: *Rainy season only started in August.

	Mahabubnagar			Hyderabad		
YR	DR	P_{rs}	P_{rs}typ.	DR	P_{rs}	P_{rs}typ.
1911	5	536	-	9	451	- -
1912	7	624	-	5	500	-
1913	10	422	- -	6	369	- -
1914	2	896	++	0	999	++
1915	3	683	o	0	1079	++
1916	2	1017	++	6	1141	++
1917	2	931	++	3	607	o
1918	10	284	- -	7	397	- -
1919	6	475	- -	6	690	+
1920	10	373	- -	7	383	- -
Av.	5.7	624	o	4.9	662	o

	Mahabubnagar			Hyderabad		
YR	DR	P_{rs}	P_{rs}typ.	DR	P_{rs}	P_{rs}typ.
1921	8	477	- -	4	546	o
1922	9	389	- -	5	456	-
1923	7	553	-	7	526	-
1924	5	560	-	3	609	o
1925	5	680	o	7	677	+
1926	5	589	o	7	436	- -
1927	2	669	o	2	671	+
1928	6	672	o	3	513	-
1929	7	395	- -	9 *	473	-
1930	2	649	o	2	604	o
Av.	5.7	563	-	4.9	551	o

Note: *Rainy season started only in August.

	Mahabubnagar			Hyderabad		
YR	DR	P_{rs}	P_{rs}typ.	DR	P_{rs}	P_{rs}typ.
1931	9	588	-	6	813	++
1932	5	728	+	9	431	- -
1933	0	687	o	4	737	+
1934	6	644	o	5	484	-
1935	5	751	+	9	577	o
1936	4	470	- -	3	463	-
1937	7	508	-	7	447	- -
1938	0	1087	++	3	617	o
1939	3	504	-	7	429	- -
1940	2	703	+	7	475	-
Av.	4.1	667	o	6.0	547	-

	Mahabubnagar			Hyderabad		
YR	DR	P_{rs}	P_{rs}typ.	DR	P_{rs}	P_{rs}typ.
1941	10	404	- -	9 *	369	- -
1942	6	401	- -	4	550	o
1943	9	622	-	6	690	+
1944	8	692	+	6	597	o
1945	2	617	-	2	628	o
1946	4	629	o	7	417	- -
1947	0	1007	++	5	799	++
1948	5	549	-	5	534	-
1949	6	680	o	6	553	o
1950	2	824	+	5	793	++
Av.	5.1	643	o	5.5	593	o

	Mahabubnagar			Hyderabad		
YR	DR	P_{rs}	P_{rs}typ.	DR	P_{rs}	P_{rs}typ.
1951	no data			7	534	-
1952	7	560	-	7	452	- -
1953	1	853	++	3	717	+
1954	2	908	++	3	713	+
1955	0	967	++	o	990	++
1956	2	1040	++	2	635	+
1957	2	779	+	4	631	+
1958	5	654	o	5	713	+
1959	2	875	++	2	700	+
1960	6	721	+	10	545	o
Av.	3.0	817	+	4.3	663	+

	Mahabubnagar			Hyderabad		
YR	DR	P_{rs}	P_{rs}typ.	DR	P_{rs}	P_{rs}typ.
1961	4	898	++	7	618	o
1962	2	859	++	5	878	++
1963	10	152?	- -	2	651	+
1964	2	1154	++	8	528	-
1965	6	666	o	7	657	+
1966	5	734	+	3	539	o
1967	2	814	+	2	785	++
1968	9	653	o	9	517	-
1969	2	647	o	7	462	-
1970	2	755	+	2	985	++
Av.	4.4	733	+	5.2	662	+

	Mahabubnagar			Hyderabad		
YR	DR	P_{rs}	P_{rs}typ.	DR	P_{rs}	P_{rs}typ.
1971	6	480	- -	7*	446	- -
1972	8	449	- -	10	307	- -
1973	6	670	o	4	678	+
1974	5	746	+	6	714	+
1975	1	864	++	3	930	++
1976	8	688	+	7	678	+
1977	4	610	-	9	484	-
1978	5	767	+	4	1007	++
1979	4 **	642	o	5	611	o
1980	5	673	o	2	727	+
Av.	5.2	659	o	5.7	658	o

Notes: * Three wet months, but rainy season only started in August.
** Rainy season started in May, followed by five wet months.

	Mahabubnagar			Hyderabad		
YR	DR	P_{rs}	P_{rs}typ.	DR	P_{rs}	P_{rs}typ.
1981	2	799	+	0	917	++
1982	6	426	- -	6	597	o
1983	10	609	-	3	934	++
1984	(9)	(426)		5	511	-
1985	(9)	(349)		9	388	- -
1986	(7)	(348)		4	538	-
1987	(7)	(488)		4	453	- -
1988	(5)	(775)		2	898	++
1989	(6)	(679)z		5	910	++
1990	(3)	(528)		5*	577	o
Av.	- -	- -	- -	4.3	672	o
1991	(2)	(706)		2	661	+
1992				4	569	o

Notes: * Rainy season only started in August.

(Mahabubnagar: town; 1951 is lacking; the figures for 1984-91 are averages for Mahabubnagar district as a whole; in earlier years the town figures were generally 15-20 per cent higher); DR = Drought Risk (see text).

P_{rs} = Rainfall in the rainy season (either June-September or July-October, whichever is higher).

P_{rs}typ = See P_{rs}.

Typology:	Mahabubnagar	Hyderabad
++	> 852 mm (1154 in 1964)	> 785 (1141 in 1916)
+	688-852 mm	631-784
o	629-87 mm	539-630
-	499-628 mm	454-538
- -	< 499 mm (284 in 1918)	< 454 (307 in 1972)
?	152 in 1963, but doubtful.	

These figures are based on the rainy season, not on the annual total; therefore they are lower than the figures given in footnote 2, which are annual figures.

4. Project Interventions in Dryland Agriculture: The Examples of the Government of India/World Bank-supported Maheswaram Project and of AWARE's Activities in Amangal *Mandal*

M. Put and A.J. Dietz

In Ranga Reddy and Mahabubnagar, and maybe in the whole of Telangana, the two most important projects with an intention to improve dryland agriculture were the World Bank-supported Maheswaram project in the south of Ranga Reddy district on the one hand, and on the other, AWARE, one of the largest NGOs in Andhra Pradesh (if not India), in one of its pioneer areas, the Amangal *mandal* in the north-east of Mahabubnagar district.

The World Bank-supported 'Pilot Project for Watershed Development in Rainfed Areas' in Maheswaram *mandal* (in short: the Maheswaram project) in the south of Ranga Reddy district was one of the most important projects in the whole of Telangana. Located 15 to 40 km south of Hyderabad, the project initially covered 23 villages and 25,000 ha. After the 1990/1 revision it was increased to 38 villages and 39,000 ha, comprising 50,000 people in 10,800 households, of which 6,200 were farming and 4,600 were landless household (GoAP, various years: 2-3).[1] Of this 'enlarged' project area, around 29,000 ha (73 per cent) was arable land, of which most (27,000 ha or 93 per cent) was dryland. Soil and water conservation activities on arable and non-arable lands (expected costs Rs 1,100/ha) together with the introduction of improved crop management technologies (expected costs Rs 500/ha) were some of its activities to increase the production of dryland crops. While most soil and water conservation activities were financed from a fund specially created for

* This chapter is largely based on two chapters included in Put (1998). Courtesy of Thesis/Thela Publishers, Amsterdam.

this project, most crop management activities were financed by ongoing schemes of the Department of Agriculture and the Department of Rural Development (through the Integrated Rural Development Programme). For extension the Maheswaram project depended on the T&V system, introduced in Andhra Pradesh only a couple of years before it started its own activities. Although the project was supposed to run for a period of 7 years (from 1983/4 to 1989/90) it ran for more than 9 years and came to a close in December 1993.

The Maheswaram project was one of the four World Bank-sponsored and one of the many other watershed projects implemented in India in the second half of the 1980s. It was in a period that the Indian government had adopted the watershed approach as one of its more important intervention strategies in rural areas. Stressing the need for integrated area development, the watershed programme deviated from many other government initiatives. First of all, having an integrated approach, the programme not only aimed at increasing and stabilizing agricultural production in the watershed, but it also tried to improve the ecological balance in the area. For the Indian bureaucracy it meant that various departments were involved. Secondly, having an area approach, the programme did not necessarily cater for the poor. In fact, having an important agricultural component, landless households were excluded from direct project activities from the start. In case the project wanted to give financial assistance to marginal and small, ST and SC farmers (e.g. in the form of subsidies for inputs), other programmes had to be asked for (financial) support.

AWARE (Action for Welfare and Awakening in Rural Environment) began its pioneering activities in 1975 in Amangal *mandal* (Mahabubnagar district) situated some 60 to 80 km south of Hyderabad. Although initially working with *Girijans* (Scheduled Tribes, STs) only, it quickly included *Harijans* (Scheduled Castes, SCs) as its target group as well. In the initial years, AWARE's activities covered some 37 villages in this *mandal*, working with a target group of more than 1,000 households and more than 6,000 people. Within twenty years the organization had managed to built an empire, covering millions of people in thousands of villages located not only in Andhra Pradesh but also in neighbouring states. In addition to AWARE's central office in Hyderabad, it had established eight regional offices, a couple of large farms, hospitals, vocational training centres for handicrafts and cottage industries, and many more buildings had just left the drawing-board.

Having an empowerment approach, AWARE focused on the

strengthening of its target groups by creating awareness about the root causes of their poverty and by organizing them in local and supra-local *sanghams*. Moreover, following an integrated approach (both welfare and awakening are included in its name), it not only implemented social motivation programmes but basic needs and economic development programmes as well. Advocating a 'withdrawal strategy', AWARE worked in each and every village for not more than 10 years, after which sustainable *sanghams* were supposed to cater to the needs of their members, while AWARE limited its activities to moral support. For agricultural extension, for instance, this means that strengthened by AWARE's involvement and backing, the *sanghams*, acting as countervailing power, were expected to press government T&V staff to visit their village.

4.1 The Maheshwaram Project

4.1.1 HISTORY

Except for a few measures taken during earlier Plans, it was not until the Seventh Plan (1985-90) that the Indian government began to promote the development of dryland agriculture in semi-arid areas on a somewhat larger scale. In fact, the first discussions in 1976 with the World Bank on a 'project for watershed development in rainfed areas' did not produce any significant results. According to the World Bank and the Indian government (1983: 1; GoI, 1988: 2) the formulation of the project was retarded because of the complexities of dryland agriculture, a lack of experience for the systematic tackling of dryland areas, a lack of agreement on the technologies to be recommended, the fluctuating priorities of the Indian government in the development of this type of agriculture, the inappropriate institutional structures and because of problems with the selection of participating states. Only after its announcement of the New Twenty Points Programme in 1982, which included the development of dryland agriculture as well as the adoption of the watershed as a national strategy for planning, did the government allocate more funds to promote this type of agriculture. The development of dryland agriculture was given a further impetus in 1983 when the Indian government and the World Bank agreed to implement such a pilot project for rainfed areas in selected watersheds in Andhra Pradesh and three other states. For Andhra Pradesh, the Indian National Watershed Development Policy Committee approved an outlay of Rs 58.2 million ($ 5.8 million) for a period of seven years.

Several guidelines for the selection of watersheds were accepted by both partners, of which the most important were (World Bank, 1983: Project File, Item 4): (1) being a representative of a major agro-climatic region with more than 750 mm of reasonably reliable annual rainfall; (2) facing resource conservation and management problems; (3) being served by an effective agricultural extension service and a feeder road system; (4) people's willingness to co-operate and strengthen local organizations; and (5) located at a place within two hours' travelling distance from an agricultural research centre capable of technical back-up, adaptive research, evaluation and training. On the basis of preliminary research undertaken by the Andhra Pradesh government on three watersheds (in Ranga Reddy, Vishakhapatnam and Chittoor district), the Maheswaram watershed in the former district was selected. However, there are doubts whether it meets the rainfall criterion. Although all project documents (e.g. GoAP, 1981: 2) report an average of 770 mm/year, the calculations in Chapter 3 indicate that the average for the area was 638 mm/year.

The preliminary research undertaken by the Andhra Pradesh government on the three watersheds dealt extensively with the agro-climatological situation in each area, the problems dryland farmers encountered and the activities to be undertaken by the government. In contrast, it hardly analysed the socio-economic characteristics of the people living in the watersheds and the ways they should be involved in project planning and implementation (GoAP, 1981). Moreover, although discussions were held with farmers not much thought was given to their needs and priorities or how innovations had to be transferred to them. In fact, civil servants together with experts on dryland agriculture proposed the agricultural recommendations (Head State Watershed Cell, personal communication, March 1994).

The main reason for the initial teething problems of the Maheswaram project were financial (APAU, nd). Initiated in 1983, seven years after the first discussions between the Indian government and the World Bank, the project started in June 1984 when an area of 32 ha was taken up on a pilot basis. Only after the Seventh Plan had become operational, financial problems had more or less been solved and from 1985/6 onwards the project was implemented on a larger scale. Afraid that it would not be able to spend the funds provided, the project management tried to catch up by implementing different activities before any kind of survey (i.e. an engineering survey on slopes and contours, gullies and hydrology; a soil survey; and a socio-economic bench-mark survey) had been completed.

The Maheswaram project did not work in all villages simultaneously. Instead, the watershed was split up in sub-watersheds and the activities were implemented in a phased manner, moving from one sub-watershed to another. Thus, villages located in a certain sub-watershed were included in the project for a couple of years only. The first activities were carried out in 1985/6 in five villages located in the first sub-watershed, including Mansanpally, one of the study villages. In subsequent years, while continuing in the first sub-watershed, the second and third sub-watershed were included in the programme. From 1987/8 onwards, the programme started its activities in another three watersheds, while slowly withdrawing its activities from the first.

Up to 1987/8, the emphasis of the soil and water conservation component of the project was on contour and graded bunding with grassy waterways and the rehabilitation of existing bunds. These expensive and complicated activities were in fact a mere continuation of already existing programmes of the Department of Agriculture. It also meant a continuation of the engineering approach present in so many government projects of the past. Major revisions for this component were proposed by a supervisory team of Indian and foreign (including World Bank) consultants in 1988 (GoI, 1988: 22; Sivamohan et al., 1993: 299), showing not only the flexibility of the project but also that decision-making was neither placed at farmers' nor at project level. Based on their advice the project slowly shifted its focus to more affordable, simple and replicable technologies (GoI, 1988: I). Thus, activities like graded bunding with grassy waterways were reduced, while land smoothening/levelling and farm ponds were completely dropped. Instead, vegetative barriers (*khus*) planted on the contour and *in-situ* soil and moisture conservation practices (like ploughing and sowing on the contour), which had been initiated on a small scale a couple of years earlier, were implemented on a much larger scale to become the main activity of the soil and water conservation component in later years (Sitapathi Rao et al. 1990: 75).

Although the component on improved crop management for dryland crops was included in the project, it had not received due attention (GoI, 1988: I). This was largely because it was envisaged that it would be financed from ongoing programmes in the state. It meant that, instead of having its own agricultural extension system, the project had to rely on the existing T&V system. Besides, for the supply of subsidized inputs like improved seeds, chemical fertilizers, plant protection chemicals, iron ploughs and the like, it had to turn to schemes like the promotional programmes for HYVs, National Oil Seed Development Programme and the District Rural Development Agency (DRDA) of Ranga Reddy district. However, the project management could not always rely on these non-project funds every year and was never sure about the inputs provided. On the whole, the money contributed by these non-project schemes was rather marginal, especially when compared to the total project budget (team leader Maheswaram project, personal communication, November 1991).

The revisions proposed by the supervision team of consultants concerning improved dryland crop management were less spectacular than for the soil and water conservation component. Crop demonstrations and the development of horticulture on dryland, launched on a small scale in earlier years, were taken up on a larger scale after 1988. More attention was also given to organic farming by digging compost pits and multi-purpose water points for fruit trees. In an attempt to increase farmers' awareness about the dryland agriculture component, the project also decided to improve its communication structure and to train village youth as its aides. A last addition was the inclusion of a rural energy management section, i.e. the distribution of more fuel efficient stoves (smokeless *chullas*) (GoI, 1988: II).

A final revision was made in 1990/91. As a consequence of the devaluation of the Indian rupee against the US dollar, the project outlay was increased by more than 40 per cent to Rs 84.2 million ($ 4.7 million). It was also decided that since the project was behind schedule, the planned closing date was extended from December 1991 to December 1993. But instead of using a part of the increased budget for its dryland agriculture component, the project area was extended to the south in 1990/91 with an additional 15 villages and an area of almost 14,000 ha, of which 83 per cent was under cultivation. One of the new villages was Debbadaguda, the second study village, situated in Kandukur *mandal.*[2]

4.1.2 IDEOLOGY AND OBJECTIVES

The idea of the Maheswaram project was to increase agricultural production on drylands on a sustained basis through a technocratic approach, using a comprehensive design (GoAP, 1985c: 2-3). This idea was clearly a mixture of Gandhian ideology (using concepts like self-reliance and self-sufficiency) and ideas prevalent in the Indian state in the 1980s (increased food grains production through a modernization approach, bridging the gap between the well-endowed irrigated areas and the less-endowed dryland areas, and an increasing environmental concern. These were combined in a holistic approach.

The concept of the watershed was the holistic approach looked for.

Here, planning and implementation of project activities were supposed to be undertaken in an integrated and comprehensive manner. Although the policy of the Indian state was based on bridging the gap between the rich and the poor, implicitly the watershed approach favoured environmental management above the socialist inspired principle of equity or re-distribution of economic growth. In contrast to many rural development programmes which focused especially on the poor, be they farmers, agricultural labourers or artisans, watershed projects did not necessarily concentrate their activities on specially defined target groups. Instead, these projects followed an area approach and tried to combine the seemingly conflicting objectives to improve the productive circumstances in the entire watershed, thereby trying to stop environmental degradation and to enhance the asset position and income earning potential of all farming households (irrespective of class and caste) in the area. The Maheswaram project, not having a separate fund to supply subsidized inputs to small and marginal, ST and SC farmers, had to appeal to other government programmes like the IRDP for financial assistance, for example.

Although the various ultimate and intermediate objectives of the Maheswaram project were often mixed up (World Bank, 1983: 7, 13; GoAP, 1985c: 4-5; GoI, 1988: 1, 4, 41; Sitapathi Rao et al., 1990; 1, 40-4), three main objectives stand out clearly. The first concerns the integrated and comprehensive development of both arable and non-arable drylands in order to increase and stabilize production of dryland crops, forage, fruit, timber and firewood. This had to be achieved through the widespread adoption of agricultural innovations suited to the existing environmental and socio-economic conditions of the area. Whereas this objective was operationalized into specific defined targets for the soil and water conservation component, such targets for the component on improved crop management for dryland crops (such as production and productivity targets) were not indicated (GoI, 1988: 5).

The second, rather peculiar, objective was the large scale verification of already developed technological packages (World Bank, 1983: 7). It was included because at the time the project started its activities some recommendations were of rather recent origin and conclusive evidence about their relative advantage was not available. It meant that the designers of the project introduced all kinds of innovations without knowing what their effects on production and the environment would be. Hence, one may argue that, in a way, the farmers in the watershed were part of an experiment.

The third objective was to restore the ecological balance in the watershed. It had to be achieved through various activities on arable and non-arable lands like soil and water conservation, social forestry, afforestation and by providing *chullas* to households.

4.1.3 ACTIVITIES AND APPROACHES

Unlike earlier soil and water conservation projects, the comprehensive and integrated design of the Maheswaram project entailed that not only arable but also non-arable lands were part and parcel of project activities.

TABLE 4.1. AGRICULTURAL RECOMMENDATIONS OF THE MAHESWARAM PROJECT FOR DRYLAND CROPS

	(Adapted) indigenous practices	Innovations introduced before Maheswaram project	Innovations introduced by Maheswaram project
Soil and water conservation structures:		Graded bunding (up to 1986/7) Diversion drains Water ways Gully control works	Contour bunds with waste weirs (1986/7 up to 1990) Vegetative barriers of Vetiver grass (*khus*) (after 1987/8) Percolation dams
In-situ soil and water conservation:	Off-season (preparatory) tillage, i.e. summer tillage for *jowar* Interculturing for all crops	Contour cultivation (preparatory tillage, sowing and interculturing) but watered down to sowing and interculturing across the major slope	Off-season (preparatory) tillage: i.e. post-harvest tillage for all crops Key lines Dead furrows
Improved dryland crop management:	Early sowing of *jowar* Intercropping *jowar* with redgram but in 2:1 ratio Application of FYM Interculturing for all crops Manual weeding	Improved seeds Optimum plant population Seed-cum-fertilizer drill Iron plough Chemical fertilizers Pesticides	Improved seeds Manure pits Integrated pest management

Sources: World Bank, 1983; GoAP, 1985c; GoI, 1988; Sitapathi Rao et al., 1990; Various Action Plans.

However, the project did not tackle issues like animal husbandry, basic needs or off-farm employment. On the other hand, this comprehensiveness meant that the project slowly shifted focus. As a consequence of greater ecological concern of the state government in New Delhi, dryland development slowly had to give way to arrest environmental degradation. In the section on project funds it will be shown that although initially most allocations were for the improvement of dryland agriculture, yet most of it was spent on forestry.

The main components of the project concerned soil and moisture conservation, the management of dryland crops, pasture and fodder development, farm forestry and afforestation. Activities like horticulture development in drylands and organic farming gained more importance after the 1988 Mid Term Review, a review that brought alterations in the other components as well. Since this study concentrates on dryland food crops like *jowar* and redgram and on the cash crop castor, the next subsections deal with activities meant to increase and stabilize the production of these crops.

Looking for affordable, simple and replicable technologies, the Maheswaram project not automatically favoured all kinds of 'real' innovations. Consequently, the recommendations were a combination of the following:

(1) Stimulating the use of (adapted) viable indigenous technologies.
(2) Stimulating the use of agricultural technologies that had been introduced by government extension services in the area already before the start of the project.
(3) Stimulating the use of more recent 'real' innovations for dryland agriculture which had not yet been introduced in the area.

These three dimensions will be used in the following introduction of the activities meant to increase and stabilize the production of dryland crops, i.e. the component on soil and water conservation structures first, followed by the component on *in-situ* soil and water conservation and finally the component on improved dryland crop management (see Table 4.1). If necessary, details of the recommendations will be given in Chapter 6.

Soil and Water Conservation Structures

The rationale for the soil and water conservation component in arable and non-arable lands of the project was formulated as follows (GoAP, 1985c: 15):

The topography is generally rolling and undulating. Major area falling under 2 to 3 per cent slope.... The agricultural lands are generally neither protected by mechanical soil conservation measures nor by regular agronomical measures, which is leading to erosion of topsoil even to exposure of subsoil, formation of rills and gullies, sand casting of low lying areas and siltation of irrigation resources. The productivity of these lands is going down year after year.

Although included in initial plans, engineering works like land smoothening and levelling as well as farm ponds (for 'life-saving irrigation' of dryland crops) were not undertaken because of the relatively high costs involved and because *in-situ* soil and moisture conservation practices (see following section) were deemed to be more appropriate. Another reason why farm ponds were not dug was because they occupied space on arable land and suffered from seepage. Thus in its first phase the project concentrated its activities on bunding and the strengthening of existing bunds, on diversion drains and water ways, on gully control works and percolation dams (GoAP, 1981: 77-8). Except for the percolation dams, these were part and parcel of the recommendations made in the 1930s under the name of the Bombay, Hyderabad and Madras Dry Farming Practices. Activities implemented on communal land were to be paid from project funds. Where the construction of graded/contour bunds was taken up on private land, farmers had to pay 75 per cent of the cost involved. Since the project expected that banks would not finance any loans for this purpose, it provided loans carrying a 6 per cent interest rate, to be repaid in 15 annual equal instalments, starting three years after the completion of the works. Bunds would be maintained by the project for one year, thereafter farmers would be responsible for their maintenance.

Not so much because of farmers' involvement but more because of outside interventions, the plans were constantly adjusted. In the early years of project implementation the activities consisted mainly of graded bunds with waterways, the standard formula of the Department of Agriculture at that time. Although the project had planned to use manual labour for their construction, yet after only a couple of months, the project management concluded that the delay in the work was caused by a shortage of labour and decided to use more expensive bulldozers instead.[3] Another change came not much later when based on reviews provided by CRIDA and a team of consultants, the bund cross sections were reduced from 0.500 to 0.334 sq m and graded bunds were replaced with contour bunds with waste weirs located at prominent depression points (Sitapathi Rao et al., 1990: 7-8; Sivamohan, 1993: 299).

More fundamental revisions came in 1986 and especially in 1988. A team of Indian and foreign (including World Bank) consultants, stressing the need for affordable, simple and replicable technology, advised stopping constructing the more expensive works. Instead of contour bunds they advised planting a vegetative barrier of Vetiver grass (*khus*). Since Vetiver was not indigenous to the area, the first 25,000 slips were brought from the World Bank-financed Kabbalnala watershed project in Karnataka in 1986. The slips were multiplied in nurseries and initial planting started in 1987/8, when *khus* was planted along with the construction of contour bunds, as was also the case in Mansanpally. In subsequent years, this technique was changed in favour of planting *khus* on the contour without constructing bunds. After the Mid Term Review of 1988, regular bunding operations were considerably reduced. They were completely discontinued after the visit of a team of consultants in September 1990. Instead, although the maintenance of the existing bunds would be attended to, the planting of *khus* hedges became the most important element of the soil and water conservation programme (GoI, 1988: II; Sitapathi Rao et al., 1990: 30-1). Hence, while the project constructed bunds (occasionally with *khus*) in Mansanpally (a village included in the project's first phase), it merely planted *khus* in Debbadaguda (a village included in its second phase).

By 1988, halfway through the project, the achievements of the soil and water conservation programme were rather disappointing. Against a target of 16,500 ha, only some 4,800 ha or 29 per cent of the area was treated with bunds, which was far less than the expected 50 per cent (GoI, 1988: 6). Slow progress was said to be due to financial constraints and teething problems. By August 1993, things had improved somewhat, at least for the *khus* hedges. Some 5,300 ha had been treated with bunds against a decreased target of 7,600 ha, and 7,400 ha with *khus* (target: 7,200 ha) (GoAP, supervision mission 1993: 16). In addition, the project constructed some 80 km of drains or *nalas* to divert excess water from the plots, plugged some 1,800 *gullies* with stones, sand bags and agave, and had dug 7 percolation tanks meant to recharge the ground water and to stabilize or improve the water level in nearby wells.

In-situ Soil and Water Conservation

The inclusion of an *in-situ* soil and moisture conservation component in the Maheswaram project was a kind of novelty. On the basis of the latest findings by CRIDA and ICRISAT and because Indian agricultural policy advocated an integrated and holistic approach, together with the changeover from conventional engineering measures to *khus*, the Mid Term Review decided to give more emphasis to *in-situ* soil and moisture conservation (GoI, 1988: 16). Similar to most other alterations, this change came about after the team of Indian and foreign (including World Bank) consultants stressed the need for affordable, simple and replicable technology; in itself this was laudable of course. However, the change was not a response to farmers' demands and it was not known whether the recommendations suited their needs.

The *in-situ* soil and moisture conservation component was carried out on private plots in arable lands in addition to construction of graded/contour bunds or planting of *khus*. More than the component on soil and water conservation structures, it consisted of a mixture of indigenous practices and innovations. Off-season (summer) preparatory tillage for *jowar* and harrowing between a standing crop (interculturing) were indigenous to the area (see Chapter 6). Knowing that not all farmers followed these indigenous practices (and if they did, it was not always to the extent the project considered optimal), the project set itself to stimulating farmers to do so. The novelty for farmers was that the project recommended that they perform these tillage and other operations along the contour. In fact, contour cultivation became one of the more important elements of the Maheswaram project. The resultant ridges, furrows and crop rows along the contour would serve as barriers to prevent run-off, would provide greater opportunity for the soil to absorb water and would arrest erosion as well. It was assumed, that through contour cultivation, crops would benefit more from the available rainwater, soil nutrients, farm yard manure (FYM) and chemical fertilizers, which would result in higher yields.

Contour cultivation, being part and parcel of the Bombay, Hyderabad and Madras Dry Farming Practices, was no 'real' innovation. 'Real' innovations introduced by the Maheswaram project numbered three:

(1) Post-harvest tillage along the contour to increase the infiltration of rainwater from *rabi*, winter and summer rains.
(2) Key lines along the contour to facilitate contour cultivation (after the change-over to *khus*, these hedges serve as key lines).
(3) A 'dead furrow' along the contour (a furrow ploughed in a standing crop which does not carry a crop itself) meant to increase the infiltration of rainwater into the root profiles of crops.

Unfortunately, the notion about some of the recommendations for *in-situ* soil and moisture conservation practices was not always very definite.

This applies to both the Plans of the projects (see various Action Plans) and to the interpretations by the staff. First of all, the concept of contour cultivation was often interchanged with cultivation across the major slope. This is a kind of half-way house concept, meaning that operations were neither undertaken along the contour nor up and down the slope. Moreover, contour/across cultivation was watered down to sowing and harrowing between a standing crop (interculturing) only. Neither the documents nor the staff have ever referred to preparatory tillage along the contour/across, for instance. Second, knowing that post-harvest tillage would be hampered in case long duration crops like redgram and castor were cultivated, project staff were not sure what to recommend. Third, the intervals between two dead furrows were not very clear to all staff either.

Improved Management of Dryland Crops

The component on improved management of dryland crops was carried out in farmers' fields only. To increase and stabilize yields of dryland crops, the Maheswaram project recommended a combination of (adapted) indigenous practices and innovations that had been developed over the years at dryland research stations (see Table 4.1). Initially, the project focused its activities for crop husbandry on the main dryland crops grown in the area: sorghum (*jowar*) with its intercrop redgram, and castor. After the Mid Term Review, following the advice of the team of Indian and foreign (including World Bank) consultants, it emphasized horticulture (vegetables and fruit trees) (GoI, 1988). The recommendations for *jowar*, redgram and castor will be introduced first, those for horticulture will briefly be discussed later (GoAP, 1981: 54-8; GoAP, various action plans; supplemented by personal communication with project staff and staff at CRIDA).

The recommendations for *jowar*, redgram and castor were as follows:

- To sow *jowar* early (or dry sowing as it is sometimes called). It is recommended for *jowar* and, if intercropped, for *jowar*/redgram. Aware that this was an indigenous practice, the Maheswaram project followed largely farmers' preferences.
- To intercrop *jowar* with redgram. Again the project was aware that it was an indigenous practice. However, the project advised the planting of lesser rows with *jowar* and more rows with redgram.
- To cultivate improved varieties of all kinds of dryland crops. Initially,

the Maheswaram project propagated improved seeds which had been released before the commencement of the project, but in later periods some of the latest improved varieties were also propagated.

- To sow a prescribed number of plants per hectare. This recommendation falls in two parts: the seed rate, normally expressed in kg/ha, and the distance between plant rows. A low seed rate and wide spacing of plants had already been advised by the Bombay, Hyderabad and Madras Dry Farming Practices.
- To use a seed drill or a seed-cum-fertilizer drill. Although seed drills were not indigenous to the area, they were indigenous in other semi-arid dryland areas, as in Karnataka. The seed-cum-fertilizer drill was an adaptation of the seed drill, but was not the result of latest research.
- To plough with iron ploughs. Like the drills, iron ploughs were not the result of latest research either.
- To apply relatively low doses of chemical fertilizers at a specific time and at a prescribed place. Whatever these recommendations, chemical fertilizers were already known in the area before the start of the project.
- To apply organic fertilizers. Observing that limited financial resources prevented many farmers from applying the recommended doses of chemical fertilizers, after the Mid Term Review the project put more emphasis on organic farming (GoI, 1988: 16). However, manuring was indigenous to the area and as such farmers did not need to be told about its benefits.
- To make compost in special pits. Realizing that insufficient and low quality Farm Yard Manure (FYM) was produced, in its final phase, the project advised farmers to dig compost pits. Although farmers were acquainted with dung hills, pits were new in the area.
- To interculture by indigenous blade harrows. Well aware that it was an indigenous practice but that farmers often limited the operation to one to two interculturings, the Maheswaram project advised farmers to remove weeds in *jowar* with the *dante* and in castor with the *guntaka* as frequently as possible.
- To weed by hand in between the crop rows, an indigenous practice as well. Initially, several kinds of herbicides were also included in the plans (GoAP, 1981: 52). Because of the high costs involved and their labour replacing nature, they were never recommended by the project, though.
- To fight pests and diseases in crops in an integrated manner, combining indigenous practices and innovations (some new, some already introduced in the area before the start of the project). For *jowar*, farmers

were either advised to sow early, or to use improved, short duration seed varieties, or to use chemicals for heavy invested plots. To control attacks of the pod-borer on redgram, the project recommended several kinds of pesticides. For castor, the project advised farmers to sow late, to remove and destroy eggs and caterpillars, to remove and burn infested plants, and to apply chemicals. In case of the semilooper, the project also advised the utilization of the presence of two important natural parasites which were generally in abundance in the second half of August as a biological control.

Inducing Farmers to Adopt the Recommendations

Convinced as it was of the incremental benefits of the recommendations, the benevolent state through the Maheswaram project used the adapted Transfer-Of-Technology (TOT) approach to communicate with farmers and to convince them about the usefulness of the recommendations. Depending on the nature of the recommendations, the project availed itself of certain specific methods to speed-up adoption, i.e. mass media (radio and leaflets), mass meetings, face to face contacts (T&V system, village volunteers), trainings and visits by farmers to agricultural research centres, subsidies of inputs and the like, and the construction of certain structures. Mass media and mass meetings were especially used to inform the farmers about the project and the activities planned, while the other means of communication were used to inform them more specifically about the nature and the advantages of the recommendations.

Concerning the soil and water conservation component, the project did not limit its activities to informing farmers about the advantages of the various elements; instead of waiting until farmers adopted the recommendations, the project constructed bunds and planted *khus* on farmers' plots itself. Although it used labourers from the villages during the first couple of months, the project turned to heavy machinery soon.

It is surprising to find that although adoption of recommended agricultural innovations for dryland crops was one of the main objectives of the project, relatively little attention was paid to the strategy of stimulating adoption. In contrast to the soil and water conservation structures, the recommendations concerning *in-situ* soil and water conservation and improved crop management were popularized by more indirect means, costing less money. The most important element consisted of informing farmers about the recommendations through the T&V system of agricultural extension, pre-seasonal trainings and demonstrations, farmers visits to agricultural research centres, and through the help of specially trained village volunteers. But except for staff who were involved in the (*in-situ*) soil and water conservation components, the project lacked its own extension service for the component on improved crop management. In addition to the supply of information, the project used subsidies to popularize various agricultural inputs. However, these subsidies were not paid from its own budget.

As the Maheswaram project had to use the services of the existing T&V system, early project documents emphasized the need to strengthen the system. Hence, the Andhra Pradesh government agreed to temporarily increase the number of Village Extension Officers (VEOs) in the area, thereby reducing the existing ratio of about 800 farming households per VEO to 400 households (World Bank, 1983: 18). In a later document this was reformulated by the Andhra Pradesh government into 'strengthening will be attended to when felt necessary' (GoAP, 1985c: 28). Nevertheless, and although the Mid Term Review repeated this need for a more intensive T&V system in the project area, the reduction in the number of households per VEO did not come about.

In addition to the T&V system, pre-monsoon training was given in the month of April. So, demonstration plots for *jowar* and castor were laid out by project staff and by scientists from CRIDA, ICRISAT and the Andhra Pradesh Agricultural University (APAU), in plots belonging to large, progressive farmers. Training and demonstration were undertaken to give farmers the opportunity to observe the recommendations, to get an idea of the possible effects it would have in their own plots, and to give feedback to agricultural researchers. However, compared to the T&V system, that provided for fortnightly visits, training and demonstration were less intensive. In fact, demonstrations were only taken up after the Mid Term Review of 1988 (GoI, 1988: 42). This review expressed its concern about the insufficient project staff available for these demonstrations and warned that inadequately done demonstrations would lead to negative results. In 1994, of the targeted 34,000 ha of farm demonstrations, 12,000 ha were laid out (GoAP, supervision mission 1993: 16; GoAP, 1995: 16).

Farmers were also taken to the research stations of CRIDA and ICRISAT, located not far from the project area. On a voluntary basis they could participate in a one or two days training course and were paid a honorarium of Rs 10 for each day of training. Over the years, 574 farmers, representing less than 10 per cent of the farming households in the watershed, participated in these trainings (GoAP, 1993: 16).

Because the project was probably not very satisfied with the progress made by above mentioned extension activities, it decided to appoint (paid) young village volunteers. From selected villages one or two young farmers were chosen, totalling some 70 youngsters, who were trained at CRIDA in *in-situ* soil and water conservation practices and in improved crop management of dryland crops. Hoping that these aides would adopt these recommendations, the project used their services to demonstrate their newly acquired skills to fellow farmers in their village. Such a village volunteer has been active in Debbadaguda, not in Mansanpally.

Earlier project documents show that those who designed the Maheswaram project had expected most effects on adoption, not from bridging the information gap, but from bridging the financial gap. Project designers were quite aware that farmers were familiar with the recommendations which were based on indigenous technologies. They were also quite aware that most farmers had at least a faint idea about the recommendations that had been introduced in the area before the start of the project. Knowing that informing farmers was a necessary but not a sufficient condition, and aware that non-adoption was more related to the relatively high costs of some recommendations, project designers decided to bridge the financial gap by subsidizing agricultural inputs (GoAP, 1981: 4-5, 53). They advised, for instance, that the difference between the costs of improved and local seed varieties (by seed exchange) be met by the project and to give subsidies of 25 to 33.3 per cent to marginal and small farmers in the case of chemical fertilizers and of 50 per cent in the case of implements and pesticides. Hence, the second largest entry in the proposed budget was that on agricultural subsidies, i.e. some Rs 21 million or 23 per cent of the budget (GoAP, 1981: 11).

However, the World Bank was not in favour of subsidies (GoI, 1988: 43). Hence, before the commencement of the project the governments of India and Andhra Pradesh decided that if subsidies were given for the component on improved management of dryland crops, they had to be financed by ongoing programmes.[4] During *kharif* 1987/8, for instance, outlets in various villages were opened which sold improved seeds of *jowar*, *bajra*, castor and maize, chemical fertilizers, pesticides and implements like iron ploughs and sprayers at subsidized prices to small and marginal farmers and SC and ST farming households. However, funds were obtained for only a couple of years and from 1991/2 onwards the project had to discontinue with the supply of subsidized inputs. It meant that, like the T&V system, the project had little control over an activity which it had thought to be very relevant. It is not surprising then that the

Mid Term Review stated that crop production did not receive due attention (GoI, 1988: I).

4.1.4 INTERNAL ORGANIZATION

Being one of four World Bank sponsored and one of many watershed projects implemented in India, the Maheswaram project was embedded in a strict hierarchy of committees, councils and cells, from Union to state to district and to project level. While some of these organizations formulated policies and laid down policy guidelines, others worked as the executive arm of these committees and were responsible for implementation of the projects. But staff of the Maheswaram project were not only accountable to people at state and national level, but also accountable to the World Bank office in New Delhi. On the basis of the advice given by the supervision teams of Indian and foreign consultants that were sent by this office to the project area regularly, project staff had to adapt its activities.

At the national level, the National Watershed Development Policy Committee (NWDPC) designed the watershed policy of the Indian government, laid down guidelines and decided about funding allocations among states. It had the powers to approve or reject the project proposals of the all watershed projects in the country, including that of the Maheswaram project. The National Watershed Development Council (NWDC) was responsible for the appraisal of the technical and economic feasibility studies included in the watershed plans made by the different states and submitted these plans to the NWDPC. In addition, the WDC had the task to supervise project implementation by the states.

At state level, two almost identical organizations were created: the state Watershed Development Policy Committee (SWDPC) and the State Watershed Development Cell (SWDC). In Andhra Pradesh, the SWDPC laid down guidelines at state level and was responsible for the approval of project plans at that level, prior to their submission to the NWDC. In Andhra Pradesh, the main tasks of the SWDC were to select suitable watersheds, prepare watershed plans, supervise the implementation of the plans and co-ordinate with all line departments involved in the projects.

At the lower level the implementing organizations were named the District Watershed Development Committee (DWDC) at the district level and the Watershed Development Team (WDT) at the project level. The DWDC, headed by the district collector, approved all watershed plans in its area, reviewed progress of the activities, co-ordinated with all line-departments and resolved local problems associated with the implementation of the watershed plans. The WDT, headed by a Joint Director of Agriculture as team leader, was responsible for detailed planning and implementation of a watershed project.

Although the original plans of the Maheswaram project conceived of a multidisciplinary team, with staff drawn from the departments of agriculture (including soil and water conservation, crop production and horticulture), animal husbandry, extension and forestry working under one umbrella, in reality the team of the project consisted mainly of staff from agriculture and forestry. For remaining activities, the team had to seek co-operation of the other concerned line-departments. This applied, for instance, to the T&V system for agricultural extension, to the Department of Rural Development and *Panchayat Raj* for subsidized inputs to special groups of farmers, the Co-operative Department to enroll all farming households as members of primary credit societies and to arrange short-term crop loans to farmers, and also to the Department of Animal Husbandry which was responsible for pasture development. Other organizations that rendered special services to the project were APAU and CRIDA both located not far from the project area. They undertook socio-economic and (adaptive) agronomic research, devised viable agricultural innovations to be recommended to farmers in the watershed, trained project personnel in these innovations and farmers who visited their premises. Finally, the Administrative Staff College of India (ASCI) was asked to monitor and evaluate the progress of the project.

Over the years, the organizational set-up at field level changed somewhat, as did the staffing pattern. Most changes meant a setback for the improved crop management component. While initially a separate Assistant Director of Agriculture and a Subject Matter Specialist (SMS) were appointed by the project to supervise the T&V system and the other elements of above mentioned component, after 1988/9, supervision became the responsibility of the staff managing the (*in-situ*) soil and water conservation components (Sitapathi Rao et al., 1990: 47-8). This meant that, below the level of the team leader, two Assistant Directors of Agriculture (ADA) managed the three components. According to plans, they were supposed to be supported by two Agricultural Officers and 30 to 50 sub-assistants who supervised the (*in-situ*) soil and water conservation activities at field level in those sub-watersheds where the project was active. In reality, the number of sub-assistants oscillated between 6 assistants in 1986/7, 30 assistants in 1989/90 and 20 assistants in 1990/1(GoAP, supervision mission 1990: 14; GoAP, supervision mission

1993: 13). The same documents illustrate that this understaffing was limited to field staff only and that it did not pertain to the office of the team leader. At the most, the project employed 103 people while the minimum number of posts sanctioned was 112, the ultimate target being 192. Nevertheless, since the project did not work in all villages simultaneously, at least 5 project staff were available for every village where the project was active, of which 2 were sub-assistants.

Not a word is uttered in the project documents on the desired qualities of the field level staff or the method by which they had to be recruited. All that is known is that, although all staff had passed the regular exams necessary to join the Department of Agriculture and that they had some basic knowledge in agriculture and extension, it was found advisable to train them in the improved technologies they were to recommend to the farmers. However, key persons and farmers complained that many field level staff lacked fundamental knowledge of dryland agriculture. These informants often stated that many staff members questioned the usefulness of indigenous agricultural technologies. In fact, it was found that quite a few field level staff lacked the dedication to improve dryland agriculture and that most were not very motivated to help socially and economically backward farmers. For most field level staff the Maheswaram project was a project like so many others. Like a large number of civil servants, their motivation to perform their duties in the project was more inspired by their salaries, the various fringe benefits (legal or not) and the fact that it was a regular job, than by some environmentally or socially inspired cause to improve the watershed or the resource-poor farmers. In fact, disrespect of some staff towards these farmers and the way they cultivated their dryland plots casts doubts about the suitability of these people for this particular job. Especially poorer farmers were displeased with the behaviour of project staff. Also, informants said that field staff favoured larger farmers, that they visited the villages very irregularly and they also had doubts about the reliability of some staff.

Although the staff of the Maheswaram project was not supposed to be transferred to other projects, during the nine years it was in operation it had several team leaders and Assistant Directors. Except for a couple of Reddy farmers in the project area who knew some top managers, most farmers were only familiar with the field staff; the only exception being one of the Assistant Directors of Agriculture, who was located in the Tuggaguda field office, located 18 km south of Hyderabad. He was known to many farmers not so much because farmers came to this office for information, inputs or to lodge complaints, but more because he was regularly found in the field, passionately trying to implement project activities and to persuade farmers to adopt the agricultural recommendations.

The Maheswaram project leaders were not very visible to farmers. They held office in Hyderabad and visited the project area very irregularly. If they did, they travelled into the field mostly during *kharif.* They were more dedicated to the organization and the state than to the beneficiaries of the project. In fact, they were more managers who tried to implement a project the best way they could, who tried to reach the targets set, and who felt accountable to those higher in the hierarchy, than they did to the beneficiaries. Also, they were more bureaucrats who implemented the activities as planned than visionary and charismatic persons. Because they sat in Hyderabad while field staff moved around in the project area, bureaucratic control and internal hierarchy was not very strict.

In a way, field staff and management felt trapped between their superiors on the one hand, who changed plans and wanted them to reach the set targets, and the farmers in the project area on the other, who were not always convinced about the necessity and the benefits of the recommendations. That the project was formulated by outsiders having their own interests and intentions with minimal involvement of the project management, shows, for instance, from comparing the original project plan (GoAP, 1981), the World Bank's Staff Appraisal Report (World Bank, 1983), the Master Plan (GoAP, 1985c) and the Mid Term Review (1988). As a result of Indian agricultural policy, but more so by the wishes of the World Bank and its consultants, elements like the subsidization of inputs and (graded and contour) bunding were removed from the project while *in-situ* soil and water conservation and *khus* planting were included. Even though flexibility was stimulated (GoI, 1988: II), the management felt that their superiors ignored the little advice they gave on activities that needed to be included or excluded from the plans. More or less the same feeling was prevalent among field staff that complained about the little room for alterations and about the fact that ideas from the field were not always supported by the 'bureaucrats'. Both the management and the field staff thought it very unwise to continue with the planting of *khus* lines, for instance. However, since discontinuation was out of the question, with indifference they merely tried to reach the set targets. It made them feel that they were executors of unnecessary activities designed by outsiders.

The relationship between field staff and management of the Maheswaram project on the one hand and farmers in the area on the other, became strained too. As will be discussed in depth in later chapters, while project staff 'merely' tried to reach the formulated targets, farmers were not always willing to co-operate. Believing that the recommendations were beneficial for farmers and that indigenous knowledge was more or less obsolete, many field staff were surprised to see that many farmers reacted with indifference to many recommendations. As a consequence, occasionally staff resorted to illegal actions or used force to implement the activities planned by the project. Their loyalty was geared more towards their superiors than towards the farmers they were supposed to assist.

4.1.5 PROJECT MEANS

The various budgets and the ultimate expenditure of the Maheswaram project clearly show that the project shifted its focus from the development of dryland agriculture to arresting ecological degradation. Initially, in 1981, the proposed budget was Rs 91.1 million ($ 10.8 million) for a period of 5 years. By far the largest part (Rs 54.3 million of which the subsidy element amounted to Rs 20.9 million) was allocated to dryland agriculture (Table 4.2). The budget was thoroughly revised in the 1985 Master Plan. Not only was the approved outlay decreased to Rs 58.2 million

TABLE 4.2. BUDGETS (1981 AND 1985) AND EXPENDITURE (1983/4-1992/3), MAHESWARAM PROJECT (*Rs million*)

Item	Proposal 1981		Master Plan 1985		Expenditure 1983-93	
	Rs	%	Rs	%	Rs	%
Dryland agriculture	54.3	59.6	4.5	7.7	2.9	3.6
Soil conservation	9.9	10.9	22.9	39.3	7.8	9.6
Animal husbandry	8.1	8.9	4.5	7.7	2.4	3.0
Forestry	3.5	3.8	14.7	25.3	35.2	43.3
Sericulture	3.6	4.0	0.0	0.0	0.0	0.0
Fisheries	0.6	0.6	0.0	0.0	0.0	0.0
Salaries and equipment	11.2	12.3	11.6	19.9	22.9	28.2
Others	0.0	0.0	0.0	0.0	10.1	12.4
Total	91.2	100.0	58.2	100.0	81.3	100.0

Note: The figures on dryland agriculture in the Master Plan and the Expenditure include horticulture; the figures on animal husbandry in the Master Plan and the Expenditure include forage development; Expenditure on 'others' is on smokeless chullas, multiple water-points, percolation tanks, compost pits, training of farmers, research, and monitoring and evaluation.

Sources: GoAP, 1981: 11; GoAP, Supervision Mission 1990: 29; GoAP, Action Plan 1993/4: 6.

($ 5.8 million) for a period of seven years (1983/4 to 1989/90), the amount allo-cated to dryland agriculture was also cut considerably to the advantage of the soil conservation and particularly the forestry programme. It is here that the interventions by the Indian government and particularly the World Bank, its funding agency, are clearly visible. Consequently, the component to improve dryland agriculture became largely dependent of other on-going agricultural programmes in the state and subsidizing inputs was not encouraged, to say the least.

The primacy of the forestry programme manifests itself even more when looking at the total expenditure of the project. Because of the devaluation of the rupee, the project outlay was increased by the World Bank to Rs 84.2 million ($ 4.7 million), of which Rs 81.3 million was spent. Most expenditure (Rs 35.2 million or 43.3 per cent) was on forestry. Only Rs 9.6 million was spent on soil and water conservation, a result of the recommendations made by the Mid Term Review of 1988 to discontinue the construction of very expensive earthen bunds and to plant *khus* hedges and to focus on *in-situ* soil and water conservation measures instead. What is conspicuous too, is the tremendous increase in the money spent on salaries and equipment. Another remarkable point is that only Rs 90,600 (0.1 per cent) was spent on 'people's participation' (coming under the heading 'others').

Table 4.2 only pertains to direct project expenditure financed by the World Bank and does not include activities financed from ongoing agricultural programmes in the state for the component on the improvement of dryland agriculture. The amount contributed by these government programmes was not known to the project management and it was not possible to obtain data from the departments financing these special schemes. However, the management of the Maheswaram project said that although expenditure must have been considerable it was less than the money financed by the World Bank on this particular component (personal communication, February 1994).[5]

Because the project management was neither able to calculate the exact expenditure per village nor the expenditure per beneficiary, all that can be calculated are very rough averages. Starting from the fact that the Maheswaram project covered 38 villages, 38.317 ha and 50,000 people in 10,800 households of which 6.200 were farming households, average expenditure totalled Rs 2.1 million per village, Rs 2,114 per ha, Rs 1,620 per capita, Rs 7,500 per household and Rs 13,065 per farming household. Focusing on expenditure made for the components on (*in-situ*) soil and water conservation and dryland agriculture, average expenditure was

Rs 0.3 million per village and Rs 1,726 per farming household. The magnitude of these amounts becomes evident when it is realized that income from daily labour averaged Rs 15/day and that the poverty line was set at an annual household income of Rs 6,400.

4.1.6 PROJECT PLANNING AND DECISION-MAKING

The government and project staff of the Maheswaram project were quite aware that the success of the project depended largely on farmers' participation (Firdausi et al., 1991: 13-16). However, the various project documents show that the project followed a top-down approach during planning and implementation and that local participation was regarded to be more or less synonymous with accepting and adopting what the project considered to be necessary measures. This rather passive form of participation is evident from the fact that farmers were not involved during the earlier phases of project formulation and design and, therefore, could hardly influence the type of activities which were to be undertaken. Although Sitapathi Rao et al. (1990: 40-4) stated that the acceptance, involvement, adoption and sustained interest of the farmers in the various activities were seen as important and that their interest and involvement could be expected as long as they were confident that their needs were met, farmers here were never asked about their needs, problems, priorities and possible solutions (Head State Watershed Cell, personal communication, March 1994).

Even during the implementation phase, the farmers' influence was very limited. The top-down approach of the project is very clear from a statement made in one of the Action Plans, which indicates that farmers were told that the project's solutions were the best (GoAP, 1986/7: 24). Thus, it is no surprise that without farmers' involvement, the planting of *khus* replaced the construction of graded and contour bunds. According to reports by APAU (n.d.: 5) and by Sivamohan et al. (1994: 137) local participation was only regarded as a tool considered essential for the successful implementation of the project and necessary in providing feedback to scientists.

Various activities were envisaged by the project designers to ensure farmers' participation (read: adoption of the recommendations) during the implementation phase, i.e. establishing Watershed Development Committees at district and village level; organizing meetings with farmers, project staff and sometimes scientists in Maheswaram and occasionally in the villages; and signing by farmers of an 'agreement bond'. However, these were not a great success:

- How the few 'progressive' farmers were selected for the District Watershed Development Committee, is not known to me, but their involvement in implementation (as in planning) of project activities was limited (GoI, 1988: 20).
- To facilitate communication between the project and the village, thereby improving the implementation of project activities, the project advised farmers to form Village Committees. These were supposed to inform project staff about the priorities of the farmers and the farmers about the activities which had been planned and the terms under which farmers were eligible to receive support. However, none of these Village Committees took off (Sivamohan et al., 1994: 137).
- Before the execution of the soil and water conservation works in a village a plan had to be made by project staff. This plan had to be discussed with the concerned farmers and, to the extent possible, any of their suggestions and modifications had to be incorporated in the plan before its implementation. Very few farmers, though, have ever seen these plans (Sitapathi Rao et al., 1990: 19-20).
- Only after a farmer had signed an 'agreement bond' could the project execute the soil and water conservation works in his land. Signing such a bond meant that the beneficiary agreed to repay the construction costs and to maintain the structures. However, we were told that many illiterate farmers did not know the contents of these bonds. Furthermore, Sitapathi Rao et al. (1990: 18) stated that against the 54 per cent who had signed such an agreement bond, 34 per cent had orally agreed while 12 per cent of their respondents had not given consent at all. They were convinced that the project staff deviated from this prescription in an attempt to achieve set targets. The same authors were inclined to believe that since repayment was not enforced, farmers believed that everything was done 'for free'. This, they argued, resulted in marginal involvement of the farmers (Sitapathi Rao et al., 1990: 40).

There are two main reasons for the lack of success. First of all, for many project officials (especially field staff) participation was a 'foreign' concept, thought up by government officials and World Bank staff in New Delhi. Being trained in a (government) culture known for its hierarchy and being familiar with the top-down approach followed in government projects, they were not accustomed to a participatory approach. Many staff had an authoritarian attitude, were paternalistic and not used to treat farmers as equals, especially if these were small and marginal, ST and SC farmers. They felt that since they knew what was best for them, farmers had to listen to their advice. They felt that involvement of 'ignorant' farmers was only counterproductive. They believed that 'socially and economically backward farmers' were not supposed to know what was best for them. Besides, they felt that participation would only result in delays. In fact, the project management, being afraid that set targets would not be reached, did not push ahead with the Village Committees. Secondly, many farmers were not very interested in the project. Having experienced the way government services normally functioned, poor farmers were convinced that the project was only beneficial for large Reddy farmers. Hence, many kept aloof when project officials visited the villages or when the project asked farmers to meet in Maheswaram. In fact, many were willing to spend their precious time for the cause of the project only when subsidized inputs were distributed.

4.2 The AWARE Approach

4.2.1 HISTORY

AWARE is clearly an exponent of the more radical organizations which emerged in the mid-1970s in India, an example of an organization opposing the service and welfare oriented approach of the Indian state and many NGOs. The foundation stones for AWARE were laid when its Keralite chairman came to Andhra Pradesh for an anthropological study on tribals. Not much later, as a civil servant implementing various poverty alleviation programmes targeted at these tribals, he concluded that these government initiatives failed to give an adequate answer to their problems. He found that since too many people were 'eating money' only a very small percentage of the government funds were used for the poor. Believing that the conditions of the poor could only be improved by NGOs, he left the civil service and joined an NGO in Anantapur district in the south of Andhra Pradesh. Some differences of opinion on ideology, approach and activities made him leave and, together with some friends and seed money from this NGO, he established an organization called the Chenmaya mission trust in Hyderabad (Chairman of RDT, personal communication, July 1993). Later, the name was changed to AWARE, he was elected chairman and some of his friends became members of the governing board. AWARE was registered under the Public Societies Act in 1975 and, on receiving $ 5,000 (Rs 40,000) from Community Aid Abroad (CAA, Australia), started its first activities in Amangal *mandal* (then Amangal block). Since its chairman believed that *Harijans* had a 'slavish attitude', initially AWARE focused on *Girijans* (Arens, 1979: 70). It was his idea that once *Harijans* saw the successes of AWARE's work with *Girijans*, they would approach the organization for help automatically. When this happened around 1979, AWARE started with programmes for Harijans in Amangal *mandal* (Chairman of AWARE, personal communication, July 1993). AWARE most probably started its activities in Ekvaipally around 1977.[6]

AWARE was founded at the right time; dissociating itself from the merely welfare oriented NGOs, it labelled itself an organization aimed at awakening and empowering the poor. Thus, AWARE not only claimed their rightful share of government funds, but according to Sohoni and Singh (1990: 20), it also intended to 'Psycho-social mainstreaming STs and SCs in national development'. These objectives corresponded more or less with the ideas prevalent in a number of progressive foreign funding agencies. In 1976, the Dutch funding agency ICCO started financing AWARE, followed by a second Dutch organization (Novib) in 1977. Within three years after its first $ 1,00,000 (Rs 2.6 million), Novib sanctioned another $ 2.9 million (Rs 47.9 million) for three years.

Made possible by funds from foreign donors and from various rural development programmes of the Indian and Andhra Pradesh government, by loans from banks and by money received from the target group, AWARE's growth has been phenomenal. Starting from the least developed Telangana region, AWARE spread its activities to the whole of Andhra Pradesh and beyond. By 1993, AWARE's activities covered 1.3 million people living in 3,344 villages located in Andhra Pradesh, Orissa, Karnataka, Tamil Nadu, Uttar Pradesh and Kerala. Having 'withdrawn' from 1,620 villages (with 0.8 million people), this means that over a period of almost 20 years, some 2.1 million people in 4,964 villages benefited from AWARE's activities (AWARE, 1994a: 112). In addition, AWARE managed to increase its programme of activities as well, which included all kinds of social motivational trainings, programmes on basic needs and on economic development.

According to AWARE's own documents its agricultural programme had the following impressive effects (AWARE, 1988?; AWARE, 1994a): from 1976 up to 1993, as a result of the almost 10,000 open wells which had been dug or improved, of the more than 7,000 pump sets which had been distributed and of the 72 minor and 74 lift irrigation projects which had been implemented, 36,000 ha of dryland had been converted into wetland. In addition, almost 12,000 ha of dryland had been covered under its soil conservation and land levelling programme. Besides which AWARE gave loans to buy nearly 25,000 plough bullocks. Moreover, from 1988 up to 1993, AWARE provided loans to sow more than 22,000 ha with improved seeds, gave loans to buy chemical fertilizers and pesticides, provided loans to almost 1,000 farming households to buy improved implements and trained almost 10,000 farmers on all kinds of agricultural recommendations.[7] As a consequence of the agricultural programme, some 75,000 tonnes of additional food grains must have been produced per year.

Not unlike other quickly expanding organizations and the Maheswaram project, AWARE experienced growing pains, some small, some large, which it succeeded in overcoming. AWARE's organizational structure could not keep pace with the increase in its geographical scale and activities, for instance. After a decade of tremendous growth, the organization had lost its flexibility, had turned into a bureaucratic organization, more or less resembling that of the government. A first storm broke out in 1985, when members of the target group, dissatisfied with the functioning of staff at various levels in the organization and the way it had implemented some projects, questioned the integrity of the staff and the effectiveness of the organization (Kumar et al., 1986: 20; Sohoni and Singh, 1990: 83; Wils et al., 1994: 2). Displeased with the way AWARE reported the progress of its activities and having heard about the problems in the field, Novib also pressed for improvement (Novib, 1987: 2-3). Convinced that the social and economic progress at village level was considerable and that only a minority of the staff was to be criticized, AWARE changed its organizational set-up from a pyramid to a strongly decentralized and bottom-up system in which Cluster Development Officers (CDOs, later renamed Cluster Development Coordinators) worked closely with VAs (Village Associations or *sanghams*). However, this new set-up only created a new pyramid. Also, the management style of its chairman did not reduce the organization's red tape.

Another problem AWARE managed to counter occurred in 1988. In the view of AWARE's chairman it was a well-planned undercover operation of the communist-inspired political opposition that wanted to destroy his organization (Wils et al., 1994: 2). In the opinion of others it was just a reaction of dissatisfied employees who had formed what might be called a trade union, demanding better wages, fringe benefits and employment security (Prasad, 1992: 76). A couple of *sanghams* supported, in their view, the legitimate demands of the staff, while others opposed these actions. The establishment of this union created a lot of tension among the staff itself, between a part of the staff and AWARE's chairman and management, and between various *sanghams*. It culminated in the dismissal of many employees and AWARE's withdrawal from the area where the disturbance had originated. Nearly 1,000 villages were declared 'independent' by AWARE's chairman even though the intervention cycle had not yet been fully completed. Nevertheless, in due course the labour situation of the staff improved and for those who had worked in the organization for more than five years, AWARE introduced gratuity, a savings fund and medical and educational benefits.

The discontent of the staff coincided with the strained relationship which had evolved between AWARE and the Indian government; alleging that AWARE worked against 'public interest', the Indian government prohibited AWARE from receiving money from foreign funding agencies and blocked all bank accounts by the end of 1988. Since the government failed to provide justification for its action, AWARE won various lawsuits. Nevertheless, it took up to March 1990, before the final, positive verdict was announced, a period during which AWARE had no other option but to scale down many of its activities (AWARE, 1990a: 10; AWARE, 1994a: 63-71). Although AWARE and its foreign funding agencies had a rather strained relationship too (quarrelling about, for instance, AWARE's organizational structure and staffing, its reporting on both financial matters and on progress made in the villages, and its unwillingness to install a monitoring and evaluation system) (see, for instance, Nothdruft and Reddy, 1991: 31-2, 36), these agencies kept supporting AWARE, both morally and, after the ban on receiving foreign money was lifted, financially.

In the twenty years of its functioning and in spite of these upheavals, AWARE has often been regarded as a very big success, as a 'best practice'. Although some critical comments have been made, most (evaluation) studies have reported very favourably about the organization (Kumar et al., 1986; Beets et al., 1988; Sohoni and Singh, 1990; Nothdruft and Reddy, 1991; Wils et al., 1994)[9] and many Indian newspapers and magazines reported approvingly about the organization and its chairman. *The Week*, for instance, made him 'Man of the Year' in 1989, and the famous Indian cartoonist and writer Kushwant Singh wrote in the *Hindustan Times* of 24 August 1992 that AWARE 'had become one of the most powerful revolutionary forces in the country'. On the basis of these reports and articles, AWARE served as an example for other NGOs. Foreign funding agencies, like the Dutch Novib and the Belgian NCOS, used AWARE in many publicity and fund raising campaigns in their country.[9]

By the beginning of 1994, AWARE's most important foreign donors agreed to support its newest programme, which allowed it to embark upon new activities and extend its actions to many new villages in various states.[10] AWARE expected that as a result it would have expanded its direct and indirect activities to 10,000 villages and 10 million people by the end of the century (AWARE, 1990b: 15). Some months after the start of this programme, foreign funding agencies were informed about huge financial irregularities by AWARE since the late 1980s. Instead of using the financial contributions of foreign donors for its various programmes, a major part had been put aside in a secret fund comprising about Rs 400 million ($ 15 million) in 1994 (Novib India desk, personal communication, 20 June 1994). After the decision of the foreign funding agencies to withdraw with immediate effect their financial support, AWARE fired many staff, others left of their own accord and many activities were suspended. Although at the time of writing AWARE was still functioning, there is every reason to believe that it had lost much of its credibility, not only in the donor community but also among the members of its target groups.[11] Actually, when its chairman stood as an independent candidate in the 1994 general election for the Legislative Assembly of Andhra Pradesh for a constituency near Amangal, he lost.[12]

4.2.2 IDEOLOGY AND OBJECTIVES

AWARE claimed to be a non-political, secular, social service organization meant to strengthen the social and economic position of the rural poor (Nothdruft and Reddy, 1991: 12). In contrast to the Indian government and many welfare and development NGOs, AWARE did not solely want to tackle the symptoms of poverty or improve the economic position of the poor, but wanted to tackle the root causes of poverty. It was convinced that rural poverty was preserved by the rural elite, a combination of landlords, politicians and government servants, who oppressed and exploited the poor. It argued that because the oppressed lacked moral courage to fight against the various offences caused by this elite, social and economic improvement did not take place or was limited in scale. AWARE felt that by awakening its target groups of SCs and STs about this exploitation and informing them about their socio-economic and political rights, they could be motivated to unite and confront their oppressors. Therefore, as important and primary steps for rural development, AWARE used social education and motivation campaigns to awaken the poor, to disclose to them the root causes of their misery. Once awakened, the rural poor could be trained in organizational and management skills and could be helped to improve their economic position by providing various basic needs and economic programmes.

That AWARE's ideology (or perhaps better: the ideology of its chairman) oscillated between the ideas of the social action groups and that of the Gandhians is evident from various phrases it used in its documents (see, for instance, Sohoni and Singh, 1990). Influenced by the ideas of radical groups which emerged in the 1970s, AWARE analysed the problems of the rural poor as follows: it argued that for centuries *Harijans* as well as *Girijans* were marginalized by the powerful sections of society and suffered from an identity defined in negative terms. Both communities had been the victims of social abuse, subjected to humiliating treatment and discrimination, and also faced economic exploitation. In addition, Harijans had been exposed to practices of exclusion and untouchability, which were sanctioned by religion. AWARE stated that twenty years of welfare and relief, as well as economic development programmes, whether implemented by the Indian government or by service or development NGOs, had hardly made any progress on the path of social and economic improvement for the poorest of the poor. It was AWARE's firm belief that poverty was a structural phenomenon, rooted in an unequal distribution of power, wealth and assets. It was convinced that the only way to change it was to restructure society and to liberate the poorest of the poor from social and economic bondage. However, AWARE wanted to work within the confines of the Indian constitution and to use legitimate democratic channels and procedures. Arising out of the ideas of the social action groups, AWARE used concepts like empowerment of the people, awakening, awareness-building and mobilization of the poor, and the need to establish people's organizations.

Although AWARE used rather radical terminology, now and then, and portrayed itself as a radical organization (AWARE, 1994a: 82), AWARE's chairman (being a dedicated Hindu of Brahmin descent and loyal to all religious rites) fully subscribed to the Gandhian principle of non-violence. Although Gandhi opposed untouchability and the backward position of the poor stoutly, he was never in favour of a complete change in the social system. He thought that by improving the conditions of the entire village community, all inhabitants, irrespective of caste, would benefit. Health, education and shelter became important ingredients of the community approach, developed by his followers and included in the various government programmes after Independence. Gandhian concepts like non-violence, developing harmony among the various sections of society, self-reliance, self-sufficiency at village level, emancipation and the decentralization of power, were accepted by AWARE without copying what it calls the misinterpretations (i.e. the community approach) of Gandhianism (AWARE, 1994a: 78).

It is our impression that over the years AWARE became more and more inspired by Gandhian ideology and Hindu culture. The increasing guru-like attitude of its chairman is a case in point. The Leftist rhetoric found in many documents written by AWARE in the early years of its existence was hardly ever put into action, not in the early years and most certainly not in later years. While it was written in those early years that the target groups might have to face confrontations and might be encouraged to revolt against injustice (Kumar et al., 1986: 36), the organization's actions were never meant to radically overhaul the existing social order. Thus, already in late 1970s Arens (1979: 10, epilogue: 1-3) reported about the tension between the more radical staff and AWARE's chairman, writing that the moment they wanted to confront the landed elite, he ordered them to withdraw. While the ideas of these people were more in line with the ideas of the social action groups (confrontation with landlords and a radical change in the social structure at village level), AWARE's chairman was more inclined to use funds available in various government programmes targeted towards SCs and STs in the best possible, most efficient, way. In short, he was an efficiency enhancer using an empowerment approach to improve the access of the poor to government and private institutions. As a consequence of this difference in opinion most radicals left AWARE over the years.

Further research may indicate why this difference between rhetoric and reality has occurred and why this change in ideology has come about. Nevertheless, some possible explanations can be advanced. First of all, AWARE never intended to be 'radical'. It is quite possible that the radical concepts included in its documents were only meant to please foreign funding agencies. Second, AWARE may have become afraid that by supporting activities which were against the vested interests of the rural elite (which it in fact occasionally did), it would be banned by the government (which almost happened in 1988). Third, since AWARE intended to become a large organization and, for that reason, was (partly) in need of government funds, it was not willing to attack either the state or the rural elite.

This difference between rhetoric and reality as well as this change in ideology makes it somewhat difficult to pigeon-hole the organization. Prasad (1992: 79), not convinced about AWARE's empowerment approach, argues that AWARE is a welfare oriented organization, giving prominence to activities meant to improve the economic condition of the target groups rather than to social transformation activities. It is true that AWARE implemented all kinds of welfare and economic development projects (see next section). Nevertheless, AWARE can be typified as a moderate empowerment NGO which not only tried to get access to existing government programmes and services, but also induced the government to execute existing laws and to legislate new ones and organized its target groups in *sanghams* which had to act as countervailing power.

That AWARE was less than the radical NGO it pretended to be, is also shown by its objectives. Although various documents report of ultimate, long-term and short-term objectives, and mix up objectives and activities undertaken to reach those objectives (Arens, 1979: 5-6, 29; Kumar et al., 1986, 36-7; Novib, 1987: 5-6; AWARE, 1994a: 1-2, 14, 28; Sohoni, 1994: 220-1), four main objectives clearly stand out. In an attempt to reach 'equality of mankind and universal brotherhood through welfare and awakening', the first objective was to 'socially uplift the depressed classes to the zone suitable for development', that is to say 'psychological placement of the depressed classes in the mainstream of existence'. In 1994, this objective was reformulated into the integral upliftment of depressed minorities like the *Girijans* and *Harijans*, including their gradual integration into the wider society, based on equity and full participation in society, economy, and polity (Wils et al., 1994: 81). These formulations indicate that it was AWARE's objective to integrate the members of the target groups into the mainstream of society (a concept it did not define).

AWARE's second objective was to increase production so that *Girijans* and *Harijans* could benefit directly from economic growth. Its third objective was the even distribution of resources for better co-existence of people. Its fourth objective was to create alert, sustainable and autonomous communities of *Girijans* and *Harijans* and to assist these to a point where they could function on their own without further help from AWARE. In line with its ideas that it did not want to work in a 'blue-print' fashion, AWARE did not operationalize these objectives into clearly defined targets. Nevertheless, from various documents it may be concluded that AWARE envisioned that, within a period of ten years after it had entered a village, the target group should have become independent from the organization and 50 per cent of the poor should have been brought above the poverty line.

4.2.3 ACTIVITIES AND APPROACHES

Much more than in any government programme (or in the Maheswaram project), AWARE followed an integrated approach consisting of all kinds of activities, agricultural development being one. The interrelatedness of AWARE's social motivation and agricultural programme necessitates a comprehensive review of these two elements.

At no stage did AWARE perceive itself as an alternative to the state (Sohoni and Singh, 1990: 19), running only parallel development programmes. Thus, AWARE combined a 'parallel strategy' with an 'access strategy', meaning that, on the one hand, it assisted the target groups through programmes running parallel to the government (on health and education) and, on the other hand, it assisted them in organizing themselves to get access to resources, to government programmes, to institutions like banks and to decision-making bodies, to claim from society, from the landed elite and from the government what AWARE considered to be their rightful and equitable share in development.

In AWARE's programmes, both strategies were combined in a comprehensive way enabling the organization to reach its objectives of social upliftment and economic development. For analytical purposes these activities can de divided into a social programme, a basic needs programme and an income and employment generation programme.[13] Although these three programmes grew considerably over the years, they were visible right from the beginning of AWARE's interventions.

Social Mobilization

AWARE's social programme formed the foundation of the other two programmes. It had two interlinked components, i.e. an organization and an awareness building programme for Girijans and Harijans. To attack the root causes of poverty AWARE felt that an active mobilization campaign of the rural poor had to be launched. Hence, one of its major activities was the organization of the target groups at village level in *sanghams* like a Village Association, a Women's Organization and a Youth Organization. These *sanghams* were considered to be vehicles for progress, for change, as a necessary condition for the successful introduction and implementation of various social, basic needs and economic projects. From the outset AWARE indicated that as a catalyst it could only assist and guide the target groups. It stipulated that the members of the target group were themselves responsible for their own development and the way it would take and that they had to bring about the necessary changes themselves. AWARE felt that it could only help them in this process by motivating the target groups to organize themselves and by assisting these organizations to design and implement their own programmes.

Arguing that the members of the target groups were not aware of the causes of their misery, had an inferiority complex and feared the members of the higher castes and the bureaucracy, AWARE started with a programme of motivational training in which awareness-building and empowering the poor were the most important elements. This training, also containing psychological elements, was not only meant to change the attitude of the members of the *sangham*, to attack their 'culture of submissiveness', but also to make them identify their own problems and to prepare their own answers and plan of action. The training had to create a feeling of confidence, so that they could stand up for their rights, would no longer be afraid to face higher castes and ultimately would start various developmental activities. Moreover, through the training, the members of the *sangham* were informed about government programmes and the ways to get access to public and private institutions. Once aware of these programmes, the *sangham* had to lay its hands on the funds and improve the social and economic conditions of its members. In AWARE's ideology, *sanghams* were a prerequisite to rural development, which would not succeed if people did not associate in planning for their own development and participate consciously in the implementation of the planned activities.

Depending on the subject taught, motivational training was given by AWARE's staff or, at the invitation of AWARE, by government or bank officials or others involved in rural development. This was mainly done during group meetings in one or two villages, on a regular basis, if possible once a month. On average, these *gramasabhas* took two to three days. Initial focus during the training was on all kinds of social topics like alcoholism, indebtedness, wife beating, inactivity, bonded labour, untouchability and social abuses by higher castes and the rights and entitlements of Girijans and Harijans. It was also meant to stimulate the target groups to effectively use their vote. Second, it focused on basic needs and economic topics like community health, minimum wages, marketing, agriculture, animal husbandry and cottage industry to mention a few.

In addition to the *gramasabhas*, AWARE used *mahasabhas* or mass meetings to inform the members of the target groups on issues like land reforms, wages and bonded labour and to get political leverage as well. After 1983, community education centres were set up in some villages, headed by a specially trained organizer belonging to the village. These provided for a more permanent meeting place for the members of the target group. Besides these interpersonal contacts, mass media in the form of leaflets, radio talks, movies and videos, were used.

Another kind of motivational training was geared towards institutional development. Directed at the leadership of *sanghams*, at leaders of supra-local organizations created by AWARE and at AWARE staff, this training was meant to build a qualified and motivated group of workers, who could act as change agents. They were trained in the legal status and the rights of the target groups, how to get access to the bureaucracy and to banks, how to manage grass root organizations and in planning and implementation of development programmes. To get political leverage, AWARE also induced local leaders to participate in local and supra-local elections, as candidates of existing parties or as independents.

Basic Needs

From the start of its activities, AWARE felt that awareness and organization building had to be accompanied with programmes on basic needs and economic development (Sohoni and Singh, 1990: 51). As AWARE believed that community health was a prerequisite of socio-economic development, it was one of the more important elements in the basic needs programme. Hence, AWARE established various community health centres in the state. Another important element was legal assistance. For this purpose AWARE established legal cells consisting of full-time lawyers who advised the target group on topics like land reforms, minimum wages, child labour, dowry system, emancipation of women, abolition of bonded labour and human rights. Other elements of the programme were rehabilitation of bonded labourers, a community programme for common welfare (social forestry, horticulture, sericulture, village drainage systems and biogas plants) and cyclone and disaster relief.

Economic Development

Recognizing that economic benefits were of prime importance to arouse the interest of the members of the target groups, in most years AWARE spent 50 to 75 per cent of total expenditure on economic development. AWARE's economic programme consisted of three components. The first, agricultural development, was geared to small and marginal farmers while the other two, animal husbandry and cottage industries, mainly catered for the landless poor. All three had three elements in common: (1) AWARE informed the members of the *sangham* about government programmes targeted towards SCs and STs and helped them to apply for and to obtain the schemes; (2) in addition to these regular government programmes,

AWARE itself implemented some schemes; and (3) besides the distribution of assets and inputs, a training component was included.

Agriculture

AWARE considered agriculture by far the most important and reliable source of income for the rural poor (Sohoni and Singh, 1990: 51). However, a major portion of the area covered by its activities was located in the semi-arid tropics and a large proportion of the land owned by its beneficiaries was low-yielding dryland. Nevertheless, influenced by the success of the Green Revolution, stimulated by the funds available in various government programmes and knowing that the *sanghams* were greatly in favour of this solution, AWARE focused on paddy production. Based on the basic premise that only multiple cropping with an assured supply of irrigation water and other material inputs and technical services could trigger agricultural development, AWARE followed the same technical modernization approach as the Indian government. Consequently, in about 20 years 36,000 ha of dryland was converted into wetland.

Such a modernization approach was followed for dryland agriculture as well, albeit on a smaller scale. Convinced that the limited yield potential of the not very fertile drylands would gradually be reduced by the loss of top soil, AWARE started a soil and water conservation programme that resembled that of the government, i.e. making graded bunds and other structures. In addition, it started informing farmers about improved crop management technologies for dryland crops and distributed all kinds of inputs, often using government funds. This modernization and intensification approach was criticized by the 1986 evaluation study which recommended that AWARE should embark upon less demanding and less risky technologies (Kumar et al., 1986: 14). Although AWARE argued that it was in favour of more sustainable agriculture in later reports (AWARE, 1992; AWARE, 1994a), the organization continued to stimulate high-input agriculture.[14]

AWARE had developed a five tier approach for its agricultural programme:

(1) To improve the availability of water for marginal and small farmers of the target groups by digging, deepening and improving open wells, drilling tube-wells, supplying pump sets (diesel and electric), construction of minor irrigation works (tanks, check-dams), lift irrigation and percolation tanks.

(2) To conserve the soil of drylands through land levelling, bunding, stone terracing, gully plugging, bunding with pasture and stream bank protection.
(3) To train farmers in soil and water conservation on drylands and in improved crop management for both dryland and wetland crops.
(4) To provide farmers with various agricultural inputs from its own funds or to help them obtain inputs like improved seeds, chemical fertilizers, pesticides and suitable implements, provided government programmes were available. Implements were supposed to be owned by the *sangham* and shared by different people.
(5) To sell the agricultural surplus generated by higher production against remunerative prices. Hence, the importance attached by AWARE to marketing of crops by cooperatives.

More specifically, AWARE proposed many recommendations for improving dryland agriculture in Amangal *mandal* (Table 4.3). These were very similar to those made by the Maheswaram project. The main differences concern some of the 'real' innovations of the Maheswaram project which were not recommended by AWARE, i.e. contour bunds with waste weirs and *khus* (being soil and water conservation structures); key lines and dead furrows (*in-situ* soil and water conservation); and manure pits (improved dryland crop management). In addition, diversion drains, water ways and the seed-cum-fertilizer drill were also not recommended by AWARE. The large similarity of the recommendations might come as a surprise, given the different ideology, objectives and planning procedure of AWARE and the state and its Maheswaram project. However, since AWARE did not generate its own agricultural innovations (it started with adaptive research in a very small scale in 1993/4) and had no qualified staff, it was almost totally dependent on agricultural innovations generated by public research.

Because most AWARE staff lacked basic agricultural knowledge and experience, and because AWARE did not have an extension service of its own, AWARE was forced to play a subsidiary role as agricultural change agency. For instance, its farms were mainly used for production (and income) purposes, hardly for agricultural training of farmers.[15] Thus, AWARE relied largely on government T&V staff, who were invited to inform farmers during special technical training camps. But, as stated in earlier chapters, the number of Village Extension Officers in the project area was limited. Moreover, they were not very motivated to visit small and marginal farmers. However, through its empowerment strategy and through its *sanghams*, AWARE felt that T&V staff could be forced to meet poor farmers regularly. Although this dependency on the T&V system was in line with its access strategy, implicitly it meant that AWARE relied on the adapted Transfer-of-Technology (TOT) approach of agricultural extension. In spite of its own participatory approach, AWARE did not stimulate a more horizontal flow of agricultural information from farmer to farmer or asked farmers for feedback on the recommendations made by the T&V system.

To a very large extent AWARE also depended on government funds for soil and water conservation and the provision of (subsidized) inputs. Similar to the ideas of the state and its Maheswaram project, AWARE had expected most effects on adoption of these agricultural recommendations,

TABLE 4.3. AGRICULTURAL RECOMMENDATIONS BY AWARE FOR DRYLAND CROPS IN AMANGAL *MANDAL*

	(Adapted) indigenous practices	Innovations introduced before AWARE's involvement	Innovations introduced by AWARE
Soil and water conservation structures:		Graded bunding Gully control works Land levelling	
In-situ soil and water conservation:	Off-season (preparatory) tillage, i.e. summer tillage for *jowar* Interculturing for all crops	Contour cultivation (preparatory tillage, sowing and interculturing) but watered down to sowing and interculturing across the major slope	Off-season (preparatory) tillage, i.e. post-harvest tillage for all crops
Improved dryland crop management:	Early sowing of *jowar* Intercropping *jowar* with redgram but in 2:1 ratio Application of FYM Interculturing for all crops Manual weeding	Improved seeds Optimum plant population Iron plough Chemical fertilizers Pesticides	Improved seeds Integrated pest management

Source: Various documents supplemented with information from AWARE field staff and T&V extension officers.

not from bridging the information gap but from bridging the financial gap. Hence, AWARE tried to improve the access to government subsidy schemes for the members of its target groups. In Amangal *mandal*, AWARE used the government's seed exchange programme to induce farmers to use improved seed varieties, while farmers also received subsidized chemical fertilizers, implements and pesticides from government schemes. In addition, AWARE gave loans from its own funds to buy agricultural inputs, implements and farm machinery.

Animal Husbandry and Cottage Industries

Under its programme on animal husbandry, sheep, goats, dairy cattle and pigs were distributed, in particular, to landless households and women, with the objective of generating some additional income. Under the programme of petty production, trade and services, loans and a training were given to artisans for basket-making, weaving, leaf-plate-making, mat-making, footwear-making, vegetable vending and tailoring. In addition, AWARE gave loans to buy band sets and bullock carts, assisted in the establishment of brick, lime-kiln and salt cooperatives and, in 1989, implemented a savings and credit programme for women. To a very large extent AWARE used government programmes like the IRDP and the JRY to finance these elements of its own economic programme. Although perfectly in line with its access strategy, above mentioned activities resembled in almost every way those of the government, which makes me doubt the innovativeness of this element of its economic programme.

In 1977, AWARE decided to give nothing for free. Instead it gave loans to individuals which could be used to buy assets and inputs. However, since AWARE was not registered as a bank, these loans were interest free. The leadership of the *sangham*, or someone from AWARE responsible for the area, selected eligible households, distributed loans and assets and collected installments. The money thus repaid was turned into a revolving fund. Concerning government programmes like IRDP, the same selection procedure was followed. Sometimes AWARE deposited money in a bank responsible for the IRDP loans, as an insurance in case the members of the *sangham* defaulted.

It was AWARE's policy to limit its intervention in a village to ten years. Its withdrawal strategy consisted of three phases, each phase characterized by a specific programme of activities. The first phase, normally lasting about three years, focused on motivational training and organization building. During the second phase, lasting about four years, the basic needs and economic programmes were implemented. During the third phase, lasting about two years, AWARE built up a hierarchy of people's organizations at supra-village level (see next section). At the end of the eighth year, once the *sangham* had become self-reliant, AWARE gradually started withdrawing from the village. Slowly its paid staff was replaced by local people who would be responsible for further planning and the implementation of all developmental activities in the village. By the tenth year, AWARE would have withdrawn from the village and the area, and the local leaders, the *sanghams* and the various service societies would have to manage their own affairs and pool financial resources to undertake developmental activities. By that time, AWARE was no longer directly responsible for managerial and financial assistance, and gave only indirect, moral support.

Although AWARE's ideology became increasingly Gandhian thought based, it resisted the idea of community development if community meant the entire village, irrespective of caste. Hence, as only AWARE's target groups could enlist for membership, *sanghams* did not reflect the entire village society. These multi-purpose grass roots organizations had four main objectives: (1) to create a sense of unity among the target group so that they would not hesitate to make a fist against oppression; (2) to design and implement village development plans; (3) to compel government agencies to implement development programmes efficiently and effectively; (4) and, for AWARE, a way to implement its own activities.

Unfortunately, AWARE did not clearly define what its target groups were. Sometimes it referred to 'the poor' or 'the poorest of the poor', or to the landless and to marginal and small farmers, or it claimed to be working 'primarily' and then 'exclusively' for Girijans and Harijans (Anubhav, 1986: 26; Kumar et al., 1986: 37; AWARE, 1990: 4; AWARE, 1994a: 1, 80). Again on other occasions it stated that it also included the service communities (predominantly BCs) as its target groups (AWARE, 1988?: 179-80). Nevertheless, for most foreign donors it was 'obvious' that AWARE's activities were geared towards STs and SCs only (Wils et al., 1994: 1), a point of view held by us too.

4.2.4 INTERNAL ORGANIZATION

Organizationally, AWARE passed trough two distinct stages: from 1975 up to 1985 it had a pyramidal structure of a few people at the top and a large group of fieldworkers at its bulky base. Responsible for dozens of villages, these fieldworkers not only implemented projects, but also had all kinds of administrative and financial tasks. Responding to complaints by its target groups, who argued that the organization resembled the state bureaucracy, where fieldworkers, occupied with desk work, remained largely in their offices, an alternative organizational model was designed (Sohoni and Singh, 1990: 83). The flow-chart of this new 'cluster-approach', which functioned with small modifications from 1985 onwards, is given in Figure 4.1. Since AWARE created target group organizations as well, these are also included.

From 1985 onwards, AWARE's area of operation was divided into villages, clusters, areas and zones. Each cluster consisted of about 20 villages, inhabited by some 1800 households of the target group. Some four to five clusters made up one area, each area containing between 80 and 100 villages. Five areas were combined into a zone, each zone encompassing some 400 to 500 villages. The headquarters supervised

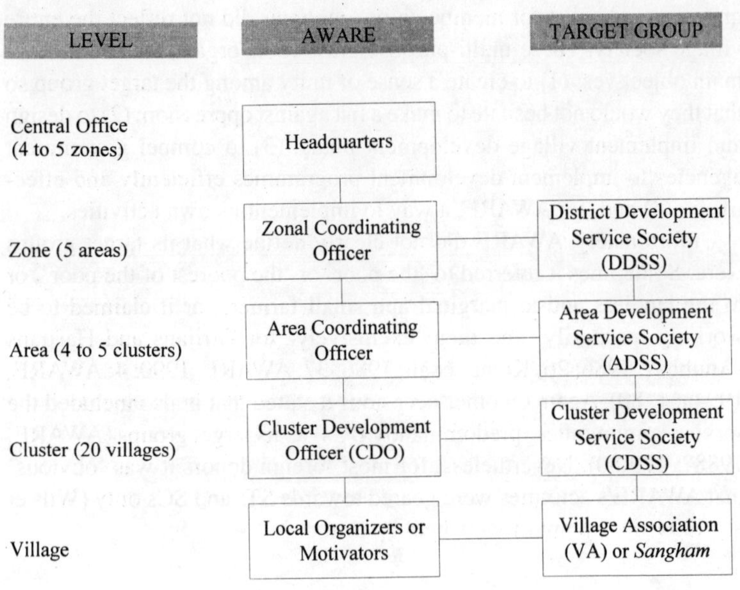

Sources: After Kumar et al. (1986: 38) and Prasad (1992: 63).

FIGURE 4.1. ORGANIZATIONAL SET-UP OF AWARE AND ITS TARGET GROUP ORGANIZATIONS AFTER 1985.

five zones or some 2,000 villages where AWARE implemented its programme. In addition to its central office, located in Hyderabad, AWARE had eight regional offices, headed by an administrative officer who looked after the administration of grants and budgets and other requirements of the various projects in that particular region. Furthermore, AWARE owned a couple of farms, hospitals and vocational training centres for handicrafts and cottage industries.[16]

In most cases AWARE entered a village on request of villagers (often leaders of the target group) or government officials, and not so much on the basis of an analysis of its socio-economic problems (Prasad, 1992: 79). After such request was honoured, AWARE appointed a staff member (the Cluster Development Officer, CDO) who was responsible for a cluster of some 20 villages. He was assisted by one or two local organizers, also called motivators, who were selected from the literate members of the target group in the cluster. These organizers were trained for three months. Being volunteers, they received an allowance for expenses only. At village level, the CDO and the organizers were supported by village volunteers.

AWARE viewed the CDOs as a cadre of grass roots officers who assisted the *sanghams* in the planning, budgeting and implementation of development programmes. They interacted with AWARE headquarters and the *sanghams* on the one hand, and local government functionaries and the *sanghams* on the other. Assisted by the organizers and village volunteers, the main task of the CDO was to set up various Village Associations or *sanghams*, to train their leaders, thereby promoting their leadership qualities, and to help and supervise the *sanghams* with the implementation of the various programmes. He was also responsible for raising local financial resources to match AWARE's loans.

Area Development Officers and Zonal Coordinating Officers supervised, coordinated and monitored the progress of the various activities undertaken by AWARE staff at lower levels. In addition, they assisted the Area Development Service Society (ADSS), respectively the District Development Service Society (DDSS), organizations set up by AWARE to promote the interests of the *sanghams* at that particular level. They also contacted civil servants responsible for rural development and interacted with bank officials. The Zonal Coordinating Officer was the most senior officer among the AWARE field staff.[17] As a member of AWARE's 'core team', he was also responsible for policy formulation. In contrast to other field staff, he worked at AWARE's central office in Hyderabad.

AWARE's day to day affairs were managed by a core group of nine people, i.e. AWARE's founder and chairman, the project director, the zonal officers, the programme coordinator, the chief accounts officer and the internal auditor. The core group not only reviewed the progress of AWARE's activities and sanctioned the necessary funds, it also decided in all policy matters (Sohoni, 1994: 238). Although under control of a general and a governing body, it was AWARE's chairman who to a large extent controlled the organization and set out its policy. This has never changed since Arens (1979: 9) reported on it. This general body consisted of 300 people who met twice a year and who, every five years, elected a governing body of 9 people from their midst, who convened every month. Until July 1989, the general body consisted mainly of educated urban intellectuals. Afterwards, following the decision that the presidents of the ADSSs were to have a seat in this body as well, the target group was represented in the general body by 30 people (Nothdruft and Reddy, 1991: 12-13). Not a single Harijan or Girijan was ever included in the governing body. AWARE staff was not represented in either of the two bodies.

Figure 4.1 shows that AWARE was very innovative in setting up various organizations at different levels which were meant to promote the interests of the target group of Harijans and Girijans. *Sanghams* were AWARE's main promoters of change at village level. In a large majority of cases separate *sanghams* were created for STs and SCs. All STs and SCs could enlist as members and pay a membership fee. From their midst, they elected a president, a secretary and some additional members of the board. These officials were unpaid, getting an allowance for expenses only. After consultations with the CDO and with AWARE headquarters, the members of the *sanghams* were supposed to identify the needs and to plan and implement all development activities, be it social awareness training, basic needs or economic programmes. *Sanghams* were also supposed to plan their budgets, identify beneficiaries for loans and arrange for their recovery. The most important selection criterion for these loans was supposed to be economic, i.e. the most needy members of the *sangham*; whether the beneficiary would be able to repay the loan was also taken into account.

At cluster level, a Cluster Development Service Society (CDSS) was created by AWARE. Its members were the presidents and secretaries of the various *sanghams* in the cluster. From their midst, they elected the members of the board of the CDSS. The CDSS was supposed to interact and communicate intensively with the CDO, who helped it with its tasks, i.e. mobilizing local resources (money from the target group as well as from the government), promoting closer inter-village relations, designing and implementing inter-village projects, and providing a marketing infrastructure. Once AWARE had withdrawn from an area, it was the CDSSs and the higher level target group organizations who were supposed to form the institutional framework through which the developmental activities were carried on. Once AWARE had withdrawn its own CDO, the 'independent' cluster had to appoint and pay a CDO from its midst. This 'local' CDO, who was accountable to the CDSS, had to obtain financial and other kind of support (from the T&V system, for instance), necessary to implement the development activities designed by the CDSS.

The elected members of the boards of the CDSSs in an area made up the ADSS. From their ranks they elected a board and these office bearers were seated in the DDS, the top level organization of the target group. The representatives of the ADSS also sat in AWARE's general body. It was the task of the ADSS and the DDS to safeguard the interests of the target population and to start different developmental activities which were beyond the capacities of individual clusters or an area. Important activities were, for instance, arranging mass rallies on various relevant issues, acquiring and renting out of expensive productive assets, like cranes for well-digging, developing a marketing infrastructure, creating credit cooperatives and organizing legal support.

As a result of the enormous growth of project activities, AWARE's staff increased considerably: from 105 full-time and 40 part-time workers in 1979 to more than 1,000 full-time and over 1,500 part-time workers in the beginning of the 1990s. Its force of unpaid volunteers increased to 35,000 (Sohoni, 1994: 230). It is our impression that in the early 1990s only a relatively small percentage of AWARE's paid staff were *Harijans* and *Girijans*, a situation that was more or less similar to the one observed by Arens (1979: 22) in the early years of AWARE's existence.[18] The requirement of a university degree clearly favoured males with a higher class and caste background. Although AWARE succeeded in recruiting more women in recent times, most staff were men. Most of the women worked in one of AWARE's offices in the lowest ranks as typists and the like, not in the field.

Most field-workers of AWARE possessed a university degree in disciplines which had little bearing on the type of work they were doing (Kumar et al., 1986: 62). Although AWARE recruited a sufficient number of new staff through advertisements in newspapers, yet quite a considerable number of employees were recommended by AWARE's own staff. This created a system of patronage, where new staff did 'favours' for those who took them in. People who responded to advertisements were hardly ever selected during an interview session but during a six weeks training, which AWARE described as really 'Spartan'. Since AWARE's chairman felt that unfit candidates would refuse hard labour, trainees had to rise early and had to do a lot of manual work. Candidates were mainly informed about AWARE's ideology, objectives and programmes, and were trained in planning. Although they were also supposed to implement AWARE's economic programme, they were hardly trained in more practical subjects like (dryland) agriculture, animal husbandry and the like. Most of the field staff we met lacked fundamental knowledge about dryland agriculture, for instance. In fact, various evaluation studies have expressed their concern on the slow and unsatisfactory recruitment of specialized staff (Arens, 1979: 17, 22; Kumar et al., 1986: 10; Nothdruft and Reddy, 1991: 16).

It is our impression that the remarks made by Bijoy et al. (1991: 15, 17) also apply to AWARE: a considerable number of staff had joined AWARE because they needed a job, not so much because they were ideologically motivated or socially committed to assist AWARE's target groups in their attempts to fight against oppression and exploitation and to overcome their poverty. In fact, quite a number of staff preferred the civil service because jobs were easier and less time-consuming, and they would receive higher salary and better fringe benefits and especially because of employment security as well.[19] Although it is not possible to substantiate it with hard figures, dissatisfaction among the workers led to a considerable turn-over of staff during the fieldwork period. Dissatisfaction also affected the motivation of the staff and led to irregular practices. Informants suggested that since CDOs had to visit the villages without proper transport facilities, paying the costs involved themselves, they met the villagers very irregularly. Other 'solutions' to supplement the salaries seemed to be in vogue too. Many key persons asserted that CDOs quite frequently had other jobs, pocketed part of the money targeted to their cluster or claimed payment on fake bills.[20]

In spite of the change in its organizational structure in 1985, AWARE resembled every inch an hierarchically structured organization in which a chairman peremptorily tells the staff what to do. Wanting a very servile staff who obeyed orders, he and his management team discouraged discussions and critical thinking. The fact that individual initiative was not appreciated, not only discouraged most staff, it also created an atmosphere of 'taking things easy'. Also, newcomers who fell to their tasks enthusiastically and 'ran too fast', were told to slow down and not to take initiatives.

Neither AWARE's hierarchical structure nor the attitude and behaviour of its chairman stimulated the internal democracy of the organization (Arens, 1979: 9). Although AWARE's approach was based on people's participation, he controlled the organization, formulated its policy and had a final say in the planning of the programmes and of individual projects.[21] To keep in control, he took all kinds of bureaucratic measures which resulted in a lot of tedious work. In that sense, AWARE's headquarters very much resembled any government office. There were piles of files with red and white tape everywhere, nearly all pending. Moreover, his management style prevented the development of good quality second-line management.[22]

AWARE's chairman was not only very visible in the organization, he was very manifest in the villages as well. In spite of the fact that AWARE was meant to be an organization established for the benefit of the members of the target group and in spite of the fact that its social motivation programme was meant to create the belief that 'we are AWARE', most villagers considered him to be AWARE in person. As a consequence of his pivotal role, many members of the *sanghams* relied on his promises only and did not attach great value to the promises of CDOs. This attitude was strengthened by the CDOs who regularly told the target group to have confidence in him. Hence, it is hardly surprising to find that beneficiaries may have thought that AWARE's chairman was god incarnate who gave away his own money.

4.2.5 PROJECT MEANS

It is difficult to give a complete overview of AWARE's income and expenditure over the years, in particular concerning the activities undertaken in Amangal *mandal*, including Ekvaipally. Besides non-availability of certain data, it is due to the fact that since the late 1980s (thus after AWARE's involvement in Ekvaipally) about Rs 400 million ($ 15 million) given by foreign donors were put aside in a secret fund. Hence, the next tables and lines have to be read cautiously.[23]

AWARE received money from various Indian and foreign sources. Together with its most important foreign partners, AWARE had established a consortium of which it held the chair. These were, Novib and ICCO of the Netherlands, NCOS of Belgium, Diakonia of Sweden and the Deutsche Welthungerhilfe (German Agro Action). Other, but from a financial point of view less important, foreign funding agencies were Community Aid Abroad (Australia), Help the Aged (Great Britain), Ford Foundation

(United States), CIDA (Canada) and the European Community. The most important Indian sources were the Central and State governments (supplying funds from programmes like the DPAP and IRDP, and from agencies meant to assist STs and SCs), banks and the Investment Cooperation of India. The members of the target groups supplied money too, in the shape of donations, payment for the services rendered and savings in revolving funds. A final source of money originated from AWARE's own enterprises like its centres for handicrafts and farms.

From 1976 to 1993, yearly average income of AWARE was some Rs 35 million. Although it was AWARE's goal not to become dependent on foreign donors and although its programme foresaw in the mobilization of government resources for about 40 per cent and of the target group of up to 30 per cent (Wils et al., 1994: 1-2; Sohoni and Singh, 1990: 93), Table 4.4 shows that in reality the organization received a major and increasing part of its money from foreign donors. This conclusion makes one wonder why AWARE more or less replicated government programmes like the IRDP and did not use the money for alternative basic needs and economic activities. The major proportion of the funds was provided by Novib. Between 1976 and 1994 it gave some fl. 38 million in current figures (against 1994 $: 20 million) (Novib India desk, personal

TABLE 4.4. AWARE'S SOURCES OF FUNDING (1976-87 AND 1988-93) (*Rs million*)

	1976-87		1988-93	
Source	Abs.	%	Abs.	%
Novib			78.2	27.7
ICCO			46.7	16.7
Diakonia			29.4	10.3
NCOS			38.6	13.8
Others			3.5	1.4
Total Foreign	164.9	48.5	196.4	69.9
Government agencies			30.7	11.0
Banks			6.6	2.5
AWARE			5.2	1.8
Revolving Fund			16.4	5.7
Target Group			21.9	7.8
Others			3.9	1.4
Total Local	175.4	51.5	84.7	30.1
Grand Total	340.3	100.0	281.1	100.0

Sources: 1976-87: Prasad (1992: 75); 1988-93: AWARE (1994a: 100-13).

TABLE 4.5 EXPENDITURE BY AWARE, BY PROGRAMME (1988-93)

Programme	Expenditure (*Rs millions*)		Beneficiaries (N)	Expenditure/ capita (*Rs*)
	Abs.	%		
Social Transformation	33.1	11.8	704,253	47
Basic Needs	36.4	13.0	na	na
Economic Development	180.0	64.1		
Agriculture	75.1	26.8	69,526	1,080
Landless Poor	29.9	10.7	57,700	518
Cottage Industry	6.3	2.3	15,608	406
Women	42.4	15.1	1,18,261	359
Bonded Labour	24.6	8.8	12,254	2,008
Others	1.7	0.6	na	na
Administration	31.1	11.1	na	na
Grand Total	280.6	100.0		

Note: Because of financial irregularities, these figures must be read with great caution.
Source: AWARE (1994a: 100-13).

communication, 10 June 1994). The contribution of the members of the target groups (7.8 per cent during 1988-93, about half in cash and half in labour) was less than expected by AWARE, as was the proportion furnished by the revolving fund (5.7 per cent), which had everything to do with the very poor reimbursement rates which stood at some 15 per cent (AWARE 1990: 16; Novib India desk, personal communication, 9 June 1993).

From 1988 to 1993, yearly average expenditure was more than Rs 45 million.[24] Kumar et al. (1986: 13), concluding that the economic programme took the biggest share in 1984 and 1985 (74.7 per cent and 79.9 per cent respectively), in which agriculture was a big spender (40.3 per cent and 53.8 per cent of total expenditure in 1984 and 1985), advised to scale down these outlays. Table 4.5 shows that AWARE only partly succeeded in this endeavour. Averaging 64.1 per cent for 1988-93, its economic programme still took the biggest share, agriculture remaining the biggest spender. Expenditure on social transformation, considered to be its most important programme of activities, averaged 11.8 per cent. The high expenditure on its agricultural programme was mainly due to AWARE's modernization and intensification policy: i.e. the construction of irrigation facilities, improvement of the quality of lands and the supply of all kinds of inputs like improved seeds, chemical fertilizers and pesticides. How much was spent on dryland agriculture is not known. Averaging Rs 12.5 million, yearly expenditure for agriculture was higher than the average Rs 9 million spent by the Maheswaram project. Of course, the region and the number of people covered by AWARE were much larger than the Maheswaram project, but it indicates the magnitude of AWARE's organization and level of operations.

Except for the information that around Rs 3.8 million (15 per cent of total expenditure) were spent in Amangal *mandal* in 1987 (Novib, 1987: 16), other data on AWARE's expenditure or on the number of its beneficiaries in this *mandal* and Ekvaipally are lacking. Hence, the following data relate to AWARE's total area of operation. Prasad (1992: 76) reported that while average per capita expenditure during 1976 to 1990 was Rs 112, average expenditure per household in the employment generating programme and the agricultural programme was much higher: Rs 2,075 and Rs 1,243 respectively. Table 4.5 shows that the highest per household expenditure for 1988 to 1993 was for bonded labour (Rs 2,008) followed by agricultural development (Rs 1,080). Compared to the Maheswaram project which spent on average Rs 1,726 per farming household on dryland development and soil and water conservation, the expenditure per household of AWARE's agricultural programme was considerably lower. Together with the fact that the proportion spent by AWARE on administration was half the proportion spent by the Maheswaram project, this might be an indication of AWARE's higher efficiency.

4.2.6 PROJECT PLANNING AND DECISION-MAKING

Unlike government-sponsored development projects, where a structured sequence of activities is emphasized to attain set targets, AWARE claimed to follow a more flexible approach in which decision-making of the target groups during all phases of the project was stimulated. After carrying out the social motivation programme and establishing a *sangham*, CDOs helped the members of the *sangham* to outline an action plan. Although this is a very active form of participation, these plans had to suit AWARE's ideology and objectives and had to be in line with its programmes. Consequently, all village plans contained programmes on organization and awareness building, on basic needs and on income and employment generation. Nevertheless, some adaptations were possible and depending on the needs and priorities of a *sangham*, some specific activities could be undertaken.

Ideally, the final decisions about development plans were taken by the members of the *sangham* who decided about priorities, keeping in view the availability of funds as well as the needs of the people. These plans were discussed with the CDO, reformulated if necessary, and then sent to AWARE's headquarters for final approval. After approval, headquarters passed funds to its field offices. At local level, these funds had to be supplemented with local money from various government programmes and banks and passed on to the *sangham*. Depending on the financial resources, it was the *sangham's* task to carry out the plan, assisted by AWARE's CDO.

This 'bottom-up' planning was based on the assumption that all members of the *sangham* participated in the discussions, that the *sangham* was internally democratic and led by effective, scrupulous and accountable leaders. While Wils et al. (1994: 81-2) seem to argue that this assumption is valid, others have their doubts. For the early years, Arens (1979: 25-6), reported that, in a few villages, not the *sangham* but the president and secretary decided about loans and that, to ensure repayment, hardly any loans were given to landless households. Prasad (1992: 79), reporting on more recent practices, was very negative, arguing that because the president and secretary in the study village played a major role in identifying beneficiaries, they became exploiters within the *sangham* and benefited considerably from funds and schemes which came their way. Also, whether the independent *sanghams* were really independent, self-sufficient and self-reliant, were still functioning and designing village development plans, was part of a debate as well (Prasad, 1992: 79; Novib India desk, personal communication, 9 June 1993; Wils et al., 1994: 96).

4.3 Village-Level Hypothesis

The staff of the Maheswaram project indicated that activities in Mansanpally were restricted to the *tanda* and hence were concentrated on the STs, who are mainly marginal and small farmers (see also Chapter 1 and 6). Activities in Debbadaguda were meant to benefit the entire farming population in the village, irrespective of caste or class background. In Ekvaipally, it was AWARE's intention to concentrate its work on SCs and STs. Sticking to the intentions of the projects and assuming that the comparative advantages of AWARE resulted in better effects (measured in adoption rates of recommended dryland technologies) for their target group of SC and ST farmers, the following hypotheses were formulated.

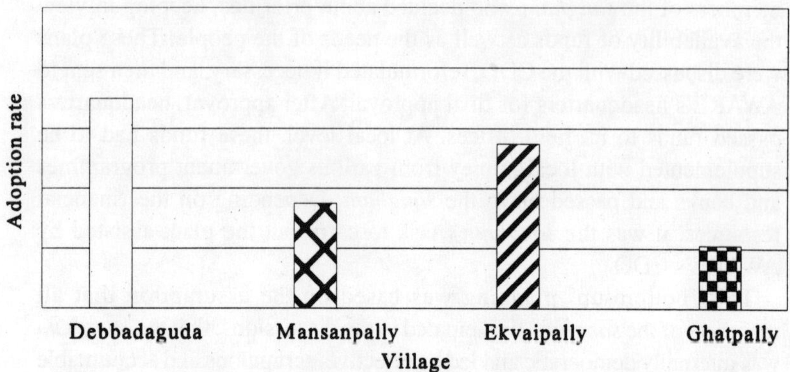

FIGURE 4.2. EXPECTED DIFFERENCES IN ADOPTION RATE OF DRYLAND AGRICULTURE RECOMMENDATIONS, BY VILLAGE.

4.3.1 CHANGES IN THE VILLAGES

Since the Maheswaram project covered all communities in Debbadaguda and only the ST community in Mansanpally and since AWARE focused on the SC and ST community in Ekvaipally, it is expected that project interventions, both in the sphere of soil and water conservation technologies and in the sphere of crop management technologies, resulted in highest adoption rates in Debbadaguda, followed by Ekvaipally and Mansanpally. Lowest adoption rates are expected in the village without project activities, Ghatpally. The expected differences are illustrated in Figure 4.2. This and subsequent figures are only illustrative: their intention is only to indicate the relative position of one village, community or class compared to the others. Hence, no exact percentages are given.

4.3.2 COMPARISON BETWEEN THE COMMUNITIES IN THE VILLAGES

In Ghatpally, the non-intervention village, a 'normal distribution' of adoption rates of recommendations for dryland agriculture is expected, meaning that relatively more FC farmers have adopted the recommendations, than BC farmers, ST farmers and SC farmers, in that order. The expectation that relatively more ST than SC farmers have adopted, while both belong to the most deprived groups, is based on the assumption that ST farmers are more land-oriented and hence more interested in improving land, water and crop management, than SC farmers who are expected to be more oriented to non-farm and off-farm employment opportunities, with lower priority for (improving) their own farming

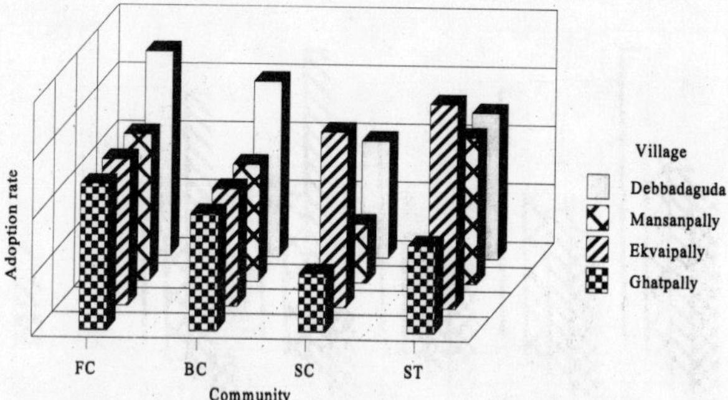

FIGURE 4.3. EXPECTED DIFFERENCES IN ADOPTION RATE OF DRYLAND AGRICULTURE RECOMMENDATIONS, BY COMMUNITY, BY VILLAGE.

activities. Since the Maheswaram project followed an area approach, a non-targeted intervention is assumed in Debbadaguda, the 'strong intervention village' of the project. Hence, the distribution in Debbadaguda is expected to be similar to that of Ghatpally, albeit at a higher level for all communities: relatively more FC farmers would have adopted the recommendations than BC farmers, followed by ST farmers, with SC farmers lagging behind. In Mansanpally, the Maheswaram village where activities were said to be restricted to the ST area, it is expected that the ranking is as follows: more or less the same number of FC and ST farmers would have adopted, followed by BC farmers, while SC farmers rank last. In Ekvaipally, because of AWARE's interventions, it is expected that relatively more ST farmers would have adopted, followed by SC farmers, FC farmers and BC farmers. The expected differences are illustrated in Figure 4.3.

COMPARISON BETWEEN THE LAND OWNERSHIP CLASSES IN THE VILLAGES

The same 'normal distribution' of adoption rates of recommendations for dryland agriculture is expected among the various land-ownership classes in Ghatpally: it is expected that more large farmers (more than 10 ha) would have adopted the recommendations for dryland agriculture than medium farmers (5-10 ha), followed by small-medium farmers (2-5 ha) and small farmers (1-2 ha). The number of marginal farmers (less than 1 ha) who would have adopted is expected to be the lowest. As a result of the non-targeted intervention of the Maheswaram project in Debbadaguda,

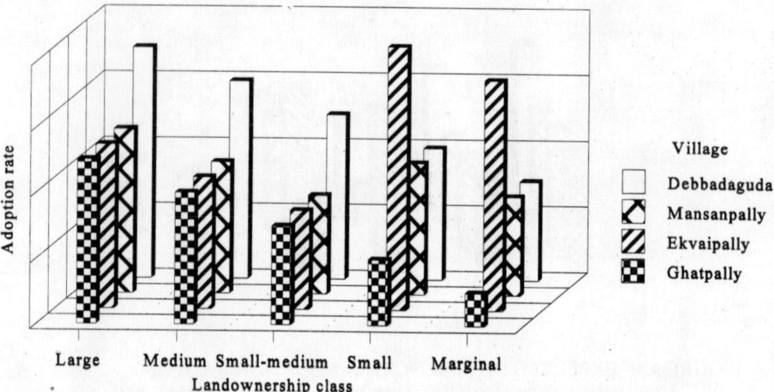

FIGURE 4.4. EXPECTED DIFFERENCES IN ADOPTION RATE OF DRYLAND AGRICULTURE RECOMMENDATIONS, BY LANDOWNERSHIP CLASS, BY VILLAGE.

the same 'normal distribution' as in Ghatpally is expected, albeit on a higher level for all classes than in this latter village. In Mansanpally the Maheswaram project, covering only the ST area, concentrated its activities on marginal and small farmers. It is expected that this intervention would result in similar adoption rates for, on the one hand, medium and small farmers, and on the other hand, but at a lower level, for small-medium and marginal farmers. Thus, the expected ranking on adoption is as follows: more large farmers would have adopted than medium and small (having a more or less similar ranking), whilst small-medium and marginal farmers come last (having a more or less similar ranking). AWARE focused its interventions on STs and SCs, communities which consist largely of small and marginal farmers. Assuming that the effects of the AWARE intervention are superior to the effects of the Maheswaram project, the expected ranking is as follows: more small farmers will have adopted the recommendations for dryland agriculture than marginal, large, medium and small-medium farmers, in that order. The expected differences are illustrated in Figure 4.4.

4.3.4 COMPARISON BETWEEN THE SAME CLASSES AND COMMUNITIES IN THE VILLAGES

As neither Ghatpally nor Ekvaipally or Mansanpally experienced specific project interventions directed at FC and BC farmers, the same adoption rates are expected for these two communities in each of these villages.

Since project interventions clearly included FC and BC farmers in Debbadaguda, the adoption rates of these two communities are expected to be higher compared to the other three villages. For SC farmers, the highest adoption rates are expected in Ekvaipally, followed by Debbadaguda. The lowest and comparable rates are expected in Ghatpally and Mansanpally. For ST farmers, the ranking is expected to be as follows: highest rates in Ekvaipally, followed by Mansanpally, Debbadaguda and Ghatpally. The rates for ST farmers in Mansanpally and Debbadaguda are expected to be more or less comparable but higher than in Ghatpally and in all four cases higher than for SC farmers. The expected differences are illustrated in Figure 4.3.

As neither Ghatpally nor Ekvaipally or Mansanpally experienced specific project interventions directed at large, medium and small-medium farmers, the same adoption rates are expected for these three classes (although Ghatpally may be lagging behind a bit). In Debbadaguda, project interventions included all farmers. Thus, the adoption rates for these three land-ownership classes are expected to be higher compared to the other three villages. For small and marginal farmers the highest rates are expected in Ekvaipally, followed by Debbadaguda and Mansanpally (with similar rates), while the lowest rates are foreseen in Ghatpally. The expected differences are illustrated in Figure 4.4.

4.4 Conclusions

The Maheswaram project is a clear example of a top-down project. Farmers had neither a say in project formulation and design nor in the nature of the activities to be undertaken. In fact, for various reasons, the measures taken by the project to increase participation during the implementation phase were not successful. For the project, participation of farmers was synonymous with accepting the recommended agricultural innovations for (*in-situ*) soil and water conservation and dryland crop management, a passive form of participation.

Top-down planning is also evident from the fact that although initially designed by civil servants of the Department of Agriculture of the government of Andhra Pradesh, many revisions were introduced by the government of India in cooperation with the World Bank. A first revision was a shift in policy. By shifting its focus from production to environmental conservation, from agriculture to the conservation of the entire watershed, the development of dryland agriculture received less attention.

This is evident from the financial resources allocated to this component. Second, it shifted its focus from expensive soil and water conservation structures to more affordable, simple and replicable technologies, notably vegetative barriers of Vetiver grass (*khus*) and *in-situ* soil and moisture conservation practices. Since the number of viable agricultural innovations for dryland agriculture was limited, the project not automatically favoured all kinds of recent innovations. Consequently, the recommendations made by the Maheswaram project were a combination of three: (1) stimulating the use of (adapted) viable indigenous technologies; (2) stimulating the use of agricultural technologies that had been introduced by government extension services in the area already before the start of the project; and (3) stimulating the use of more recent 'real' innovations for dryland agriculture which had not yet been introduced in the area.

The objective of developing dryland agriculture through the widespread adoption of these recommendations was partly hampered because the project was largely dependent on the existing T&V extension system. Although the government of Andhra Pradesh had agreed to increase the number of T&V staff in the area for the purpose of the project, this did not materialize. Moreover, the idea to speed up adoption rates by bridging the financial gap through the supply of all kinds of agricultural inputs on a subsidy, met resistance from the World Bank. Consequently, to supply subsidized inputs, the project was dependant on other ongoing schemes in the state. However, it was never sure whether and when it would be assisted by these other programmes.

On many occasions, project management and field staff felt trapped between their superiors and the farmers. Project management often regarded themselves as mere executors of activities designed by outsiders and, like field staff, felt some animosity on the part of the farmers as well. The fact that suggestions like their advice to stop the planting of *khus*, were not accepted by these outsiders, and the fact that many farmers were reluctant to accept their advice, did not stimulate them in their work either. Most staff resembled the civil servant known from many publications, i.e. not very dedicated and socially motivated to assist poor dryland farmers to improve agricultural production. For most of them this was just another project which earned them a living.

AWARE's ideology is a mixture of ideas promoted by social action groups and by Gandhians. Although Gandhian thought has become more important, particularly in its early years AWARE used a lot of leftist rhetoric. Because it never intended to radically overturn the existing social order, AWARE is best labelled a moderate empowerment NGO. To reach its objectives of social, political and economic development of its target groups, *Girijans* and *Harijans*, AWARE followed an integrated approach by initiating all kinds of social motivation (including awareness training and organization building), basic needs and economic development programmes. Financially, agricultural development, including a component on soil and water conservation and dryland crop management, was one of the most important elements of AWARE's activities. The strategy used to assist these target groups was a combination of a 'parallel' and an 'access' strategy, meaning that besides developing separate programmes without government involvement, AWARE also initiated programmes using government schemes and programmes. For agriculture it meant that AWARE relied largely on agricultural innovations designed by public research, on the T&V system of agricultural extension and on all kinds of government programmes subsidizing assets and agricultural inputs. This means that in spite of its 'radical flavour', AWARE's recommendations for dryland agriculture hardly differed from those of the Maheswaram project.

Using concepts like participation and empowerment, AWARE established different kinds of local and supra-local organizations that had to safeguard the interests of the target groups. These organizations were not only created to get access to the government and financial agencies like banks, or to act as pressure groups to force the government to execute existing laws and to legislate new ones, they were also created to design and implement development programmes at village and cluster level in a bottom-up manner. This is a very active form of participation. However, doubts were expressed whether the members of the *sanghams* could really include their needs and priorities in projects since these had to be in line with AWARE's activities and with government programmes. In addition, it was doubted whether *sanghams* and other supra-local organizations were internally democratic and led by effective, scrupulous and accountable leaders, who acted for the benefit of the members. It was also argued that more than being created to empower the target group, *sanghams* were meant to enhance the efficiency and effectiveness of projects.

Although AWARE stimulated participation of the target groups, these were only marginally represented in AWARE's governing bodies. Whatever these bodies, AWARE's policy as well as its programme of activities were designed by its chairman who controlled the organization.

His leadership style prevented the establishment of second-line management, resulted in a lot of bureaucratic work and made many observers doubt AWARE's internal democracy. Consequently, most staff turned servile, afraid to question his decisions. With a mixture of fear and contempt they were not very much motivated to work to the best of their abilities. Financial considerations and not altruistic sentiments were the basis of motivation of most of the staff. Furthermore, many were less socially motivated and had joined AWARE because they were in need of a job. In fact, many were eager to leave the organization for a government job.

NOTES

1. These figures are based on government statistics. It must be remembered that these are not necessarily in accordance with reality.
2. To get a reliable picture of the activities of the project in Mansanpally and Debbadaguda and to measure its effects, senior staff were asked for all kinds of information (when it started, the various activities undertaken by the project and in which year, the names of the beneficiaries of various schemes and the money involved, the staff working in the villages and the total money spent by the project in the villages). We received part of the information, in particular the information concerning the soil and water conservation programme. The project management claimed that they had no data concerning the crop improvement programme, more specifically, to the beneficiaries who received subsidized inputs, because these were financed from non-project funds. Unfortunately, we did not manage to get these data from these other sources either.
3. This seems to be a rather odd conclusion since the rural labour market is not tight at all, most certainly not in the slack season.
4. These inputs were financed by schemes like the High Yielding Varieties Programme (HYVP), the National Oilseeds Development Programme, the National Watershed Development Programme in Rainfed Areas (NWDPRA) and the Drought Prone Area Programme (DPAP) (Action Plan 1987/8: 10-11).
5. In 1986/7, for instance, the DRDA was requested to provide an amount of Rs 5,00,000 for arranging timely inputs for farmers (Action Plan 1986/7: 10).
6. To get a reliable picture of its activities in Ekvaipally and to measure its effects, we asked AWARE staff and its chairman for all kinds of information (when AWARE had started in Ekvaipally, which activities it had undertaken and in which year, the names of the beneficiaries, the money involved, the AWARE staff working in the village, what local organizations it had established, their members and the composition of their boards). Unfortunately, during the thirty months of the study, AWARE evaded giving

this information. Although they claimed to have written records on their interventions in Ekvaipally, except for one report (AWARE, 1994b) no written information was supplied. Unfortunately, this very general statement was useless for research purposes.

7. Sohoni and Singh (1990: 24) claim that nearly 1,00,000 people were trained.
8. Many of these studies have been undertaken by the same people and in a number of cases one of the authors belonged to the AWARE management team.
9. Novib used AWARE, for instance, in their 'Gast aan Tafel' (Guest at your Table) campaign. Together with NCOS, it published a book written by M. Van Bouwel (1993) on AWARE. In 1989, NCOS published a comic strip drawn by the famous Belgian Vandersteen studios on AWARE's bonded labour programme: Bessy Natuurkommando: De Loonslaven (Bessy Environmental Command: Bonded Labour).
10. In defiance of its name, AWARE started a cancer hospital in Hyderabad in the early 1990s.
11. The latest news is that Indian auditors, although confirming the existence of such a secret fund, did not find proof of large scale corruption by the top management. Hence, ICCO is considering resuming its financing of AWARE once again. Novib, lacking confidence in AWARE, has decided not to take such a step (Trouw, 11 Nov. 1997).
12. Pointing at AWARE's chairman, constituents are reported to have said: 'You cannot eat from a bare tree'.
13. The section on these programmes is primarily based on various AWARE documents as well as on reports by Arens (1979), Kumar et al. (1986), Sohoni and Singh (1990), Prasad (1992), Sohoni (1994) and Wils et al. (1994).
14. As a matter of fact AWARE had its own rig to drill bore holes. Moreover, it intensified the agricultural production on its own farms by converting dryland into wetland and continued to use pesticides, for instance.
15. The Human Resource Development Centre was a farm-cum-training complex (see also note 16). Except for a training given for (new) staff, and contrary to AWARE's claims that the centre was meant to train farmers as well, not a single agricultural training was given during the fieldwork period (covering some 30 months).
16. Located some 24 km south of Hyderabad, in the Maheswaram project area, AWARE had established a 30 ha farm-cum-training centre named after a former chief justice of India and a member of AWARE's governing body. In the centre, there were three blocks, each having six very basic rooms for the AWARE caretakers, those without toilet meant for AWARE's permanent labourers. There was a luxurious sixty room guest house for AWARE staff, to be used during training sessions. M. Put had the privilege to be the only guest in this guest house for most of the fieldwork period. Furthermore, the centre had two lecture halls, an amphitheatre and a large, posh, shrine for the deceased mother of AWARE's chairman. Under construction were a library to accommodate hundreds of thousands of books, an administration building

and a luxurious 16 room hotel with swimming pool for foreign guests. AWARE expressed its gratitude to the Dutch funding agency Novib for its generous cooperation and financial assistance towards the building of the centre (AWARE, 1991).

During the fieldwork period, all kinds of irrigated crops like fruit, vegetables, mulberry, paddy, seed maize and seed *bajra* were grown. These were irrigated by more than a dozen tube wells (fruit trees by drip irrigation), even to such an extent that water was often lacking. Except for a very small plot cultivated with castor (which failed), dryland crops were not produced during this period. In fact, more than a demonstration farm for (dryland) crops, AWARE regarded it as a source of institutional income.

17. On paper, from 1991 to 1994, from the beginning to the end of the fieldwork, only three zonal officers had been appointed. The two vacancies were never filled.
18. AWARE did not give any particulars (e.g. educational level, caste and tasks) about their staff. This kind of information was mostly obtained from (former) staff.
19. Although salaries were raised over the years, they remained rather low. In 1993, a CDO received Rs 1,500 ($ 57) per month. Travelling expenses within the cluster of villages and expenses for accommodation had to be paid out of his own pocket. An Area Coordinating Officer received Rs 1,800 ($ 68) per month plus actual travelling expenses outside the villages. The average salaries paid to graduates in government jobs or in the private sector were at least Rs 2,500 ($ 94) per month excluding expenses. Most fringe benefits were paid to staff that worked for AWARE for five years of more.
20. Quite often beneficiaries were told by AWARE staff to give them a certain percentage of the loans to cover their expenses and to compensate for the trouble they had faced.
21. *India Today* (15 Sept. 1992) reported as follows: 'he (AWARE's chairman) has a streak of the autocrat in him. He may delegate authority but the ultimate decision is always his and he is intolerant of dissent.'
22. Prasad (1992: 133-5) asserted that one could clearly find '*maa-bap*' relations in AWARE, i.e. both members of the staff and of the target group looked for guidance and direction to AWARE's chairman who was considered to be a 'divine' person. Over the years many observers have raised questions about his style of leadership (lack of second-line management and internal democracy), irregular practices, behaviour towards women and intentions with AWARE (Novib, 1991: 10; Groteman et al., 1993: 23; Novib India desk, personal communication, 9 June 1993; Trouw, 21 Oct. 1995). In fact, the earliest doubts originated from 1979 (Arens, 1979: 7).
23. Comparing the total receipt in current figures during 1976 and 1994 (Rs 620 million) with the secret fund of around Rs 400 million, shows the sheer magnitude of AWARE's financial irregularities.
24. Information about AWARE's expenditure per programme before 1984 is not available.

5. An Introduction to the Study Villages

M. Put

The objective of this introductory chapter to the study villages is to describe some of their basic features; to demonstrate that the procedure discussed in Chapter 1 has resulted in the selection of villages that are indeed more or less identical to the chosen characteristics; and to demonstrate that village society and economy has changed enormously since independence.

5.1 Basic Features

5.1.1 LOCATION

The study villages, inhabited by 1,200 to 2,200 people, are situated on both sides of a bitumen road connecting Hyderabad with Sri Sailam, a pilgrim centre some 210 km to the south of the capital. Mansanpally, one of the villages of the Maheswaram project, is situated on one of its metalled (sand mixed with small rocks) branches, at a distance of some 30 km from Hyderabad (see Map 5.1). From Mansanpally, travelling to the south for another 5 km, one touches upon Ghatpally, the non-intervention village. The bus from Hyderabad takes about two hours to reach both villages. Debbadaguda, the second village incorporated in the Maheswaram project, is situated on another branch of the Sri Sailam road, at a distance of some 40 km south of Hyderabad. Ekvaipally, the village included in AWARE's activities, is situated further south, some 55 km from the capital. As with the other two villages, both Debbadaguda and Ekvaipally were accessible by metalled road. By bus, the journey from Hyderabad to Debbadaguda took two hours and it took another 30 minutes to reach Ekvaipally.

Although the connectivity between the study villages and Hyderabad has improved over the last decade or so and although Hyderabad's presence was felt in all study villages, the psychological distance was still rather

* This chapter is included in Put (1998). Courtesy of Thesis/Thela Publishers, Amsterdam.

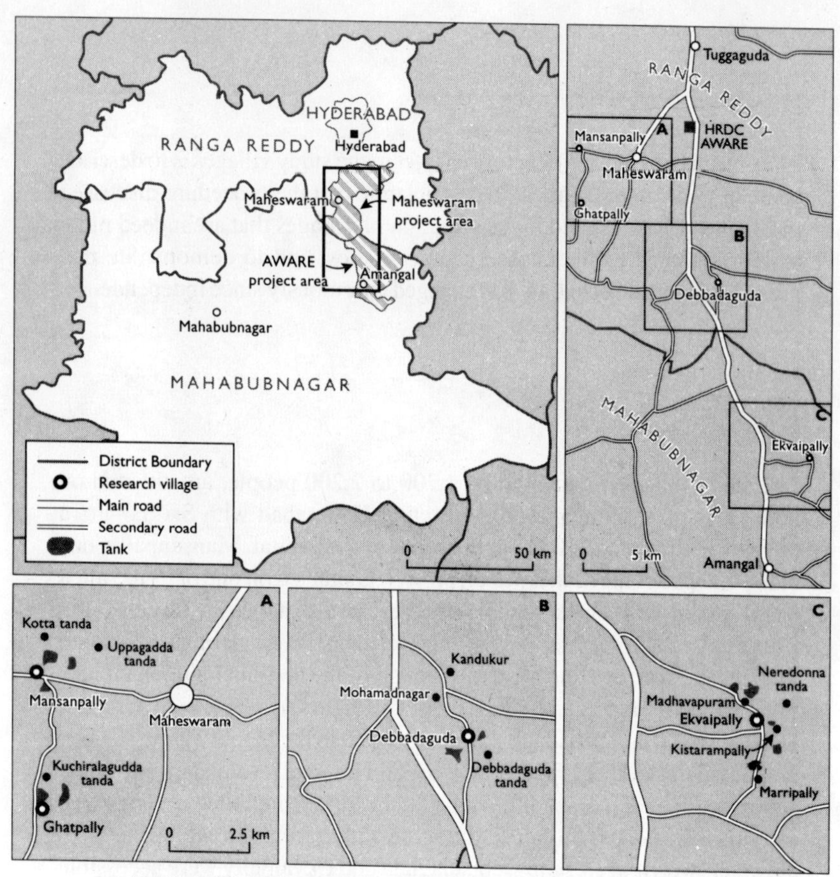

MAP 5.1. LOCATION OF THE PROJECTS AND STUDY VILLAGES.

great, especially in the minds of the economically poor and socially backward farmers and agricultural labourers. For communication purposes, the few television sets in each village were more important than the Sri Sailam road. In fact, mutual contacts between capital and village, although increasing, were limited. A bus trip not only took a long time, a return ticket, costing between Rs 12 and Rs 18, more than equalled daily wages.[1] Of the economically poor and socially backward, only a few commuted to the capital on a more or less regular basis, for instance, to look for work. Most villagers, however, travelled to the city on special occasions only, to sell agricultural products or to buy goods which were not available in their own and surrounding villages. It was predominantly the richer villagers who were the more frequent visitors to the capital. Some of them had relatives living in Hyderabad or had children attending a school there or they travelled to Hyderabad on business, to buy agricultural inputs or to meet civil servants and employees of banks.

In spite of being located in the relative vicinity to Hyderabad, the area was still very rural. Hyderabad's industrial areas were primarily located on the northern and western side of the capital and only a few small factories were situated along the Sri Sailam road. None of these were located in the study villages. Hyderabad's presence made itself felt more in agriculture than in industry. Since the late 1980s, being situated along the same road, some wealthy people of rural and especially urban origin had established orchards and poultry farms which catered to the Hyderabad market. In addition, influenced by an increase in demand, Hyderabadi merchants had begun to buy milk and flowers from the area. More interior villages and dryland agriculture were hardly touched upon by these new commercial activities. Except for some poultry farms in Debbadaguda, some small patches of land cultivated with flowers in Mansanpally and milk collected in all four study villages, this also applied to the study villages.

All four study villages consisted of a main village, the residence of most people, and one or more hamlets. Only with some difficulty, was it possible to identify the quarters of the Forward Castes (FCs), Backward Castes (BCs) and Scheduled Castes (SCs). A strict spatial separation was still the rule for the Scheduled Tribes (STs) who lived in their own hamlets, called *tandas*, situated between 2 to 4 km from the main village. Ghatpally had one, Mansanpally had two *tandas*, all inhabited by Lambadas. These could be reached on foot only. Debbadaguda had two hamlets, one *tanda*, the residence of Lambadas, and a hamlet inhabited by FC, BC and SC households. Both could be reached by dirt road.

Ekvaipally had four hamlets, two *tandas* and two hamlets, that could be reached on foot only. Both hamlets were inhabited by FC, BC and SC households, one *tanda* by Lambadas and the other (named after AWARE's chairman) by Chenchus.

5.1.2 NATURAL ENVIRONMENT

The four study villages are located in the same agro-ecological zone, which is a low rainfall zone (National Commission on Agriculture, 1976). The south-west monsoon usually becomes active around mid-June and recedes by mid-October. In this period the area receives on average 500 mm of rain. Since the north-east monsoon adds only little to this average, as do the other two seasons, with 638 mm the yearly average is only relatively higher. However, these are only averages and yearly rainfall variability is large (see Chapter 3). Walker and Ryan (1990: 34) report for Aurepally, an ICRISAT study village not far from the study area, that a year of normal rainfall within the standard deviation of a ten-year (1976 to 1986) annual mean, was a rare event. To reassure the readers they add that although four out of ten years were unusually dry, a severe and prolonged rainfall deficit did not occur.

The topography of the study area is mostly rolling and undulating, the slope generally being less than 2 per cent (GoAP, 1985c: 15). The few hillocks in the south (also near Ekvaipally) are sparsely covered with trees, giving the tribal population of the area an opportunity to collect forest produce like wood, bark (for tanning), fruits, gum and honey. Tree-like vegetation in the other parts of the study area is scarce. Because of its forestry programme, the Maheswaram project succeeded in greening parts of the project area. With the exception of these plantations, the hillocks near Ekvaipally and some commercial plantations with fruit trees (like the toddy palms creating a romantic scenery in Ekvaipally), trees are scarce. On the whole the vegetation in the region is more of a shrubby type, being grass with isolated bushes. These common property resources and the permanent fallows are used by small and big ruminants for grazing.

Alfisols (red soils) dominate in the four villages although some patches of Vertisols (black soils) are present as well. Normally Alfisols are sub-divided into *chalka* soils (sandy loams) and *dubba* soils (loamy sands). However, a considerable heterogeneity exists in dryland soils in the study villages [see also Dvorak (1988) for Aurepally, the ICRISAT study village]. Using characteristics like colour, 'temperature' (hot and cold soils), taste (saltiness), depth, occurrence of gravel in the profile, topsoil and subsoil texture, and parent material, farmers classify village soils into many subgroups. It is not uncommon to find that the soil of a dryland plot is made up of more than one subgroup. This soil heterogeneity makes it difficult to correlate soil type with the yield of any dryland crop grown in a certain plot.

Since the village territories are not traversed by rivers, alluvial soils are not present. Soils found in wetland, although looking darker and having a finer texture than dryland soils, originate from the same parent material as the latter. Irrigation, however, has modified the soil characteristics. To what extent this has happened depends on the number of years the plot has been irrigated. Over the years, wetland under tanks may have accumulated silt or farmers may have spread silt on their irrigated fields themselves. Improved crop management like the application of relatively higher quantities of organic and inorganic fertilizers, has resulted in a better or even much better texture and fertility of these wetland soils compared to the soils on dryland plots.

5.1.3 INFRASTRUCTURE

Government intervention makes itself felt, in particular, concerning the infrastructure. It has not only changed the outlook of all four study villages, but has had a tremendous effect on village life and on agricultural development as well. The improved infrastructure, for instance, enhanced the connectivity of all four villages with Hyderabad. As it was one of the selection criteria (see Chapter 1), it was almost identical in the four study villages. Since the beginning of the 1970s, all were connected to the electricity grid. Although not always legal, electricity was used to illuminate rooms in some (particularly FC) houses, a few streets, to run flour mills and more importantly to operate water pumps to irrigate crops. Villagers faced relatively large problems with the electricity supply, though. Heavy fluctuations in voltage blew up water pumps and other equipment regularly. Power cuts, in summer and at the beginning of *kharif*, sometimes lasting more than five hours per day, made even wetland cultivation a risky undertaking.

A private tempo service (a goods carrier carrying people as well) was started in the area around 1975 and the public bus service was started in the late 1980s. Tempos and buses link the four study villages with Hyderabad and ply two to three times per day. Although rarely punctual, they are a very good means of transporting people and agricultural products to the markets and shops in Hyderabad.

With the arrival of government financed hand-pumps in the 1980s, unprotected open wells became obsolete as a source for drinking water. Every village has at least seven hand-pumps (Ekvaipally had thirteen of these devices), meaning that all communities had access to at least one protected source of drinking water.[2] Financed by the government, some drinking water tanks made of stone and closed off to prevent pollution had also been constructed in Mansanpally and Debbadaguda after 1990. Unfortunately, they did not work properly.

All four villages have government sponsored primary schools.[3] All schools are up to fifth standard, except for Ekvaipally, where the school started a sixth grade in 1993. Understaffing, absence of teachers (most do not live in the villages and some have supplementary sources of income) and mediocre teaching facilities (various classes in one room, lack of furniture and books), result in sub-optimal tuition, though. Because most adults were not interested, evening classes meant for adult education were defunct in all villages.

The government sponsored Public Distribution System operate in all four study villages through the fair price or ration shop, where inhabitants having a ration card can buy subsidized rice, wheat (if available), sugar, cooking oil and kerosene. In addition, various *kirana* (private provision) shops cater to the daily needs of the inhabitants, selling food grains, vegetables, cooking oil, spices, sweets and non-food items like *beedies* (country cigarettes), stationery, soap and the like. There is a flour mill in every village as well. For banks or cooperative societies farmers have to travel less than 10 km.

5.1.4 POPULATION

The 1991 census indicates that the number of inhabitants in the four study villages ranged between 1,182 persons in Ghatpally (213 households) and 2,209 persons (367 households) in Ekvaipally. Most inhabitants, also the Lambadas and the Chenchus, were Hindus. Some Muslim households were living in every village too, which is not remarkable for a region that was under the reign of the Nizam of Hyderabad State. Next to Hindus and Muslims, both Ghatpally and Debbadaguda are also inhabited by a few Christian households. The same census shows that for 94 per cent of the households in Debbadaguda, 87 per cent of the households in Mansanpally and 83 per cent of the households in Ghatpally, agriculture (as farmer or/ and as agricultural labourer) was the most important source of income. Reliable figures for Ekvaipally are missing.

Every village is inhabited by some twenty different castes. Measured by the expanse of their landed property, by their wealth and their political power, the Reddies and the Vaisyas (both belonging to the FC community) dominate the social, political and economic landscape. In Ghatpally, the Muslim community (most were FCs as well), was important too. Barring a few, most FC households owned considerable stretches of land and are very influential and respected farmers. Besides farming, most Vaisya households are engaged in shopkeeping, in trading and acted as moneylenders. Absolutely speaking, the BC community is the dominant community. Most of them fit into the service-caste community of laundry men, pottery makers, blacksmiths, goldsmiths, barbers, carpenters and the like. Some have still retained their traditional jobs, but for most held that they got all or most of their income from other sources than their traditional occupation. The *Harijans* (Scheduled Castes or SCs), situated at the bottom of the social hierarchy, consisted of two groups, the *Mala* and *Madiga*. Most *Harijan* households had various sources of income, from agricultural labour, their traditional occupation, agriculture or wage labour outside agriculture. Most *Girijans* (Scheduled Tribes, STs) got their income from agriculture, animal husbandry, agricultural labour or from wage labour outside agriculture. Although the ST community, together with the SC and BC community, is labelled a 'deprived group' by the government, yet the Lambadas differed very much in their attitude from the other two communities. That they were less submissive and obedient than most SCs and BCs, may be related to the fact that, under the Nizam's reign, they were less subjected to severe exploitation by the landed elite and were economically more advanced than BCs and especially SCs and other tribes.

The findings by observers like Walker and Ryan (1990: 28), Robinson (1991: 9) and Drèze (1997) that the caste hierarchy is slowly eroding, also pertains to the four study villages. This is true for both the social and the economic dimensions. SC and ST households reported that although they were still discriminated against, it was much less now than in the past. This change was evident, for instance, from the fact that STs and SCs talked more freely with FCs, that they were allowed to enter the houses of the Reddies and other FCs and that irrespective of caste, children went to the same village school.[4] Low caste people claimed that these changes were the result of overall economic development which has taken place in the last decades. Nevertheless, some of the old practices and prejudices are still alive. FCs often use foul language against SCs and

STs and hardly ever enter their localities in the evening, accusing them of bad habits such as drinking. The commanding tone which is often used by FCs made SCs in particular assume an humble attitude.

The slowly eroding caste hierarchy applies to landownership and wealth as well. It is no longer true that invariably all FC households are wealthy, having the largest farms, while all SCs, without exception, are poor landless labourers. Larger FC farmers have lost land, if only because of subdivision to sons, while SC households have bought (little) land. In all of the four study villages people pointed at Reddy households which were nearly or completely landless and spoke with respect of those Harijan households which had become relatively well-to-do (medium) farmers. In the sample, the FC household having the smallest holding lived in Ghatpally. Owning only 0.6 ha of land he was just a marginal farmer. The largest SC farmer of the sample lived in Mansanpally, owning 6.2 ha, while the largest ST farmer of the sample lived in Ekvaipally, owning 6.4 ha. Still, this area is a pittance compared to the area owned by the largest farmer in the sample. Living in Debbadaguda this FC respondent owned 44.5 ha. Thus, irrespective of these changes, caste is still a good predictor and indicator of land-ownership in each of the villages, meaning that while most of the larger landowners belonged to the FC community, most of the marginal and small farmers belonged to the SC community (see next section). Irrespective to these changes, villagers indicated that a FC farmer owning 5 ha is more respected than a SC farmer owning a similar area of land.

In spite of the slowly eroding caste hierarchy, the legacy of the oppressive and exploitative Muslim-ruled Hyderabad State, with its ruling class of feudal landlords, large farmers and civil servants, can still be felt in the four villages. The various agricultural reforms promulgated after independence have not helped to change the power structure in the villages much. Although Reddies may have lost land and those in the lower ranks of the social hierarchy may have improved their economic position over the years, the former Reddy *deshmukhs* and other large farmers of the same caste still control politics in the four villages. This everlasting dominance of Reddies is clear from the fact that it is still quite common that it is from their ranks that the *sarpanch* (head of the village council) as well as the *patwari* (for keeping and maintaining accounts of village landholdings and land revenue) are elected. This had occurred in all villages except in Ekvaipally. Supported by AWARE, the former president of AWARE's village level organization (*sangham*), who was an influential BC farmer, was elected *sarpanch* in Ekvaipally.[5] Reddy dominance was also evident in other realms of daily life, for instance, from the fact that for landless labourers and smaller farmers, Reddies were the most important employers in the villages, both for casual and for permanent work. In addition, many gave credit and sometimes food and were a source of (agricultural) knowledge and information.

Being the largest landowners and having kept their key political positions in the villages, more than any other community, the Reddies and other FCs kept in touch with the bureaucratic machinery at *mandal* (the lowest bureaucratic level, below the district) and higher levels. They were quite familiar with the wheeling and dealing of politics, government agencies and banks, knew about procedures, how to play the bureaucratic games, and how to take advantage of their status. Another big advantage was that they often came from the same community as government officials and bankers, which put them on a more or less equal footing with these officers. Reddies and other FCs considered these contacts of utmost importance because it informed them about government funds, loans and subsidies to be released shortly and gave them an opportunity to influence the decision-making process at *mandal* and higher bureaucratic levels. It was these contacts which enabled them to appropriate resources for local development programmes.

Because of their contacts and knowledge about procedures of, for instance, government agencies and banks, some Reddies were highly valued intermediaries or brokers between those villagers who did not know the way around and the state. It certainly added to their status in the village community. As indicated by Drèze and Sen (1995: 105), these middlemen often used their position to further their personal interest much more than any social goal. Patronage and clientelism which are said to be among the main structural features of the Indian state (IOV, 1994: 128), was apparent in all four villages, even in Ekvaipally. It made the BC *sarpanch* in this AWARE intervention village favour his BC 'friends' (and to a lesser extent his SC 'friends'), while he was less interested in changing the dependent position of the SC and ST community.

What has improved considerably in one generation in all study villages, and what most probably has been one of the factors responsible for the slowly eroding caste structure, is the level of education. This is not only a consequence of the establishment of a primary school in every study village, it is also a result of the fact that realization has dawned on many heads of households that education is required for government jobs and for jobs in the non-agricultural sector. A steady job, supplying a regular income, was what a lot of parents, irrespective of caste, had in mind for their children, particularly for boys. Consequently, even though not all teachers were very motivated, and even though the level of education was much to be desired, children were better educated than their parents. Thus, while 286 heads of households (73 per cent) in the sample reported that they had not received any education at all, only 132 households (34 per cent) indicated that none of its members have ever been to school.[6] In Table 5.1 the level of education of heads of households is compared with the level of any of its members.[7] It indicates that the level of education has improved in all villages, more so in Debbadaguda and less so in Mansanpally. It also demonstrates that although this improvement has been largest for FCs, the level of education of SC households in Ekvaipally and ST households in Ghatpally and Debbadaguda has also improved considerably. It is remarkable, however, that while the educational improvement of SC households in Ekvaipally, AWARE's intervention village, is more than average, that of ST households lags behind.

Although parents send their children to school more often now than they did in the past, and although for that reason literacy levels are increasing, illiteracy was still prevalent in some communities. In all villages almost all FC households in the sample had at least one member who could read and write. In three out of four villages the literacy level decreased from FC to BC and SC. The only exception was Ekvaipally where the literacy level was higher among SC households compared to BC households. The literacy levels among the ST households fluctuated between the villages: in Ghatpally and Debbadaguda it was relatively high, even higher than that of the Backward and Scheduled Communities, whilst in the other two villages it was the lowest of all communities.

TABLE 5.1. HOUSEHOLDS IN WHICH THE LEVEL OF EDUCATION OF ONE OF ITS MEMBERS IS HIGHER THAN ITS HEAD, BY VILLAGE, BY COMMUNITY

	Debbadaguda		Mansanpally		Ekvaipally		Ghatpally		Total	
Community	Abs.	%	Abs.	%	Abs.	%	Abs.	%	Abs.	%
FC	17	68	19	83	21	81	16	62	73	73
BC	18	72	15	58	15	48	14	45	62	55
SC	12	48	6	25	20	69	14	48	52	49
ST	15	60	4	16	3	21	6	75	28	39
Total	62	62	44	45	59	59	50	53	215	55

Source: Own survey.

5.2 The Agricultural Economy

5.2.1 RESOURCE ENDOWMENTS

Perhaps needless to repeat, but since landless households were excluded from the sample the next section is biased towards the features of the agricultural economy of land-owning villagers.

Concerning landed property, various forms of land titles existed in the villages: land owned by an individual farmer having a *patta* (title-deed); land owned by several people (often brothers) having a *patta*; and land owned by the *poramboke* (government) but given to or encroached upon by an individual. Because these last two categories were of limited importance in all study villages, these will not be discussed separately and are included in individual land titles.[8]

The share of respondents coming under a certain holding-size class is presented in Figures 5.1 (for dryland), 5.2 (for wetland) and 5.3 (dryland and wetland combined).[9] All figures show the same unequal distribution of land. Thus, averaged for the four villages and concerning dryland only, while more than 50 per cent of all respondents owned less than 2 ha, 3 per cent had more than 10 ha.[10] Figure 5.1 shows that some differences existed between the villages. The proportion of respondents owning less than 2 ha of dryland was, for instance, somewhat less in Debbadaguda, while the proportion of those owning between 2 and 5 ha was somewhat higher.[11] As stated in Chapter 1, this was expected: while selecting the villages, Debbadaguda was considered to be somewhat more affluent than the other three villages. Nevertheless, the holding-size pattern for dryland was more or less comparable in all four villages.

The unequal distribution relates to wetland too. The proportion of respondents owning a sizable wetland holding was limited. Averaged for the four villages and concerning wetland only, 41 per cent owned no wetland at all, another 41 per cent owned less than 1 ha, while 1 per cent owned more than 10 ha.[12] Figure 5.2 indicates that the differences between the villages were small. The percentage of respondents owning more than 2 ha of wetland was somewhat higher in Ekvaipally followed by Debbadaguda.[13]

Consequently, total land (the combination of dryland and wetland) was also unequally distributed. Averaged for the four villages, 19 per cent of all respondents owned less than 1 ha (marginal farmer), 30 per cent owned between 1 and 2 ha (small farmer), 37 per cent between 2 and 5 ha (small-medium farmer), 10 per cent between 5 and 10 ha (medium farmer) and

5 per cent owned more than 10 ha of land (large farmer). Since it is said that a minimum or economic holding (which covers the cost of cultivation) ranges between 2 and 6 ha, between 49 and 85 per cent of the respondents had insufficient land. Depending on the definition used, it illustrates the limited agricultural potential of most or all marginal, small and small-medium farmers in the sample. Figure 5.3 shows that the differences between the villages were more or less similar to those found in dryland. The proportion of respondents owning less than 2 ha of land was less in Debbadaguda, while the proportion of those owning between 2 and 5 ha was somewhat more.[14] Nevertheless, the holding-size pattern was more or less comparable in all four villages.

Figures 5.4, 5.5 and 5.6 (calculated in a similar fashion as Figures 5.1 to 5.3) show the average holding-sizes by community. They confirm the earlier remark that FC households are the landed elite. While the average holding-size for the sample households was 2.5 ha of dryland and 0.7 ha of wetland, FC respondents on average owned 4.8 ha of dryland and 2.0 ha of wetland. The average holding of the other communities was much smaller, especially in regard to the irrigated area. On average, BC respondents owned 2.2 ha of dryland and 0.3 ha of wetland, SC households 1.2 ha of dryland and 0.2 ha of wetland; and ST households 1.7 ha of dryland and 0.2 ha of wetland.

Figure 5.4 shows that concerning the average dryland holding some differences existed between the villages, especially concerning FC respondents and to a lesser extent concerning BC respondents. While the average dryland holding of FC respondents was largest in Ekvaipally (6.5 ha) and Debbadaguda (6.1 ha), it was much lower in Mansanpally (2.4 ha). On average, BC respondents in Debbadaguda had the largest dryland holding (3.0 ha), followed by BC respondents in Mansanpally (2.6 ha) and those living in Ghatpally (1.5 ha).[15]

Concerning wetland, the picture is more or less the same (Figure 5.5). While the averages for BC, SC and ST respondents were more or less similar in all four villages (oscillating around 0.3 ha), the averages for FC respondents in Ekvaipally (2.8 ha) and Debbadaguda (2.7 ha) were larger than for Mansanpally (1.5 ha) and Ghatpally (1.2 ha).[16] These averages show that to a large extent, wetland farming was dominated by FC respondents. This conclusion is affirmed by Table 5.2 which demonstrates that while 41 per cent of all respondents did not own wetland, it was only 11 per cent for FC respondents. Village-wise, the proportion of households without wetland was highest in Ghatpally followed by Mansanpally, which is the result of the few, small catchment reservoirs (tanks) located in these villages and, in particular for Mansanpally, the few respondents owning

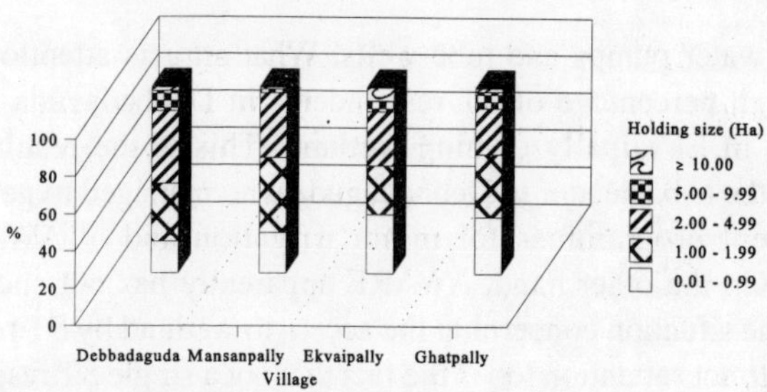

Note: The figure refers to land which is owned, land in joint ownership and to *poramboke* land.

Source: Own survey.

FIGURE 5.1. HOLDING-SIZE CLASSES OF DRYLAND (%) BY VILLAGE (1991/2-1993/4).

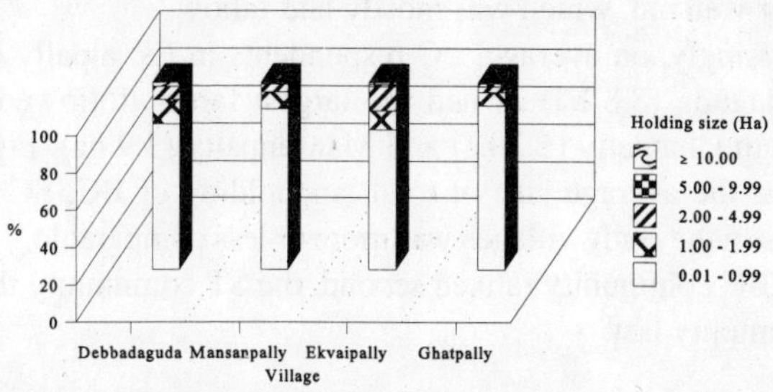

Note: The figure refers to land which is owned, land in joint ownership and to *poramboke* land.

Source: Own survey.

FIGURE 5.2. HOLDING-SIZE CLASSES OF WETLAND (%) BY VILLAGE (1991/2-1993/4).

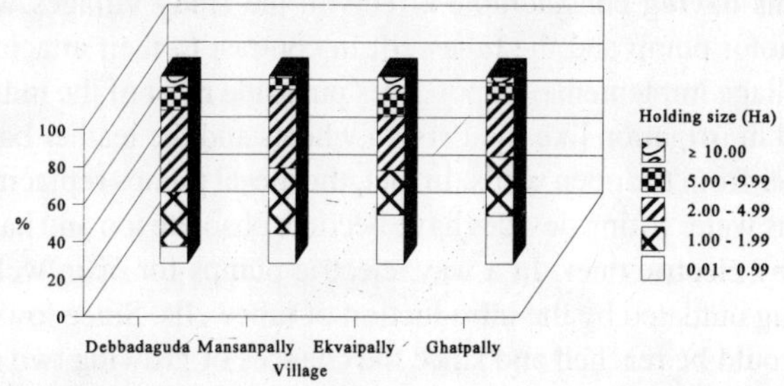

Note: The figure refers to land which is owned, land in joint ownership and to *poramboke* land.

Source: Own survey.

FIGURE 5.3. HOLDING-SIZE CLASSES OF TOTAL HOLDING (%) BY VILLAGE (1991/2-1993/4)

open wells, water pumps and tube-wells. What attracts attention is the relatively high percentage of ST respondents in Debbadaguda and SC respondents in Ekvaipally owning wetland. This is the result of the activities of the tribal leader in Debbadaguda who managed to get access to government programmes for minor irrigation and of AWARE in Ekvaipally. On the other hand, AWARE apparently has not succeed in improving the situation concerning the access to wetland by ST respondents. What attracts attention too is the fact that not a single ST respondent living in Mansanpally owned wetland, a consequence of the fact that not a single tank was located near their land (situated at a relatively high altitude) and that they did not own wells or pumps. Ownership of wetland is not necessarily synonymous with cultivation of wetland crops, though. In fact, lack of water prevented many respondents from cultivating part or all of their wetland, which was mostly laid fallow.

Not surprisingly, on average, FC respondents in Ekvaipally (9.3 ha) and Debbadaguda (8.8 ha) owned the largest farms, followed by FC respondents in Ghatpally (5.3 ha) and Mansanpally (3.9 ha). Figure 5.6 indicates that the average size of total landholding of BC, SC and ST communities in the study villages was more or less comparable.[17] In most villages the BC community ranked second, the ST community third and the SC community last.

Various trends in irrigation were discernible in the villages. In relative terms, the area under surface irrigation from tanks was declining and, both in relative and absolute terms, the area irrigated by ground water was increasing. This increase was the result of a rapid technological change in ground water pumping, stimulated through government intervention. Innovations having considerable effects in the study villages were the electric motor pump and the tubewell. In contrast to their attachment to age-old tillage implements, respondents put aside most of the indigenous tools used in irrigation like the Persian wheels and the leather bags used to lift water from the open wells. In fact, the diesel pumps replacing these indigenous water lifting devices have become obsolete too and have been replaced by electric ones. In a way, electric pumps for open wells were also getting outdated by the introduction of tubewells. Since lower lying aquifers could be reached and since the chances of growing two or three crops per year on the same plot were higher with tubewells as comapred to the combination of an open well with electric pump, most respondents preferred to have a tubewell. The increasing number of tube wells in the villages and in the region, however, caused a considerable decline in the ground water level. Consequently, open wells carried less water than in

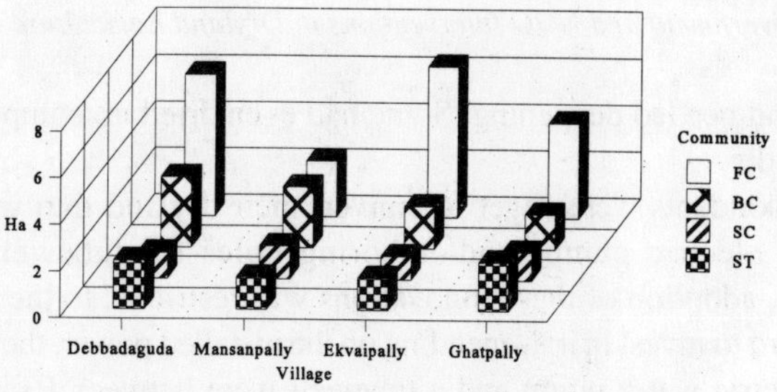

Note: The figure refers to land which is owned, land in joint ownership and to *poramboke* land.

Source: Own survey.

FIGURE 5.4. AVERAGE DRYLAND HOLDING (HA), BY COMMUNITY, BY VILLAGE (1991/2-1993/4).

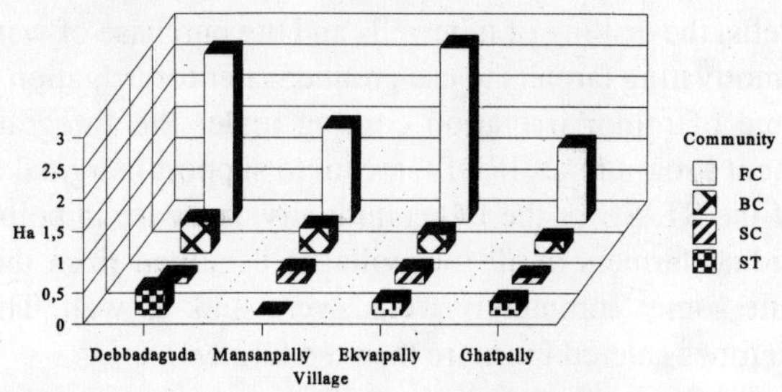

Note: The figure refers to land which is owned, land in joint ownership and to *poramboke* land.

Source: Own survey.

FIGURE 5.5. AVERAGE WETLAND HOLDING (HA), BY COMMUNITY, BY VILLAGE (1991/2-1993/4).

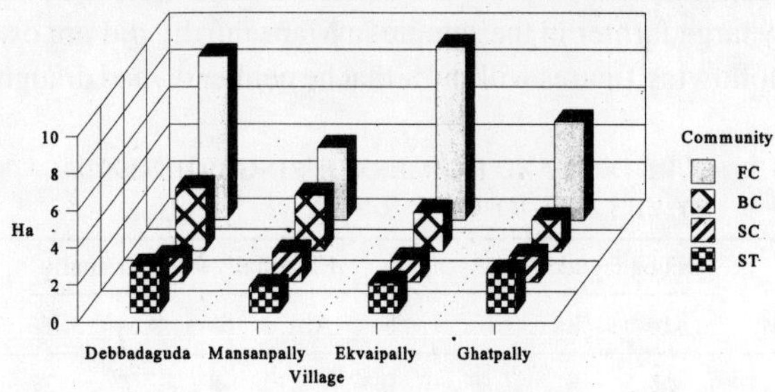

Note: The figure refers to land which is owned, land in joint ownership and to *poramboke* land.

Source: Own survey.

FIGURE 5.6. AVERAGE TOTAL HOLDING (HA), BY COMMUNITY, BY VILLAGE (1991/2-1993/4).

the past and needed deepening. Some had even dried up completely and were left idle.

All respondents were eager to convert their dryland into wetland by installing electric pumps and/or boring holes for tubewells. Being expensive, adoption of these innovations was restricted to the ones who could afford to invest in it. Depending on the installed power, the expenses for an electric water pump and a tubewell were between Rs 4,000 and Rs 10,000 and Rs 30,000 and Rs 40,000 respectively. Under certain conditions, farmers who were not able to invest these considerable sums of money were eligible to financial assistance from the Indian government. In fact, the government's rural electrification programme together with subsidies on electricity, on the deepening, rehabilitation and construction of open wells, the drilling of tubewells and the purchase of water pumps helped in motivating farmers to use ground water for irrigation purposes. The scheme of minor irrigation coming under the Integrated Rural Development Programme (IRDP), meant to support marginal and small farmers of the ST, SC or the BC community, is a case in point. Quite a few individual farmers in all four villages benefited from these IRDP loans, while some 'community wells' were sunk as well. These latter wells sometimes catered for more than ten farmers.

On average, 26 per cent of all respondents owned an open well, 26 per cent an electric water pump and 8 per cent a tubewell. Except for Mansanpally, the differences between the villages were small.[18] That relatively fewer respondents in Mansanpally owned a well or water pump than in the other villages may be related to the fact that the number of large farmers in the village and the sample was much less compared to the other villages. The only large farmer in the sample in Mansanpally did not own wetland and the following figures will show that he neither owned draught-animals,

TABLE 5.2. PURE DRYLAND FARMERS DURING 1991/2-1993/4, BY VILLAGE, BY COMMUNITY

	Debbadaguda		Mansanpally		Ekvaipally		Ghatpally		Total	
Community	Abs.	%	Abs.	%	Abs.	%	Abs.	%	Abs.	%
FC	2	8	0	0	2	8	7	27	11	11
BC	12	48	5	19	14	45	22	71	53	47
SC	16	64	11	46	9	31	15	52	51	48
ST	7	28	25	100	8	57	4	50	44	61
Total	37	37	41	42	33	33	48	51	159	41

Source: Own survey.

dairy cattle or agricultural implements. In fact, he stopped cultivating his dryland after the 1990/1 season and concentrated his economic activities on his major source of income: business.

Figures 5.7 to 5.9 indicate that in all study villages a clear positive relationship existed between holding-size and ownership of open wells, water pumps and tubewells. The highest percentages of respondents owning these were found among medium and large farmers.[19] In spite of the various government programmes, the number of marginal and small farmers owning these devices was very small. Figures 5.10 to 5.12 show that, except for open wells in Debbadaguda, relatively more FC respondents than respondents of other communities owned open wells, pumps and tubewells.[20] They also indicate that, except for Mansanpally, a relatively high percentage of ST respondents in all villages owned open wells and water pumps. In addition to private initiative, this was partly a consequence of the various government programmes. This is also evident from the open wells and water pumps in joint ownership (not included in the figures), which figured prominently in all villages except Mansanpally. It was very high in Ekvaipally, where 38 open wells and 21 water pumps were owned by more than one person. It was partly the outcome of government and, albeit on a smaller scale, AWARE interventions.

In contrast to the adoption of new irrigation equipment, most agricultural implements were age-old and of indigenous origin. The wooden plough and blade harrows like the *guntaka* and the *dante* were very simple and their design comparable to the implements made hundreds of years ago. On average, wooden ploughs were owned by 80 per cent, *guntakas* by 79 per cent and *dantes* by 77 per cent of the respondents. There was, however, some difference between the villages. While 95 per cent of the respondents in Ghatpally and 80 per cent of the respondents in Debbadaguda and Ekvaipally owned these implements, only 60 per cent of the respondents in Mansanpally possessed these tools.[21] Except for this latter village and Debbadaguda, where a relatively high percentage of marginal farmers did not own these indigenous implements, a clear positive relationship between holding-size and ownership did not exist.[22] Except for Mansanpally and Debbadaguda, where relatively more SC respondents did not own these implements, the differences between the various communities was relatively small.[23] These limited differences seem to indicate that these implements were regarded as a kind of dire necessity for each farming household. The fact that they were not very expensive (costing not more than a couple of hundred Rupees), made it relatively easy for most respondents to buy them.

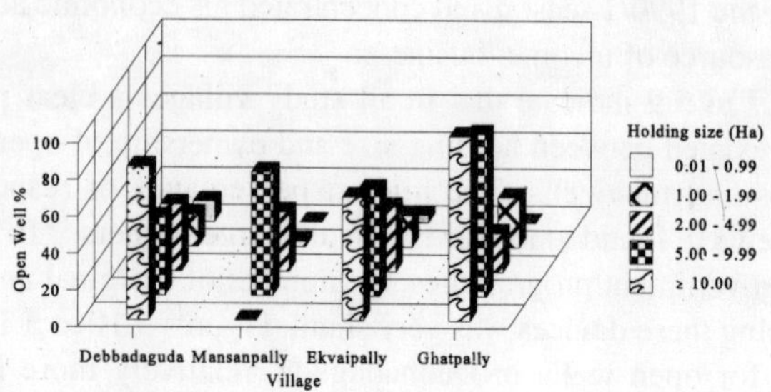

Source: Own survey.

FIGURE 5.7. PERCENTAGE FARMERS OWNING AN OPEN WELL, BY HOLDING-SIZE, BY VILLAGE (1992/3).

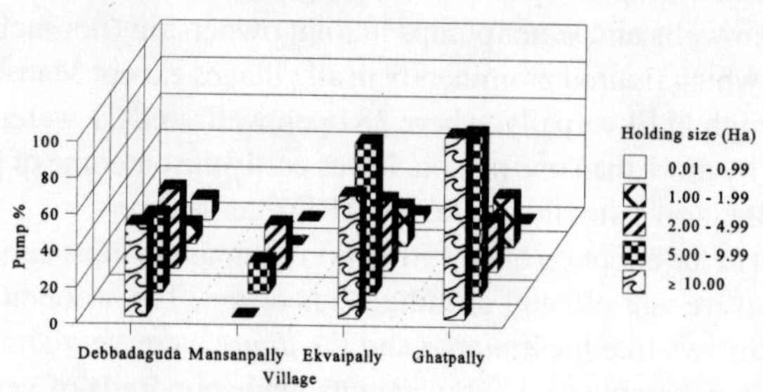

Source: Own survey.

FIGURE 5.8. PERCENTAGE FARMERS OWNING A PUMP, BY HOLDING-SIZE, BY VILLAGE (1992/3).

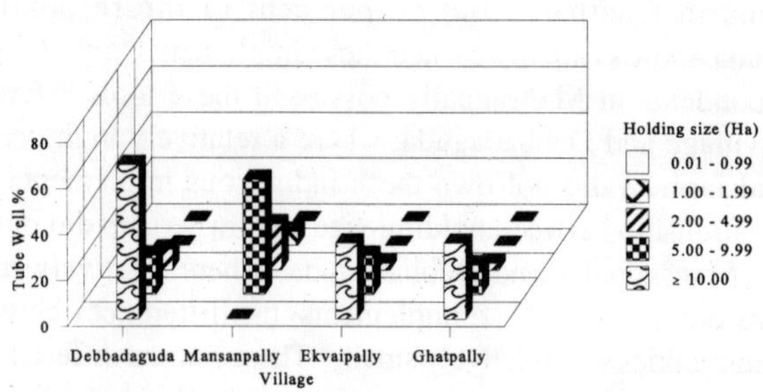

Source: Own survey.

FIGURE 5.9. PERCENTAGE FARMERS OWNING A TUBEWELL, BY HOLDING-SIZE, BY VILLAGE (1992/3).

In contrast to the rapid technological change in irrigation devices, the adoption of most other modern tools and machinery was limited. Of the modern tools, sprayers for plant protection chemicals were adopted most: by 9 per cent of the respondents. Most sprayers were bought by FC respondents owning medium to large holdings. They were used most in Ekvaipally, followed by Debbadaguda, Mansanpally and Ghatpally. The number of iron ploughs and tractors was limited to less than a handful of FC respondents in each study village, all owning rather large farms. Not a single respondent owned a thresher or a seed-cum-fertilizer drill.

All items discussed so far concern arable farming. Another, very important resource and income generating activity was animal husbandry. Arable farming and animal husbandry depend on one another. While cattle supply traction, milk and dung, crops supply by-products like fodder. This interdependency, together with the fact that most farmers did not specialize in arable farming or animal husbandry but practised both, shows that agriculture can be labelled mixed farming. Moreover, lack of sufficient common property resources and the absence of private meadows necessitated the production of crops which yielded good quality by-products in sufficient quantities.

One pair of bullocks and one piece of dairy cattle, preferably a female buffalo, is the minimal aspiration of every farming household. To till the land, two bullocks are a minimum requirement and the dairy cattle give milk for home consumption and/or for selling. A majority of 75 per cent of all respondents owned draught-animals (bulls, bullocks and male buffaloes). This percentage was highest in Ghatpally, followed by Mansanpally, Debbadaguda and Ekvaipally. Another 65 per cent of all respondents owned dairy cattle (cows and she-buffaloes). The proportion of dairy cattle-owning households was highest in Ekvaipally, followed by Ghatpally, Mansanpally and Debbadaguda.[24] Although respondents knew about superior varieties and cross-bred bullocks, buffaloes and cows, only a few large FC respondents owned a small number of such animals. Their much higher price and the fact that they needed much more care (especially higher quality fodder and medical treatment), prevented a large majority of respondents from buying these animals.

Figures 5.13 and 5.14 indicate that in all study villages a clear positive relationship existed between holding-size and ownership of draught-animals and dairy cattle.[25] Except for the large farmer in Mansanpally who gave up cultivation after the 1990/1 season, and who did not own draught-animals or dairy cattle, the highest percentage of respondents owning these animals were found among medium and large farmers. In

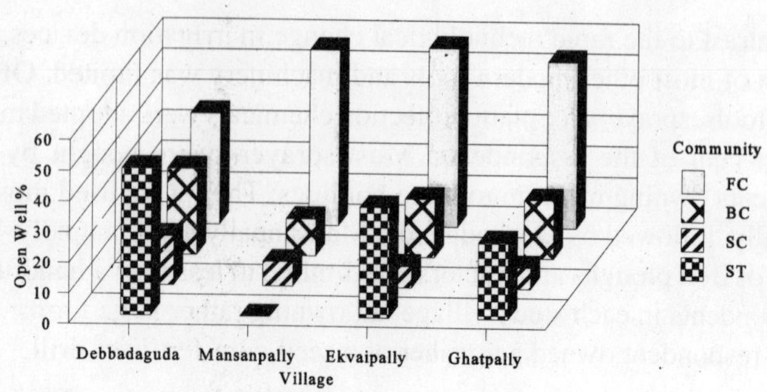

Source: Own survey.

FIGURE 5.10. PERCENTAGE FARMERS OWNING AN OPEN WELL, BY COMMUNITY, BY VILLAGE (1992/3).

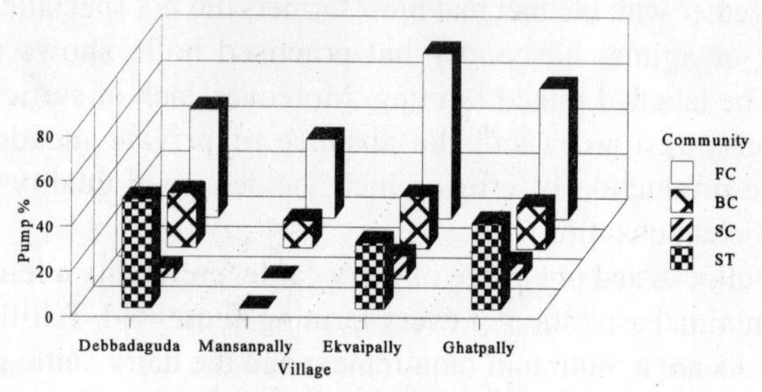

Source: Own survey.

FIGURE 5.11. PERCENTAGE FARMERS OWNING A PUMP, BY COMMUNITY, BY VILLAGE (1992/3).

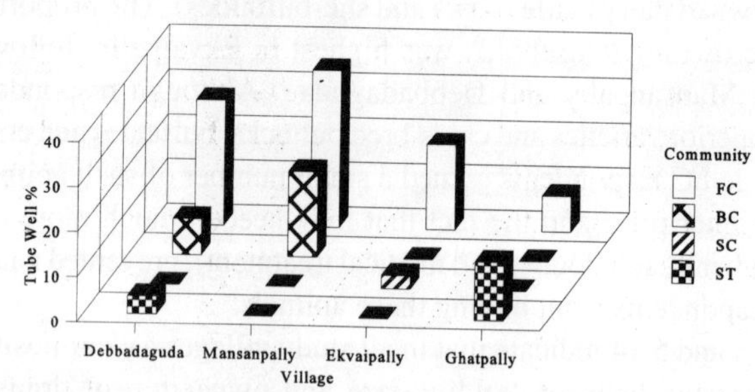

Source: Own survey.

FIGURE 5.12. PERCENTAGE FARMERS OWNING A TUBEWELL, BY COMMUNITY, BY VILLAGE (1992/3).

spite of various government programmes, the proportion of marginal and small farmers owning these animals was very low. Save for Ghatpally, this applies to all villages, and especially to Debbadaguda.

As with many other resources, except for draught-animals in Ghatpally and Debbadaguda, relatively more FC respondents than other respondents owned draught-animals and dairy cattle (Figures 5.15 and 5.16).[26] The picture concerning animal traction and the other communities is less clear. Relatively speaking, the number of respondents without draught-animals was highest for BC respondents in Ghatpally, for ST respondents in Ekvaipally and Mansanpally and for SC respondents in Debbadaguda. As far as dairy cattle are concerned, the picture is more obvious. In all four villages, the number of respondents not owning dairy cattle was highest among SC respondents. In fact, the position of SCs in Debbadaguda is striking since only a few of them owned draught-animals or dairy cattle. Trees supplied additional food (fruit) for many households and were an extra source of cash income (from bark, firewood, timber and fruits) as well. Many tribals obtained additional income by collecting forest produce from what was left of the surrounding 'forest reserves'. Outside the 'forests', but with the exception of toddy palms and trees situated on *poramboke* land, trees were in ownership meaning that the owner had access to its produce or could lease trees out to others. Most respondents, however, were petty producers owning one or two trees like *seethapal* (custard apple), tamarind or *neem*. Since these yielded hardly an abundant harvest, most respondents used the entire production for household consumption, or produced only limited quantities for the market.

Only a few respondents were large owners, owning many trees, sometimes in the form of an orchard. The only tree which provided a relatively large income for non-FC farmers was the toddy palm that produced a juice which, after fermentation, was sold as a low alcoholic drink. Although the toddy palm belongs to the owner of the land on which it grows, government regulations prescribed that toddy palms had to be leased out to the union of toddy-tappers. This union predominantly consisted of Gouds, the traditional toddy-tapper caste, which belongs to the BC community. Many Gouds living in Ekvaipally, a village characterized by the presence of many toddy palms, earned considerable sums of money from selling toddy in and outside the village.[27]

Commercial production of fruit was on the rise, predominantly because urban people had bought land in the area and had started cultivating orchards. Respondents argued that, as a consequence, land prices in the villages had gone up. The largest orchards owned by respondents, sometimes covering a couple of irrigated hectares, belonged to large FC farmers, who cultivated lemon, guava and grapes. Producing hundreds and sometimes thousands of kilograms of fruits, worth sometimes tens of thousands of Rupees, it was a very lucrative investment. In addition, and particularly because of the efforts of the Maheswaram project which promoted the production of mango, guava and pomegranate, some respondents had also taken up the cultivation of fruit trees. Most of these orchards were small, often less than 0.2 ha, and by 1994 not a single tree had begun to bear fruits.

5.2.2 WEALTH

To make a rough assessment of the average wealth of the sample households and to allow for another comparison between the study villages, the value of earlier discussed productive assets (landed property, wells, animals and implements like water pumps and hand tools) and durable consumer goods, if owned, is added.[28] In case productive assets and durables were owned by more than one person, wealth was estimated by dividing its value by the number of partners.

Average wealth of all respondents was Rs 1,62,000. With an average of Rs 2,09,000, respondents in Debbadaguda were wealthiest, which is not surprising given their better resource endowments. With an average wealth of Rs 1,83,000, Ekvaipally ranked second, Ghatpally third (Rs 1,42,000) and Mansanpally (Rs 1,14,000) ranked last. The magnitude of these figures is the result of a few very wealthy respondents living in each village. That economic inequality still reigned in the villages is clear from the difference between the poorest and the wealthiest respondent which was almost a factor 1,000 (Rs 3,930 *versus* Rs 3.8 million). In fact, two-thirds of the respondents living in Mansanpally, Ekvaipally and Ghatpally and 52 per cent of the respondents living in Debbadaguda owned less than Rs 1,00,000. The number of respondents owning more than Rs 5,00,000 was highest in Ekvaipally: they totalled 14.

With an average share of 80 per cent in total assets, land was the most important component of wealth.[29] At quite a distance from landed property, with an average share in total asset formation of 11 per cent, the second most important source of wealth were wells and implements. This was mainly due to the high costs of wells and water pumps and less so for the indigenous implements, costs of which rarely exceeded Rs 1,000. With an average share in total possessions of 8 per cent, livestock was the third component.[30] The value of the last item, consumer durables, was low, averaging 1 per cent of total wealth.

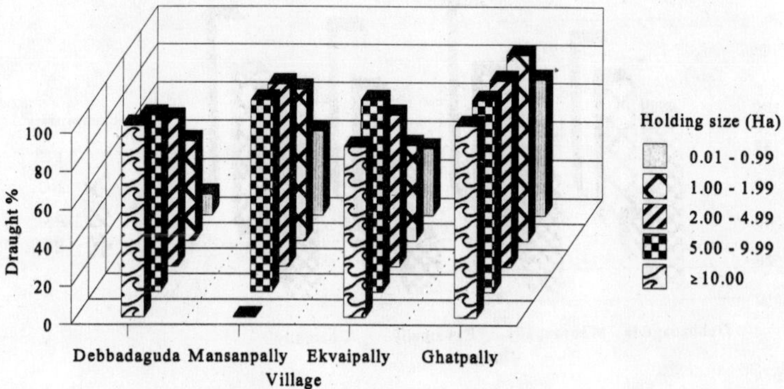

Notes: Draught animals are bulls, bullocks and male buffaloes. Percentages for Debbadaguda refer to 1992/3 only.

Source: Own survey.

FIGURE 5.13. PERCENTAGE FARMERS OWNING DRAUGHT ANIMALS, BY HOLDING-SIZE, BY VILLAGE (1991/2-1992/3).

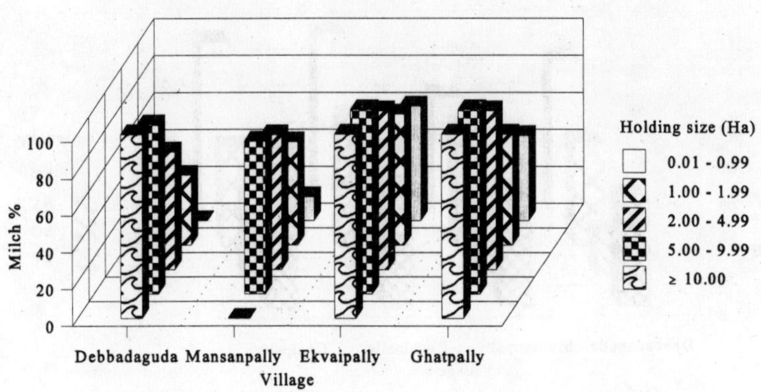

Notes: Milch animals are cows and female buffaloes. Percentages for Debbadaguda refer to 1992/3 only.

Source: Own survey.

FIGURE 5.14. PERCENTAGE FARMERS OWNING MILCH ANIMALS, BY HOLDING-SIZE, BY VILLAGE (1991/2-1992/3).

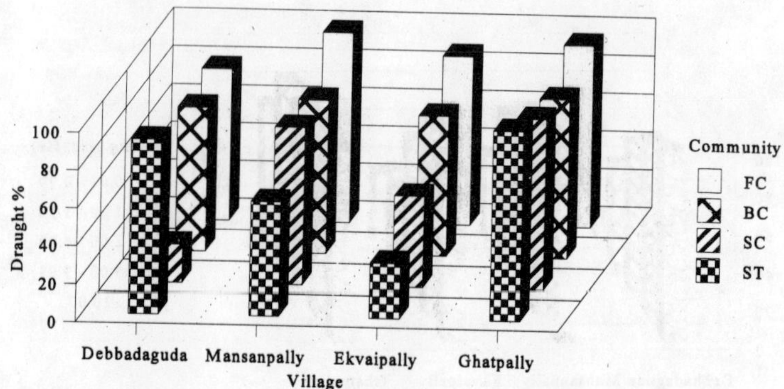

Notes: Draught animals are bulls, bullocks and male buffaloes. Percentages for Debbadaguda refer to 1992/3 only.

Source: Own survey.

FIGURE 5.15. PERCENTAGE FARMERS OWNING DRAUGHT ANIMALS, BY COMMUNITY, BY VILLAGE (1991/2-1992/3).

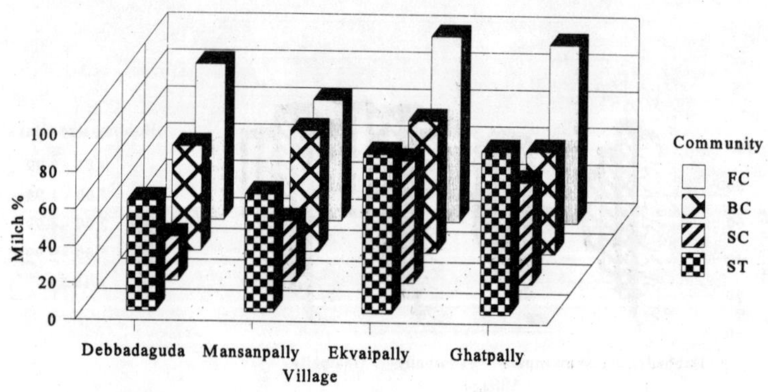

Notes: Milch animals are cows and female buffaloes. Percentages for Debbadaguda refer to 1992/3 only.

Source: Own survey.

FIGURE 5.16. PERCENTAGE FARMERS OWNING MILCH ANIMALS, BY COMMUNITY, BY VILLAGE (1991/2-1992/3).

Since landed property was the most important component of wealth it is hardly a surprise that a positive relationship existed between holding-size and wealth.[31] Large farmers living in Debbadaguda and Ghatpally had especially accumulated considerable wealth while marginal and small farmers were among the poorest in all study villages. On average, FC respondents were the wealthiest of all communities in the four villages. Except for Ghatpally, where on average ST respondents were somewhat wealthier than BC respondents, and for Mansanpally, where on average SC respondents were somewhat wealthier than ST respondents, BC respondents ranked second while ST respondents ranked third and SC respondents came last.[32] What attracts attention is that while average wealth of FC respondents was highest in Debbadaguda followed by Ekvaipally, it was lowest for SC respondents in these villages too.

5.2.3 AGRICULTURAL CREDIT

Various informal (of which moneylenders and chit funds were the most important ones) and formal (banks) sources supplied credit to the villagers. Although moneylending was veiled in secrecy some of the things observed was that in each village some five to ten moneylenders (most of them Vaisyas (business men) or large Reddy farmers) gave short-term (up to one year) and medium-term (up to two years) loans for different purposes, productive as well as for consumption. Long-term loans were not given. Whatever the loan and whoever the borrower, each loan agreement was put on paper and signed by the two parties and a witness, indicating how formalized informal credit was. Depending on his financial position, each moneylender had between ten to thirty customers and may have lent between Rs 1,00,000 and Rs 2,00,000. Totally, they must have supplied between Rs 0.5 million and Rs 1 million to the villagers each year. As mentioned by Walker and Ryan (1990: 201), moneylenders had a personalized, long-term relationship with villagers; because of this relationship and because they knew about the assets, financial position, character and reliability of the borrowers, repayment rates were high, almost always close to 100 per cent.

Most short-term or crop loans were given in cash, with an interest rate of 36 to 48 per cent per year, more reliable people paying 36 per cent. Since most large farmers could finance agricultural inputs themselves or preferred to apply for cheaper formal credit, most short-term loans were given to small-medium and medium farmers. Depending on caste, landed property and reliability, these farmers could borrow up to Rs 10,000.

However, for small-medium and medium farmers of the ST and SC community the limit was put at Rs 3,000. Since most moneylenders had put up a barrier of 2 ha of dryland, most marginal and small farmers were not eligible for cash loans. Only those who were considered reliable could get a small seasonal loan of a couple of hundred Rupees. A special type of seasonal loan was the *nagu* loan, which was given in kind. Especially marginal and small farmers, being in need of food or seeds, asked for these kind of loans. Because of their dependent position and because they were considered unreliable customers, they had to pay a high price. *Nagu* loans in food (mostly rice and *jowar*) were advanced against an interest rate of 100 to 150 per cent for six months, while *nagu* loans in seed carried an interest rate of 50 per cent for the same period. Because these high interest rates prevented many from borrowing, most moneylenders in Debbadaguda and Ghatpally had stopped supplying all *nagu* loans since the late 1980s. *Nagu* loans in seed were only supplied in Ekvaipally.

Medium-term loans were only given to the most reliable larger farmers who paid an interest rate of 36 per cent per year. Depending on caste and landed property, they could borrow up to Rs 25,000. Only on very special occasions and depending on the financial position of the moneylender and his relationship with the borrower, higher loans were given. If used for agricultural purposes, these loans were meant to buy land, to improve existing open wells and to buy water pumps. Afraid of default, moneylenders refused to finance expensive investments in dryland agriculture (e.g. the construction of bunds), a type of agriculture they considered too risky. Likewise they were reluctant to finance the digging or drilling of wells and were only interested in lending money for a water pump once water had been struck.

Another important source of informal credit in the villages were chit funds. Managed by a (often respected) villager, a chit fund consisted of a group of people who periodically contributed money to the fund, the sum periodically being earmarked for one of its members. This person could use the money for whatever purpose he/she wanted. The rotating nature of these chit funds made them less useful as a crop loan, though, since these loans are needed at a specific point in time. The periodicity, the number of participants and the amount contributed, differed per fund. Especially for landless households and for marginal and small farmers, households having no or only limited access to moneylenders and the formal credit institutions, chit funds were one of the few means to get a loan. Each year during the study period, at least two chit funds were in progress in each of the study villages. Debbadaguda even had seven funds in 1993. On the whole some 400 villagers, or approximately 25 per cent of all adults in the study villages, participated in these chit funds. While some funds catered for special groups, for boys, women or were on caste basis, participation of most groups was free, the only barrier being the periodic deposit and the reliability of the participant. The periodic deposit of most chit funds was between Rs 100 and Rs 250. But each village had its large chit fund requiring a deposit of Rs 1,000, meaning that only medium and large farmers and shopkeepers participated.

For formal credit, villagers could avail of the facility offered by branches of the State Bank of Hyderabad, the Syndicate Bank, the Canara Bank, the Farmers Cooperative Society, the Golconda Grameen Bank and the Sanghameshwara Grameena Bank, all located at a short distance from the villages. Except for Mansanpally and Ekvaipally, most respondents reported having received a loan from one of these banks at least once: in Debbadaguda 67 per cent, in Mansanpally 26 per cent, in Ekvaipally 50 per cent and in Ghatpally 79 per cent. While on average 60 per cent of the FC and BC respondents had secured loans from these formal sources, 45 per cent of the SC and ST respondents had managed to do so. These relatively high percentages may be the result of the fact that all respondents were landowning farmers and because IRDP loans had been given by the government in all study villages. Most loans were short-term crop loans, to be used to buy seeds and chemical fertilizers. Bank managers confirmed the conclusions of Walker and Ryan (1990: 215-18): repayment rates of loans were low and largely reflected expectations of government waivers on loan repayment. In fact, all farmers in Ghatpally were 'punished' by the bank because some large farmers, having a conflict with the bank on the last government waiver, refused to repay their loans. Consequently, not a single farmer in this village received formal credit in *kharif* 1992/3.

Table 5.3 demonstrates that in *kharif* 1992/3 in all study villages most

TABLE 5.3. NUMBER AND AMOUNT OF LOANS (*in Rs 1,000*) RECEIVED FOR *KHARIF* 1992/3, BY SOURCE, BY VILLAGE

	Debbadaguda		Mansanpally		Ekvaipally		Ghatpally		Total	
Source	N	Rs	N	Rs	N	Rs	N	Rs	N	Rs
Informal	55	276	65	570	58	215	41	213	219	1,274
Formal	26	431	24	278	21	381	0	0	71	1,090
Total	81	707	89	848	79	596	41	213	290	2,364

Note: For comparison reasons, the Rs 2.5 million loan given by a commercial bank to finance the poultry farm owned by a FC respondent living in Debbadaguda is not included.

Source: Own survey.

loans came from informal sources. Nevertheless, the total amount furnished by formal sources superseded the loans supplied by the informal sector in Debbadaguda and Ekvaipally, a result of the higher amounts given by the banks. It is related to the fact that, in particular, larger farmers applied for loans at banks. In fact, smaller farmers indicated that for them the informal sector was the most important source of (agricultural) credit. Flexible lending and repayment procedures, fast decision-making and the fact that no intermediaries and bribes were necessary to get loans, were important reasons why they preferred the moneylender above banks, in spite of the fact that his interest rates were up to three times higher.

5.3 The Cropping Pattern

Although indigenous agricultural technologies (together with the recommendations made by the projects) will be discussed in the next chapter, some introductory remarks concerning the cropping pattern in the villages need to be made here.

All four study villages can be characterized as dryland villages and their farmers as dryland farmers. This dominance, at least related to area, is evident from the fact that averaging for 1991/2-1993/4, the proportion of dryland to total land owned by the respondents was 77 per cent in Debbadaguda, 80 per cent in Mansanpally, 75 per cent in Ekvaipally and 82 per cent in Ghatpally (Table 5.4). Averaging 22 per cent for all study villages, the wetland area owned by the respondents was not insignificant.

Table 5.4 demonstrates that respondents left a relatively large area fallow during *kharif* 1991/2-1993/4. This applies to both dryland and wetland. On average, the proportion of dryland that was left fallow was 51 per cent in Mansanpally, 49 per cent in Debbadaguda, 43 per cent in Ekvaipally and 36 per cent in Ghatpally. It can partly be explained by the fact that

TABLE 5.4. AVERAGE AREA OF DRYLAND AND WETLAND OWNED, LAID FALLOW AND NET SOWN AREA DURING *KHARIF* 1991/2-1993/4, BY VILLAGE (*ha*)

	Debbadaguda		Mansanpally		Ekvaipally		Ghatpally		Total	
Land	Dry	Wet	Dry	Wet	Dry	Wet	Dry	Wet	Dry	Wet
Owned	307	92	191	49	273	90	207	44	978	275
Fallow	150	38	97	22	117	32	75	21	439	113
Net sown area	157	54	94	27	156	58	132	23	539	162

Source: Own survey.

large farmers had little interest in cultivating dryland and left substantial areas fallow. In *kharif* 1992/3 for instance, five large farmers from Debbadaguda did not till a notable 88 ha, one farmer from Mansanpally left some 10 ha fallow, two large farmers from Ekvaipally did not sow 25 ha and one large farmer from Ghatpally left 22 ha fallow. In the same *kharif* 1992/3 season, around 50 per cent of all respondents laid all or part of their dryland fallow. The most important reasons given for not cultivating their dryland were lack of interest, the land was needed for grazing, lack of time, low fertility of the soil, insufficient rainfall and/or insufficient capital (money, animals, seeds or chemical fertilizers) to till the land. While lack of capital was often mentioned by smaller farmers, larger farmers referred more often to the first two reasons.

The average percentage of wetland that remained uncultivated during *kharif* 1991/2-1993/4 differed only slightly from dryland: 48 per cent in Ghatpally, 45 per cent in Mansanpally, 41 per cent in Debbadaguda and 36 per cent in Ekvaipally. It was not so much a lack of interest, as insufficient water in tanks and open wells and/or insufficient capital, that prevented many farmers from cultivating all their wetland. While smaller farmers gave both reasons, larger farmers stressed the first. Restoring the fertility of the soil was an important reason to lay wetland fallow in summer, not in *kharif* or *rabi*.

Jowar (sorghum) as cereal crop, redgram (intercropped with *jowar*) and some other grams as pulse crops and castor (an oilseed) as cash crop, were the dominant *kharif* crops on dryland (Table 5.5). On average in *kharif* 1991/2-1993/4, these crops occupied between 51 per cent

TABLE 5.5. AVERAGE AREA CULTIVATED WITH DRYLAND AND WETLAND CROPS DURING *KHARIF* 1991/2-1993/4, BY VILLAGE (*ha*)

	Debbadaguda		Mansanpally		Ekvaipally		Ghatpally		Total	
Crops	Abs.	%	Abs.	%	Abs.	%	Abs.	%	Abs.	%
Jowar	61	29	54	45	58	28	63	41	236	34
Castor	71	34	26	21	48	23	47	31	192	27
Other dryland crops	25	12	14	12	50	23	22	14	111	16
Paddy	33	16	16	13	34	16	18	12	101	14
Other wetland crops	21	10	11	9	24	11	5	3	61	9
Net sown area	211	100	121	100	214	100	155	100	701	100

Note: *Jowar* includes both mono and intercropping.
Source: Own survey.

(Ekvaipally) and 72 per cent (Ghatpally) of the net sown area of the respondents. Although respondents in most villages intercropped *jowar* with redgram, respondents in Ekvaipally had a slight preference for intercropping *jowar* with *bajra* (in one furrow) and redgram (sometimes mixed with cowpea). The fact that respondents in Ekvaipally also preferred to cultivate *bajra* as a monocrop, explains to a large extent the relatively large area cultivated with other dryland crops in this village. If rainfall allowed, horsegram and safflower (both monocrops) were cultivated in *rabi* in all study villages.

The cropping pattern on dryland in the study villages was by no means static. Each year farmers adapted their cropping strategy to the whims of the monsoon. Depending on their own judgement about the sufficiency of the *kharif* rains farmers may skip the cultivation of *jowar* or castor and decide to grow a *rabi* horsegram crop instead (see section 10.1.10). Furthermore, in course of time, new crops were added or removed from the cropping pattern. Farmers pointed to two important changes in the cropping pattern on dryland that have occurred over the last 25 years or so, changes which resulted in a cropping pattern that was not as diverse as it used to be and was less diverse compared to outcome of other research (see Walker and Ryan, 1990: 46); the first being the virtual disappearance of *ragi* and small millets, as a result of which hardly any farmer kept a stock of seeds of these crops; the second being the growing importance of tomato cultivation. A change in food preferences of farmers themselves and of other (rural) people, from low status cereals like *ragi* and small millets to higher status cereals like rice, had a negative effect on the demand of these former crops and more or less forced them to stop cultivating these dryland cereals. This decision was facilitated by the fact that these coarse cereals produced less fodder than *jowar*. Facing a reduction of the pastures on the village commons, farmers gave priority to *jowar* which, as a multi-purpose crop, gives both grain and fodder. The more recent increase of tomato cultivation on dryland was related to an increase in demand from the Hyderabad consumer market which resulted in a relatively high price. A considerable number of farmers considered the good prospects of earning an additional income to outweigh the higher natural and marketing risks of this cash crop on dryland (higher than the 'traditional' dryland crops: *jowar* and castor). However, the higher risks involved as well as the higher labour input needed made most farmers plant small areas, often not exceeding 0.1 ha.

Paddy was the dominant wetland (mono)crop, particularly in *kharif*

(Table 5.5). On an average during *kharif* 1991/2-1993/4, paddy occupied between 12 per cent (Ghatpally) and 16 per cent (Debbadaguda and Ekvaipally) of the net sown area of the respondents. Vegetables (spinach, beans, tomatoes, onions, capsicum and eggplant), coriander, chilies, fruit and flowers were cultivated in *kharif* as well. Some farmers, having insufficient access to water, cultivated dryland crops like *jowar* and castor in wetland. This practice increased after *kharif* 1991/2 because the monsoon had been inadequate to refill the tanks and open wells. Inadequate water supply during *rabi* had a more dramatic effect on wetland cultivation in general and paddy cultivation in particular. First of all, compared to *kharif*, the area left fallow increased considerably during *rabi* 1991/2-1993/4. Second, while the average area cultivated with paddy in *rabi* was some 50 per cent less than in *kharif*, the area cultivated with crops that needed less water (e.g. maize, vegetables, groundnuts and flowers) increased, in particular in Mansanpally and Debbadaguda.[33]

Over the years, in contrast to dryland, the cropping pattern on wetland has become more diverse. Farmers gave two main reasons for growing new crops. First, the increasing demand from Hyderabad led to higher prices and higher benefits for crops like vegetables, flowers, fruit, spices, mulberry and wheat. Second, compared to paddy, most of these crops do not need that much water. In particular, for farmers having access to a less reliable source of water, horticulture is like a kind of safe alternative. The fact that some of the tanks and wells in Mansanpally yielded insufficient quantities of water, stimulated a growing number of especially medium and large farmers to cultivate flowers on small patches of wetland, sometimes instead of *kharif* paddy, sometimes in *rabi*, after paddy.

A very remarkable change in the cropping pattern on wetland was the result of the activities undertaken by some seed improvement companies.[34] These selected, invariably, large and almost all Reddy farmers, as contract farmers to cultivate, for instance, *bajra* and maize on wetland in *rabi*, not for food but for seed improvement. These companies gave seeds and inorganic fertilizers for free, while their agents gave advice and monitored growth regularly. Farmers indicated that this support, together with the fact that the farm gate price of the produce was settled with them in advance (meaning that a major part of the risk, which is relatively small since it concerned irrigated crops, was taken away from the farmer and put at the doorstep of the seed company), made it rather lucrative to cultivate these crops. Although contract farmers were present in all villages, they figured predominantly in Debbadaguda.

5.4 Yields, Self-sufficiency at Household Level and Cash Income from Agriculture

At the outset of this section it has to be stressed that the data on the 1989/90, 1990/1 and 1991/2 harvests was collected by survey. Except for Debbadaguda, data concerning the first two years was obtained by the end of 1991 and the beginning of 1992. In Debbadaguda, the data was obtained by the end of 1992. Data concerning 1992/3 and 1993/4 was collected by plotwise schedules during the harvest or after post harvest operations were over. Hence, all data was collected at a time the Maheswaram project and AWARE had (almost) completed their intervention strategy. Hence, it may be assumed that yields have (partly) been influenced by the activities of both projects. Realizing, however, that some of the variance between the data may be explained by the different ways of data collection, that the collection of yield data is always a hazardous affair and that yields depend on many factors outside the control of any project, the figures are not so much meant to illustrate the effect of the different intervention strategies but will only be used to illustrate the low yielding capacity of dryland crops.

That average yields of *jowar*, redgram and castor cultivated on dryland were low is illustrated in Table 5.6. Five-year averages for all study villages were 453 kg/ha for *jowar*, only 31 kg/ha for redgram (cultivated as intercrop of *jowar*) and 389 kg/ha for castor. The almost insignificant yields of redgram can partly be explained by serious attacks of the pod borer, a pest that swept the area and especially Mansanpally and Ekvaipally during 1991/2-1993/4. It made many farmers decide to decrease the

TABLE 5.6. AVERAGE YIELD OF SELECTED DRYLAND CROPS (*kg/ha*) DURING *KHARIF* 1989/90-1992/3, BY VILLAGE

	Debbadaguda			Mansanpally			Ekvaipally			Ghatpally			Total		
Year	J	R	C	J	R	C	J	R	C	J	R	C	J	R	C
1989/90	515	67	424	398	40	338	294	54	274	416	25	275	403	44	323
1990/1	512	55	396	472	42	430	281	52	265	543	50	358	449	49	356
1991/2	485	36	361	473	16	384	239	23	209	421	17	422	416	23	350
1992/3	515	21	301	497	1	382	271	4	240	633	27	777	482	13	455
1993/4	337	48	330	675	8	298	223	16	263	859	33	783	565	27	460
Average	473	45	362	503	21	366	262	30	250	574	30	523	453	31	389

Note: J = *jowar*; R = redgram; C = castor.
Source: Own survey.

number of redgram plants per ha or to stop growing redgram at all. The table points out that the averages do not differ that much between Debbadaguda, Mansanpally and Ghatpally, at least during 1989/90-1991/2. Ekvaipally clearly lags behind, though. The lower yields of *jowar* in this village can only partly be explained by the fact that a considerable number of farmers added *bajra* as an intercrop to *jowar*, the average yield of *bajra* being 72 kg/ha only. Most probably lower yields of all dryland crops in this village were also a consequence of the relatively lower rainfall in this part of the study area.

In contrast to the yields of these dryland crops, the five-year averages of *kharif* paddy for all study villages was more than 3,000 kg/ha. A minor difference existed between, on the one hand, Ghatpally and Ekvaipally, where paddy yields averaged around 3,000 kg/ha and, on the other, Mansanpally and Debbadaguda, having average yields of around 3,500 kg/ha. It meant that average paddy yields were at least five times the averages of *jowar*. In fact, the same farmer who harvested dryland crops with yields ranging from 300 to 600 kg/ha, secured a paddy harvest yielding more than 2,000 kg/ha.

The significance of wetland cultivation in the study villages is more related to the yields (also in monetary value) of wetland crops and to the investments made in wetland cultivation than to the cultivated area. Inputs of labour and capital were, for instance, much higher than in dryland cultivation. Because of its less risky nature and the higher yielding capacity of wetland crops, farmers preferred wetland above dryland and invested considerable amounts of money in the conversion of dryland into wetland. As reported in an earlier section, with the financial assistance of moneylenders (and with the support of the government) who were more interested in wetland than in dryland cultivation, a considerable number of water pumps have been bought and many new wells have been dug and drilled in each of the study villages. As a result, the area under wetland crops like paddy and vegetables has been enlarged to the detriment of the area under 'traditional' dryland crops like *jowar* and castor.

Since a relatively large portion of respondents had marginal and small holdings and produced low quantities of dryland food crops, many households were not self-sufficient in dryland food grains. Averaging for the period 1990/1-1993/4 less than 50 per cent of the *jowar* producing respondents and less than 25 per cent of the redgram producing respondents succeeded in feeding their household members for at least one year. The level of self-sufficiency of *jowar* was highest in Ghatpally (averaging more than 60 per cent) and lowest in Ekvaipally (less than 40 per cent).

For redgram, the level of self-sufficiency was highest in Debbadaguda (averaging more than 35 per cent) and lowest in Mansanpally (less than 20 per cent). In contrast, self-sufficiency of paddy producing respondents was more than 80 per cent in all villages in the same period.

In all study villages, self-sufficiency of *jowar*, redgram and paddy was highest among large FC respondents, followed in most villages by BC farmers, ST farmers and SC farmers, in that order. It was predominantly these FC farmers who marketed part of their harvest, but the number of farmers who sold *jowar* and redgram was low, as was the proportion of the production that they marketed. Averaging for the entire study period and all four villages, less than 25 per cent of the respondents sold *jowar* and less than 10 per cent sold redgram. These percentages were somewhat higher in Ghatpally and Debbadaguda. Most of those who marketed part of their produce sold less than 1,000 kg of *jowar* and less than 500 kg of redgram. In contrast, up to two-thirds of the paddy producing respondents marketed part of their crop, a proportion that was somewhat higher in Ekvaipally and Debbadaguda than in the other villages. The quantities sold were much larger than for dryland food grains. One respondent from Debbadaguda, a large FC farmer, once sold 1,00,000 kg of paddy for instance.

Low and insecure yields of dryland crops together with the fact that many respondents were not self-sufficient in staple food, forced many farmers to diversify their sources of (cash) income. In fact, a substantial number of respondents had three or more sources of (cash) income, i.e. besides arable farming also from animal husbandry and agricultural wage employment. Other important (cash) income generating sources were toddy-tapping, collecting forest produce (both, particularly, in Ekvaipally), services (traditional jobs of the BC community) and wage employment outside the agricultural sector (particularly brick-making and construction). In 1992/3, for instance, and averaging for all study villages, only 31 per cent of the cash income originated from crops which were sold. The marketing of arable surplus was more important in Ghatpally and Debbadaguda, with respectively 43 per cent and 36 per cent, than in Mansanpally (26 per cent) and Ekvaipally (20 per cent). Most probably, this difference is related to the higher yields of dryland crops in Ghatpally and the higher yields of paddy in Debbadaguda. For many respondents, and in particular for marginal and small farmers, the own farm was not an important source of cash income. Averaging for all study villages, marginal farmers received 13 per cent of their cash income from marketed crops, small farmers 22 per cent. Except for Debbadaguda, the average share of cash income from marketed crops was highest for FC respondents (55 per cent). In Debbadaguda, it was the ST community that on average received the highest share of cash income from marketed crops (56 per cent). It was the highest of all communities in all villages, clearly a consequence of the relatively better resource endowment of these Lambadas, partly a result of government support they had received over the years.

5.5 Conclusion

This chapter has demonstrated that the study villages were indeed identical on the criteria mentioned in Chapter 1. Located between 30 and 55 km south of Hyderabad, the infrastructure in the villages was more or less comparable: all four study villages were connected by a metalled road to the main Hyderabad-Sri Sailam motorway, had bus facilities linking the villages with the capital, were connected to the electricity grid, had a school, a ration shop as well as private shops and safe water supply. The availability of many of these facilities is not only an indication of the presence of the Indian government but also an indication of a changing rural society and the introduction and adoption of innovations. The fact, that tiles had replaced thatch, and that concrete slabs were in turn replacing tiles, that many people were proud to show their watches and radios, that a few had television sets, that electric water pumps had replaced diesel pumps, that drilled tubewells were slowly replacing open wells and that life insurance policies had been taken by a considerable number of respondents, shows that farmers are not as 'innocent' as many outsiders think them to be.

The agro-ecological situation in the four study villages was almost identical as well. Although most probably the yearly rainfall averages for Ekvaipally were somewhat lower than in the other study villages, all were located in the same low rainfall zone and had the same type of soil. Hence, the cropping pattern did not differ much between the villages. The most important dryland crops in *kharif* were *jowar*, intercropped with redgram and in Ekvaipally sometimes with *bajra* and cowpea, and castor. Only if soil moisture permitted, some farmers also grew *rabi* crops like horsegram and safflower on dryland. In *kharif*, irrigated agriculture on wetland was dominated by paddy cultivation. Lack of sufficient quantities of water in *rabi* and summer forced most farmers to grow crops that needed less water like vegetables, maize (cultivated for seed improvement for seed companies), fruits and flowers. More than on dryland, farmers experimented with new (varieties of) crops on wetland. An outstanding feature in all study villages was the relatively large area of both dryland and wetland that remained fallow in *kharif*. For reasons of lack of interest, lack of time and the need for pastures, larger farmers left (part of) their dryland fallow. Low fertility of the soil, insufficient rainfall and/or insufficient capital were the most important reasons why smaller farmers were reluctant or unable to till their dryland plot.

With five-year averages of 453 kg/ha for *jowar*, 389 kg/ha for castor and only 31 kg/ha for redgram (as intercrop), yields of dryland crops were low. These low and insecure yields combined with small holdings forced many smaller farmers to get additional income from other sources, both inside (like animal husbandry and as agricultural labourer) and outside agriculture (like well digging and in construction). Large farmers diversified their economic activities by cultivating not only paddy on wetland but by growing all kinds of commercial crops (like vegetables, fruit, flowers and crops for seed improvement) and got additional income from animal husbandry and business. Thus, being a risky and not the only source of income, a considerable number of farmers were reluctant to invest much time and capital in dryland agriculture. In a way this decision was supported by moneylenders who, because of higher yields and lower risks, preferred to finance improvements in wetland agriculture above improvements in dryland agriculture.

The fact that yields of dryland and wetland crops differed considerably for the same farmer indicates that from a farmer's point of view it was very rational to invest more in wetland than in dryland agriculture. The preference to convert dryland into wetland, to prefer electric pump sets above diesel pumps and tubewells above open wells, are cases in point showing that farmers did innovate. This rationality of farmers' behaviour negates the views which claim that custom or tradition constrains agricultural development in general and dryland agriculture in particular.

The social differences between the four study villages were small. Although certainly not vanished, the caste hierarchy was slowly eroding in all villages and although still present, untouchability was considered less a problem by the respondents than in the past. In all villages FC farmers were the dominant community and the landed elite. Both economically and socially, Reddies still dominated the rural scene. They owned considerable quantities of resources like dryland, wetland, livestock and implements, and this large resource endowment was expressed in social and political power at village and sometimes at regional level. It not only meant that most of them found it relatively easy to get access to all kinds of institutions, it also meant that their communication network was spread over a relatively large area. This allowed many of them to act as intermediarics for others in the village, especially for people at the lower end of the social hierarchy. Slowly, however, these latter households gained more economic and political power. Some have bought land, have been elected in official bodies and communicated more with the outside world.

Economic differences between the villages were somewhat larger. As was expected, economically Debbadaguda ranked first. Some farmers in this village not only owned a considerable area of wetland (cultivating not only paddy but also crops for seed improvement) but also owned vineyards and poultry farms. Although paddy cultivation was important in Ekvaipally, the production of toddy was also very significant, especially for BC farmers. Compared to the other three villages, Ghatpally stood out for its relatively higher yields of some dryland crops and due to the fact that its farmers owned more livestock and implements. Economically, Mansanpally lagged somewhat behind, a consequence of a less favourable resource endowment, particularly concerning irrigation facilities. Some farmers tried to overcome the limited availability of water by cultivating flowers on wetland. To a large extent these economic differences between the study villages and between the villagers can be explained by the different access they had to water, to irrigate wetland crops.

NOTES

1. Averaging for the fieldwork period, Rs 25 equalled $ 1.
2. The only household owning its own hand-pump was the Brahmin household in Mansanpally. Because hand-pumps had been dug in all quarters in the villages, disputes between FC and SC households about socially 'polluted' water did not rise during the study period.
3. The school constructed from government funds in the Lambada *tanda* of Ekvaipally in 1991 was not so much built because the tribals had asked for it but more for the benefit of the former president of AWARE's village level organization (*sangham*) who as *sarpanch* withheld part of the construction materials. The building was not used for teaching (the teacher taught in Ekvaipally), but as a shed for animals and goods.
4. It is also interesting to note that the teacher in Ghatpally was a Harijan.
5. Incidentally, it shows that in reality AWARE's target group strategy included BCs.
6. Heads of households that did not receive education were 82 in Mansanpally (84 per cent), 75 in Ekvaipally (75 per cent), 68 in Ghatpally (72 per cent) and 61 in Debbadaguda (61 per cent). Households without any educated

members were 49 in Mansanpally (50 per cent), 34 in Ekvaipally (34 per cent), 29 in Ghatpally (31 per cent) and 20 in Debbadaguda (20 per cent).

7. For these and other tables/figures, the number of respondents is 100 for Debbadaguda, 98 for Mansanpally, 100 for Ekvaipally and 94 for Ghatpally.
8. Leaseholds were present as well, but share-cropping (sharing inputs and outputs in various percentages between the owner of the land and the cultivator) and leasing (a fixed amount is agreed upon between owner and cultivator at the beginning of the agricultural season) were of minor importance and will not be discussed.
9. Size was calculated by averaging the holding-size over the entire study period.
10. Averaged for the four villages and concerning dryland only, 26 per cent of all respondents owned less than 1 ha, 33 per cent owned between 1 and 2 ha, 32 per cent owned between 2 and 5 ha, 7 per cent between 5 and 10 ha and another 3 per cent owned more than 10 ha.
11. The proportion of respondents in Debbadaguda, Mansanpally, Ekvaipally and Ghatpally owning less than 1 ha of dryland was 17 per cent, 22 per cent, 32 per cent and 31 per cent respectively; owning between 1 and 2 ha was 31 per cent, 40 per cent, 26 per cent and 34 per cent respectively; owning between 2 and 5 ha was 39 per cent, 35 per cent, 29 per cent and 25 per cent respectively; owning between 5 and 10 ha was 11 per cent, 2 per cent, 6 per cent and 9 per cent respectively; owning more than 10 ha of dryland was 2 per cent, 1 per cent, 7 per cent and 2 per cent respectively.
12. Averaged for the four villages and concerning wetland only, 41 per cent of all respondents owned no wetland at all, 41 per cent owned less than 1 ha, 10 per cent owned between 1 and 2 ha, 7 per cent between 2 and 5 ha, 1 per cent between 5 and 10 ha and another 1 per cent owned more than 10 ha.
13. The proportion of respondents in Debbadaguda, Mansanpally, Ekvaipally and Ghatpally owning no wetland was 37 per cent, 42 per cent, 33 per cent and 51 per cent respectively; owning less than 1 ha of wetland was 41 per cent, 44 per cent, 42 per cent and 36 per cent respectively; owning between 1 and 2 ha was 13 per cent, 9 per cent, 10 per cent and 7 per cent respectively; owning between 2 and 5 ha was 7 per cent, 4 per cent, 13 per cent and 3 per cent respectively; owning between 5 and 10 ha was 0 per cent, 1 per cent, 1 per cent and 2 per cent respectively; owning more than 10 ha of wetland was 2 per cent, 0 per cent, 1 per cent and 0 per cent respectively.
14. The proportion of marginal farmers (owning less than 1 ha of land) in the sample in Debbadaguda, Mansanpally, Ekvaipally and Ghatpally was 9 per cent, 16 per cent, 26 per cent and 26 per cent respectively; of small farmers (owning between 1 and 2 ha) was 29 per cent, 35 per cent, 24 per cent and 32 per cent respectively; of small-medium farmers (owning between 2 and 5 ha) was 44 per cent, 42 per cent, 29 per cent and 31 per cent respectively; of medium farmers (owning between 5 and 10 ha) was 12 per cent, 6 per cent, 12 per cent and 9 per cent respectively; and of large farmers

(owning more than 10 ha of land) was 6 per cent, 1 per cent, 9 per cent and 3 per cent respectively.

15. The average dryland holding of FC respondents in Debbadaguda, Mansanpally, Ekvaipally and Ghatpally was 6.1 ha, 2.4 ha, 6.5 ha and 4.1 ha respectively; of BC respondents 3.0 ha, 2.6 ha, 1.8 ha and 1.5 ha respectively; of SC respondents 1.1 ha, 1.4 ha, 1.0 ha and 1.2 ha respectively; of ST respondents 2.1 ha, 1.4 ha, 1.4 ha and 2.1 ha respectively.
16. The average wetland holding of FC respondents in Debbadaguda, Mansanpally, Ekvaipally and Ghatpally was 2.7 ha, 1.5 ha, 2.8 ha and 1.2 ha respectively; of BC respondents 0.5 ha, 0.4 ha, 0.3 ha and 0.2 ha respectively; of SC respondents 0.1 ha, 0.2 ha, 0.2 ha and 0.1 ha respectively; of ST respondents 0.4 ha, 0.0 ha, 0.2 ha and 0.2 ha respectively.
17. The average total landholding of FC respondents in Debbadaguda, Mansanpally, Ekvaipally and Ghatpally was 8.8 ha, 3.9 ha, 9.3 ha and 5.3 ha respectively; of BC respondents 3.4 ha, 3.0 ha, 2.1 ha and 1.7 ha respectively; of SC respondents 1.3 ha, 1.6 ha, 1.2 ha and 1.4 ha respectively; of ST respondents 2.5 ha, 1.4 ha, 1.6 ha and 2.2 ha respectively.
18. The proportion of respondents owning an open well in Debbadaguda, Mansanpally, Ekvaipally and Ghatpally was 32 per cent, 18 per cent, 28 per cent and 26 per cent respectively; owning a water pump was 31 per cent, 11 per cent, 33 per cent and 28 per cent respectively; owning a tubewell was 10 per cent, 13 per cent, 6 per cent and 3 per cent respectively.
19. The proportion of marginal farmers owning an open well in Debbadaguda, Mansanpally, Ekvaipally and Ghatpally was 11 per cent, 0 per cent, 4 per cent and 0 per cent respectively; of small farmers was 17 per cent, 3 per cent, 17 per cent and 27 per cent respectively; of small-medium farmers was 36 per cent, 32 per cent, 35 per cent and 21 per cent respectively; of medium farmers was 42 per cent, 67 per cent, 58 per cent and 88 per cent respectively; of large farmers was 83 per cent, 0 per cent, 67 per cent and 100 per cent respectively.

    The proportion of marginal farmers owning a water pump in Debbadaguda, Mansanpally, Ekvaipally and Ghatpally was 11 per cent, 0 per cent, 4 per cent and 0 per cent respectively; of small farmers was 10 per cent, 0 per cent, 21 per cent and 27 per cent respectively; of small-medium farmers was 43 per cent, 24 per cent, 38 per cent and 27 per cent respectively; of medium farmers was 42 per cent, 17 per cent, 83 per cent and 88 per cent respectively; of large farmers was 50 per cent, 0 per cent, 67 per cent and 100 per cent respectively.

    The proportion of marginal farmers owning a tubewell in Debbadaguda, Mansanpally, Ekvaipally and Ghatpally was 0 per cent, 0 per cent, 0 per cent and 0 per cent respectively; of small farmers was 0 per cent, 6 per cent, 0 per cent and 0 per cent respectively; of small-medium farmers was 9 per cent, 20 per cent, 3 per cent and 3 per cent respectively; of medium farmers was

17 per cent, 50 per cent, 17 per cent and 13 per cent respectively; of large farmers was 67 per cent, 0 per cent, 33 per cent and 33 per cent respectively.

20. The proportion of FC respondents owning an open well in Debbadaguda, Mansanpally, Ekvaipally and Ghatpally was 36 per cent, 57 per cent, 58 per cent and 54 per cent respectively; of BC respondents was 28 per cent, 12 per cent, 19 per cent and 19 per cent respectively; of SC respondents was 16 per cent, 8 per cent, 7 per cent and 7 per cent respectively; of ST respondents was 48 per cent, 0 per cent, 36 per cent and 25 per cent respectively.

    The proportion of FC respondents owning a water pump in Debbadaguda, Mansanpally, Ekvaipally and Ghatpally was 48 per cent, 35 per cent, 73 per cent and 58 per cent respectively; of BC respondents was 24 per cent, 12 per cent, 23 per cent and 19 per cent respectively; of SC respondents was 4 per cent, 0 per cent, 10 per cent and 7 per cent respectively; of ST respondents was 48 per cent, 0 per cent, 29 per cent and 38 per cent respectively.

    The proportion of FC respondents owning a tubewell in Debbadaguda, Mansanpally, Ekvaipally and Ghatpally was 28 per cent, 35 per cent, 19 per cent and 8 per cent respectively; of BC respondents was 8 per cent, 19 per cent, 0 per cent and 0 per cent respectively; of SC respondents was 0 per cent, 0 per cent, 3 per cent and 0 per cent respectively; of ST respondents was 4 per cent, 0 per cent, 0 per cent and 13 per cent respectively.

21. The proportion of respondents owning a wooden plough in Debbadaguda, Mansanpally, Ekvaipally and Ghatpally was 83 per cent, 61 per cent, 78 per cent and 100 per cent respectively; owning a *guntaka* was 82 per cent, 60 per cent, 77 per cent and 97 per cent respectively; owning a *dante* was 81 per cent, 55 per cent, 78 per cent and 94 per cent respectively.

22. The proportion of marginal farmers owning a wooden plough in Debbadaguda, Mansanpally, Ekvaipally and Ghatpally was 33 per cent, 6 per cent, 58 per cent and 100 per cent respectively; of small farmers was 69 per cent, 50 per cent, 83 per cent and 100 per cent respectively; of small-medium farmers was 96 per cent, 90 per cent, 79 per cent and 100 per cent respectively; of medium farmers was 100 per cent, 83 per cent, 100 per cent and 100 per cent respectively; of large farmers was 100 per cent, 0 per cent, 89 per cent and 100 per cent respectively.

    The proportion of marginal farmers owning a *guntaka* in Debbadaguda, Mansanpally, Ekvaipally and Ghatpally was 33 per cent, 6 per cent, 54 per cent and 92 per cent respectively; of small farmers was 66 per cent, 53 per cent, 83 per cent and 100 per cent respectively; of small-medium farmers was 96 per cent, 86 per cent, 79 per cent and 97 per cent respectively; of medium farmers was 100 per cent, 83 per cent, 100 per cent and 100 per cent respectively; of large farmers was 100 per cent, 0 per cent, 89 per cent and 100 per cent respectively.

    The proportion of marginal farmers owning a *dante* in Debbadaguda, Mansanpally, Ekvaipally and Ghatpally was 33 per cent, 6 per cent, 58 per cent and 79 per cent respectively; of small farmers was 66 per cent, 50 per cent, 83 per cent and 100 per cent respectively; of small-medium farmers was 94 per cent, 76 per cent, 79 per cent and 97 per cent respectively; of medium farmers was 100 per cent, 83 per cent, 100 per cent and 100 per cent respectively; of large farmers was 100 per cent, 0 per cent, 89 per cent and 100 per cent respectively.

23. The proportion of FC respondents owning a wooden plough in Debbadaguda, Mansanpally, Ekvaipally and Ghatpally was 88 per cent, 96 per cent, 85 per cent and 100 per cent respectively; of BC respondents was 88 per cent, 69 per cent, 74 per cent and 100 per cent respectively; of SC respondents was 56 per cent, 29 per cent, 72 per cent and 100 per cent respectively; of ST respondents was 100 per cent, 52 per cent, 86 per cent and 100 per cent respectively.

The proportion of FC respondents owning a *guntaka* in Debbadaguda, Mansanpally, Ekvaipally and Ghatpally was 88 per cent, 87 per cent, 85 per cent and 96 per cent respectively; of BC respondents was 88 per cent, 73 per cent, 74 per cent and 97 per cent respectively; of SC respondents was 52 per cent, 29 per cent, 69 per cent and 97 per cent respectively; of ST respondents was 100 per cent, 52 per cent, 86 per cent and 100 per cent respectively.

The proportion of FC respondents owning a *dante* in Debbadaguda, Mansanpally, Ekvaipally and Ghatpally was 88 per cent, 74 per cent, 85 per cent and 96 per cent respectively; of BC respondents was 88 per cent, 69 per cent, 74 per cent and 90 per cent respectively; of SC respondents was 48 per cent, 29 per cent, 72 per cent and 93 per cent respectively; of ST respondents was 100 per cent, 48 per cent, 86 per cent and 100 per cent respectively.

24. Since the number of respondents keeping share-animals was very small, these and following figures relate to owned animals only. The proportion of respondents owning draught-animals in Debbadaguda, Mansanpally, Ekvaipally and Ghatpally was 67 per cent, 81 per cent, 64 per cent and 90 per cent respectively; owning dairy cattle was 56 per cent, 57 per cent, 79 per cent and 69 per cent respectively. While almost all respondents owned a couple of chicken, only a minority of the households kept small ruminants like goats and sheep. In addition, a couple of poultry farms were operated by a few FC respondents in Debbadaguda.

25. The proportion of marginal farmers owning draught-animals in Debbadaguda, Mansanpally, Ekvaipally and Ghatpally was 11 per cent, 44 per cent, 35 per cent and 71 per cent respectively; of small farmers was 52 per cent, 79 per cent, 50 per cent and 97 per cent respectively; of small-medium farmers was 77 per cent, 95 per cent, 79 per cent and 97 per cent respectively; of medium farmers was 92 per cent, 100 per cent, 100 per cent and 100 per cent respectively; of large farmers was 100 per cent, 0 per cent, 89 per cent and 100 per cent respectively.

The proportion of marginal farmers owning dairy cattle in Debbadaguda,

Mansanpally, Ekvaipally and Ghatpally was 0 per cent, 13 per cent, 62 per cent and 46 per cent respectively; of small farmers was 38 per cent, 56 per cent, 71 per cent and 60 per cent respectively; of small-medium farmers was 64 per cent, 73 per cent, 86 per cent and 86 per cent respectively; of medium farmers was 92 per cent, 83 per cent, 100 per cent and 100 per cent respectively; of large farmers was 100 per cent, 0 per cent, 100 per cent and 100 per cent respectively.

26. The proportion of FC respondents owning draught-animals in Debbadaguda, Mansanpally, Ekvaipally and Ghatpally was 80 per cent, 100 per cent, 89 per cent and 96 per cent respectively; of BC respondents was 76 per cent, 81 per cent, 74 per cent and 84 per cent respectively; of SC respondents was 20 per cent, 83 per cent, 48 per cent and 90 per cent respectively; of ST respondents was 92 per cent, 60 per cent, 29 per cent and 100 per cent respectively.

    The proportion of FC respondents owning dairy cattle in Debbadaguda, Mansanpally, Ekvaipally and Ghatpally was 84 per cent, 65 per cent, 100 per cent and 96 per cent respectively; of BC respondents was 56 per cent, 65 per cent, 71 per cent and 55 per cent respectively; of SC respondents was 24 per cent, 33 per cent, 66 per cent and 55 per cent respectively; of ST respondents was 60 per cent, 64 per cent, 86 per cent and 88 per cent respectively.

27. Although AWARE had an anti-alcohol programme, it never tried to stop the production of toddy in Ekvaipally. In fact, the former BC president of the AWARE *sangham* was president of the toddy-tappers' union in the area.

28. Durables are items like television sets, radios, refrigerators, bicycles, mopeds, motor-cycles and scooters. For reasons of lack of reliability, gold is excluded from the calculation.

29. The average value of dryland was Rs 35,000/ha against an average Rs 65,000/ha for wetland. Dryland of good quality was worth Rs 60,000/ha against Rs 120,00/ha for good quality wetland.

30. The most important elements of livestock were draught-animals and dairy cattle. For a bullock or a female buffalo of average quality, a farmer had to pay Rs 2,500, better breeds costing up to Rs 5,000 a piece.

31. Average wealth of marginal farmers in Debbadaguda, Mansanpally, Ekvaipally and Ghatpally was Rs 17,000, Rs 26,000, Rs 23,000 and Rs 37,000 respectively; of small farmers was Rs 63,000, Rs 61,000, Rs 47,000 and Rs 71,000 respectively; of small-medium farmers was Rs 1,26,000, Rs 1,44,000, Rs 1,15,000 and Rs 1,16,000 respectively; of medium farmers was Rs 3,13,000, Rs 4,01,000, Rs 4,18,000 and Rs 3,39,000 respectively; of large farmers was Rs 15,97,000, Rs 3,00,000, Rs 9,07,000 and Rs 14,26,000 respectively.

32. Average wealth of FC respondents in Debbadaguda, Mansanpally, Ekvaipally and Ghatpally was Rs 5,70,000, Rs 2,32,000, Rs 5,31,000 and Rs 3,29,000

respectively; of BC respondents was Rs 1,28,000, Rs 1,16,000, Rs 82,000 and Rs 81,000 respectively; of SC respondents was Rs 38,000, Rs 60,000, Rs 37,000 and Rs 56,000 respectively; of ST respondents was Rs 98,000, Rs 54,000, Rs 60,000 and Rs 84,000 respectively.

33. The average area cultivated with paddy in *rabi* 1992/3-1993/4 was 8 ha (4 per cent of the net sown area of dry and wetland) in Debbadaguda, 4 ha (3 per cent) in Mansanpally, 10 ha (5 per cent) in Ekvaipally and 9 ha (6 per cent) in Ghatpally.
34. One seed company started a seed improvement farm in the vicinity of Ghatpally.

6. Indigenous and Recommended Technologies for Dryland Agriculture in the Study Villages

M. Put

Are farmers victims or beneficiaries of project interventions? Assessing how farmers have reacted upon the interventions of the Maheswaram project and AWARE is the focus of this and forthcoming chapters. However, before analysing in detail farmers' response upon the recommended agricultural technologies, these are compared with the indigenous technologies first. This is the first of the three objectives in this chapter. The second objective is to describe the project interventions in Mansanpally, the village representing the early activities of the Maheswaram project, in Debbadaguda, the village representing its second phase and in Ekvaipally, the village representing the early activities of AWARE. It discusses how beneficiaries have reacted to project interventions and upon actions undertaken by project staff. The third objective is to analyse to what extent the two projects have succeeded in becoming an important source of information about innovations for dryland agriculture.

6.1 Soil and Water Conservation Practices: Bunds and other Structures

6.1.1 INDIGENOUS

Most respondents were aware of soil and water losses in their dryland plots. For instance, they often complained that heavy rains created gullies and washed away fertile topsoil, seeds, Farm Yard Manure (FYM) and costly chemical fertilizers. Nevertheless, only a few respondents were eager to invest in measures to restrict these losses, reasoning that the

* This chapter is included in Put (1988). Courtesy of Thesis/Thela Publishers, Amsterdam.

benefits of these measures were insufficient to cover the costs and to make up for the efforts. That farmers were convinced of the benefits of soil and water conservation measures in wetland is apparent from the fact that while converting dryland into wetland they levelled the plot, constructed border bunds and made bunds within the plot itself, dividing it into smaller segments. Although these activities forced farmers to spend considerable sums of money on labour and often heavy machinery (tractors with levellers) and although the cultivable area of the plot was diminished to some extent, farmers were very eager to invest in soil and water conservation in wetland. In fact, during the fieldwork period, quite a few dryland plots were treated this way and were converted into wetland. It shows that it was not a lack of information or knowledge that influenced farmers' behaviour towards soil and water conservation measures in dryland, but sound economic reasoning.

Small and low earthen bunds constructed on the borders of a dryland plot were by far the most important indigenous structures to control soil and water loss. Often a waste weir in one of the bunds allowed excess water to flow downwards easily without damaging the standing crop. The fact that most bunds deviated quite markedly from the contour indicates that they served multiple objectives of which soil and water conservation was only one. Perhaps their most important purpose was to demarcate the land, to protect the dryland plot from encroachment by a neighbouring farmer. It shows from the fact that only a few farmers constructed bunds inside the plot to decrease its length, although these additional bunds would have helped to check the velocity of the run-off and to prevent soil and water loss. Another reason why farmers preferred straight border bunds was their functioning as key lines, making it more convenient to perform all kinds of tillage operations in a straight line. Most of these indigenous bunds had been constructed many decades ago. Nowadays, most farmers were not eager to invest time, money and labour in the construction of new or the maintenance of existing dryland bunds. Most maintenance in dryland was done automatically during tillage, when some soil was put against it. This is again in sharp contrast to most wetland bunds which were repaired every year.

Only a minority of the respondents had made other structures than these border bunds. For instance, only a handful of farmers did what was mentioned earlier and had supplemented these border bunds with bunds inside the plot, making it look like a wetland plot. Some others, instead of constructing border bunds, made a bund at the top of a plot, and/or one at the bottom end, and/or constructed one or more bunds somewhere in between. Besides bunds, stone checks or gully plugs in depressions, rills or gullies were made inside the plots. In addition, to slow down the velocity of the run-off, most respondents cultivating two or more crops in one plot left a grass strip in between these crops.

6.1.2 RECOMMENDED

Following the ideas generated by public research stations, the Maheswaram project promoted the construction of graded/contour bunds in Mansanpally (a first phase village). AWARE advised graded bunds in Ekvaipally. Vegetative barriers of Vetiver grass (*khus*) were promoted by the Maheswaram project only, especially after 1988. The project planted these on a very small scale in Mansanpally, while dryland plots in Debbadaguda (a second phase village) were treated with these vegetative barriers only. Another difference between the two projects was that in addition to informing farmers about the advantages of the recommendation, the Maheswaram project built/planted these as well.

While the indigenous bunds satisfy the multiple objectives of the farmers, the single concern of graded/contour bunds and *khus* is the conservation of water and soil. It is said that they perform this task better than indigenous border bunds because they are constructed on or slightly deviating from the contour and because they are higher and more robust. In addition, their layout is such that they break up plots into smaller segments, intercepting run-off before its volume and velocity reach such magnitude which causes serious erosion and allowing rainwater more time to infiltrate into the soil. However, compared with indigenous bunds, graded/contour bunds have a couple of disadvantages. First of all, they are much more expensive to construct. Being much bigger in width and height than border bunds, the area occupied is also larger, sometimes covering up to 5 per cent of the plot. Often, additional area is also lost for cultivation due to stagnation of water in front of the bund and because fertile topsoil is used to construct it. Inherent in their nature, graded/contour bunds might cut across farm boundaries and might also divide a plot in many pieces. Consequently, a farmer is confronted with a reduction in the efficiency of operations like ploughing and harrowing.

Since the Maheswaram project was not able to reach its targets for bunding and since the per hectare costs of the bunds were much higher than budgeted, to speed up the work and to cut down on expenses, *khus* lines were introduced in 1986/7. Stressing the need for affordable, simple and replicable technology, the Mid Term Review stated that the estimated costs of *khus* lines were much lower than those of contour bunds: Rs 400 *versus* Rs 1,100/ha (GoI, 1988: II, 42). Other advantages of *khus* above bunds were said to be the less time needed for planting and the fact that not much arable land was lost. Other claims were that it would quickly form a dense, permanent hedge; that it was perennial and required minimal maintenance; that it was practically sterile and would not become a weed; that it was generally unpalatable to livestock although the soft young leaves could be used for fodder; that it could withstand drought; that it would not compete with crops; and that it would grow in all kinds of soil (World Bank, 1990: 46-7).

6.2 *In-situ* Soil and Moisture Conservation Activities

6.2.1 INDIGENOUS

Summer tillage, as an indigenous *in-situ* soil and moisture conservation operation, was limited to plots sown with castor, to prepare it for the next *jowar* crop. It was limited to castor plots because the soil (Alfisol) was still relatively soft after harvest in February/March. This softness had everything to do with the rather short three months period between the last tillage operation in October/November and the harvest. Summer tillage was undertaken in March or April, a couple of weeks before the sowing of *jowar*. Farmers preferred to prepare the plot with a blade harrow (the *guntaka*), a light implement which did not penetrate the soil very deeply. They did not use a plough because the soil was not sufficiently soft for the bullocks to open it with ease.

In contrast, summer tillage was not an indigenous practice after a *jowar* crop, to prepare it for castor. The eight-month period between the last tillage operation in *jowar* plots in June/July and the first pre-monsoon showers in March/April the following year, allowed the soil to form a crust and to become very hard. Since the feeble bullocks were not able to pull neither a blade harrow nor a plough through such a crust, farmers postponed preparatory tillage until the soil had become sufficiently wet and soft, after the start of the monsoon, by the end of June. The softness of the soil allowed tillage with a plough. It was then that castor, one of the few crops which can be sown rather late in *kharif*, was sown.

Although post-harvest tillage is possible after a *jowar* harvest (in September/October the soil is still rather soft), it was not undertaken. The most important reason was that since it was never sure whether a new crop would be sown in that particular plot, farmers did not want to run the risk of losing the investments made. Besides, farmers who intercropped *jowar* with redgram opposed post-harvest tillage because the indigenous redgram variety, which ripened only in December or January, was still in the field. Farmers who monocropped *jowar* felt that post-harvest tillage was not worthwhile since the next crop, almost invariably a castor crop, was sown rather late, after the monsoon had already softened the soil. Moreover, post-harvest tillage more or less clashed with the paddy harvest and other income-generating activities. The few farmers who tilled the soil after a *jowar* harvest, did so for a *rabi* crop (horsegram or safflower). Although beneficial for soil and moisture conservation, such a preparatory tillage is not equivalent to post-harvest tillage.

Most preparatory tillage operations, whether undertaken by *guntaka* for *jowar* or by plough for castor, consisted of two turns, once up and down and once across the slope, with the first turn always against last year's sowing direction. Only a few farmers ploughed or harrowed diagonally as well. Farmers preferred this crosswise tillage because it provided the best possible tilth, opening the soil and breaking its crust and lumps better than one-directional tillage operations. Although farmers argued that the number of preparatory tillage operations depended on the type of soil and the presence of weeds (not so much on the quantity of rain fallen and the type of crop) most *jowar* and castor plots were tilled once in such a crosswise manner.

More important for *in-situ* soil and moisture conservation than the direction of preparatory tillage operations is the direction of sowing. The furrows and the resultant crop rows last longer than the furrows or other micro structures made while preparing the plot. Most farmers argued that a new plant would flourish only if sown in a spot not utilized by the previous crop. They considered that the chance of seeds falling in a place once occupied by a previous crop, would be much lower if a plot was sown up and down the slope and across the slope in alternate years. The idea that different crops did not necessarily use the same nutrients and could be sown in the same spot without any drawbacks and thus in the same direction every year, was not considered valid. Although farmers were very aware of the detrimental effects of sowing against the gradient and were aware of the practice of ploughing across the slope every year, many adhered to this crosswise sowing. Most of the plots owned by these farmers were located in a more or less flat terrain and loss of fertile topsoil, water, seeds, FYM and chemical fertilizers was of minor concern to them. Farmers having plots located in an undulating terrain were more concerned with this loss of topsoil, water and inputs. They stated that since sowing up and down the slope decreased soil fertility, which was detrimental to plant performance and yields, they sowed across the slope every year. Many of these farmers said that if their plots had been more or less level, they would also have sown in alternate directions every year.

Interculturing with a blade harrow (*dante* or *guntaka*) between a standing crop (and occasionally with a plough in castor) was the last indigenous element of *in-situ* soil and moisture conservation operations in dryland agriculture. Because leaves of longer plants would get damaged, most farmers considered more than two harrowings unnecessary and harmful for *jowar.* Castor, a more sturdy plant, could be intercultured up to four times, and in a rough manner. However, when intercultured with a plough, it was limited to a single operation. Farmers were quite aware of the advantages of interculturing with a blade harrow: i.e. since the softened soil would improve infiltration of rainwater, plants would benefit more from the moisture available in the root profile; weeds were removed; plants were somewhat uprooted stimulating them to improve their root system; and farmers believed that branches would proliferate in uprooted and damaged castor plants, which would improve yield. The blade harrows produced small heaps of soil every 10 m or so which acted as temporary soil and water conservation structures, as a kind of tied-ridges. Because most interculturing operations were undertaken between the plant rows, its direction was determined by the sowing direction of the crop, which could be up and down or across the slope, depending on the preferences of the farmer. The exception was interculturing with a plough in castor, which was always done against the sowing direction.

6.2.2 RECOMMENDED

As indicated in Chapters 7 and 9, the recommendations for *in-situ* soil and moisture conservation were a mixture of indigenous technologies (summer tillage for *jowar* and interculturing), technologies devised in the 1930s (contour cultivation) and 'real' innovations (post-harvest tillage, key lines and dead furrows). Except for key lines and dead furrows which were recommended by the Maheswaram project only, both projects advised similar technologies.

Concerning the recommendations for preparatory tillage of dryland plots, two problems arose. First of all, the summer period was not clearly defined by the Maheswaram project. If its documents referred to a specified period at all, it was often March to May (GoAP, 1981: 55). Including the month of May is rather peculiar, though, because most farmers preferred to sow *jowar* in the last week of May. The second problem was that staff of the two projects as well as the T&V staff knew that post harvest tillage would be hampered in case long duration crops like redgram and castor were still in the plot.

One of the most important recommendations of *in-situ* soil and moisture conservation was contour cultivation. The resultant ridges, furrows and crop rows would serve as barriers to prevent run-off and would provide a greater opportunity for the soil to absorb water and to arrest erosion as well. Through contour cultivation, crops would benefit more from the available rainwater while soil nutrients, FYM and chemical fertilizers would not wash away to lower levels in the plot. All this would result in higher yields. To facilitate contour cultivation the Maheswaram project made so-called key lines on the contour. After the changeover to *khus*, these hedges served as key lines.

Unfortunately, two problems arose with this recommendation. First of all, both projects and the T&V system interchanged contour cultivation with cultivation across the major slope. This is a kind of in-between concept, meaning that operations are neither undertaken on the contour nor up and down the slope. Secondly, while all kinds of tillage operations (preparatory ploughing, sowing and interculturing) can be undertaken along the contour line, it was watered down to sowing only. Neither documents nor staff ever referred to preparatory tillage.

The Maheswaram project also advised farmers to plough a so-called dead furrow on the contour. This furrow had to be ploughed in a standing crop some 15 to 20 days after sowing and did not carry a crop itself. It was meant to minimize soil and water loss and to increase the infiltration of rainwater into the root profiles of crops. Being a rather recently developed practice, this novelty created ambiguity among project documents and staff. Intervals between two dead furrows ranged between 2 m and 5 m, while five to six rows were also mentioned (Action Plan 1989/90: 4; Krishna Rao, 1991: 108-114; Sivamohan et al., 1994: 128; GoAP, 199?: 9).

6.3 Crop Management Practices: Rotation and Intercropping

6.3.1 INDIGENOUS

Once sown, a dryland crop was more or less left to its own devices. In fact it is this easiness of cultivation, an aspect which has received little attention in the literature so far, which is an important criterion for farmers.

Dryland crops like *jowar*, *bajra*, all pulses and castor, were not only considered drought resistant and suitable to the agro-ecological situation of the area, they were also regarded easy crops. Know-how or experience was hardly needed to grow these crops, once sown they did not need too much labour or other inputs, and they did not need a lot of care until harvest.

Rotation of crops was one of the practices which prevented soil exhaustion and which contributed in getting high yields on dryland. Most farmers preferred a two-year rotation, cultivating one part of the holding with *jowar* or a *jowar*-redgram intercrop and the other part with castor, changing the pattern every year, meaning that castor followed *jowar* and vice versa. A minority, mostly because their holdings were small, preferred to cultivate their entire dryland plot with one crop, one year *jowar* and the next year castor. Another minority preferred a three-year rotation, *jowar*-castor-fallow or *jowar*-castor-horsegram or *jowar*-castor-safflower, where *jowar* was either intercropped or not.

In spite of these preferences, farmers did not always maintain their rotation schedule. The erratic and unpredictable monsoon often compelled them to grow another crop rather than the planned one. On quite a number of occasions, farmers did not cultivate at all and let the plot lie fallow. In addition, factors like insufficient inputs had the same effect.

Intercropping was another way to prevent soil exhaustion of dryland. In one and the same plot one main crop was sown together with one or more subsidiary crops or intercrops. Intercrops could be sown mixed with the main crop or between the main crop in parallel rows. Often, intercrops were leguminous crops like pulses which are nitrogen fixing, enriching the soil. Since pulses were rich in protein, intercropping with these crops also enabled farmers to improve the diet of the members of their household. In fact, asked why they intercropped *jowar* with redgram, most farmers referred to the advantage of one plot supplying two different food crops. Only a minority of farmers knew about the beneficial effect of redgram on yields of *jowar*.

The indigenous intercropping pattern on dryland consisted in all villages of a couple of rows sown with *jowar* followed by one row of redgram. Although most farmers preferred a 4:1 ratio, five rows of *jowar* interchanged with four or six or sometimes seven, was a rather regular practice. It indicates that farmers did not always take that much care. Some farmers in Ekvaipally preferred to intercrop *jowar* not only with redgram but also with *bajra* and sometimes added cowpea as well. They sowed (a higher quantity of) *jowar* seeds mixed with (a lower quantity of) *bajra* seeds in a couple of rows, followed by one row containing the redgram/cowpea mix. It is not exactly known why these Ekvaipally farmers preferred this type of intercropping. A possible explanation might be that this village received less rain than the other three villages and that this type of intercropping gave farmers a reasonable chance that at least one crop would yield a suitable harvest.

6.3.2 RECOMMENDED

Both the Maheswaram project and AWARE had no intentions to radically change the existing cropping pattern, the existing rotation of crops or the intercropping pattern on dryland. However, both projects knew that not all farmers intercropped *jowar*. In addition, since public research had concluded that the indigenous intercropping practices had best be adapted, both projects set themselves motivating farmers to do so. However, the adaptations promoted by the Maheswaram project drifted now and then. Sometimes, all they advised was intercropping *jowar* with redgram; sometimes they advised sowing two rows of *jowar* followed by one row of redgram (a 2:1 ratio); on other occasions they mentioned a 2:1 or 3:1 ratio; and in still other documents they incorporated *bajra* and maize in the intercropping pattern (see the various Action Plans). In other leaflets they recommended intercropping castor and redgram in a 5:1 ratio, a recommendation against the advise given by CRIDA, which recommended growing castor as a monocrop. To my knowledge, AWARE was not very specific about intercropping either. The documents recommended intercropping *jowar* with redgram without giving a certain ratio. However, the T&V extension system which was used by AWARE recommended a 2:1 ratio.

6.4 Crop Management Practices: Choice of Suitable Seed Varieties

6.4.1 INDIGENOUS

Farmers were very particular about seeds. These were judged not only in terms of yield of produce (grain, kernels, seeds or fruits), but also in terms of the quantity of fodder produced, inputs needed, farm-gate price of the produce, easiness of cultivation and resistance against drought, floods, pests and diseases. It meant that whenever there was a perceptible decline in yields or a perceptible change in any other property, farmers changed to another seed variety. In fact, the same demonstration effect described by Sunitha Raju (1990) for farmers in Coastal Andhra also happened in the study villages. When farmers used improved seeds which had superior properties above the old ones, other farmers wanted to buy these from them.

Some of the most important criteria to be met by any *jowar* variety were related to it being a multi-purpose crop. As a food crop, *jowar* was valued on grain yield potential, palatability of the food, and the period it kept the feeling of hunger away. As a fodder crop, it was judged on criteria like yield potential of fodder and the palatability of its stalks and leaves. For two reasons its fodder crop characteristics had become more important than its food crop properties. First, the increased consumption of rice, in particular the result of the establishment of ration shops which supplied rice at subsidized prices, had led to a decrease in the demand for grain of *jowar* (also by farmers themselves). Second, the reduction in communal and other grazing lands had led to an increase in the demand for fodder produced by *jowar*.

Although its leaves were eaten by goats and sheep and its stalks were used as fuel, castor was not considered a multi-purpose crop. Since it was grown for the production of oilseeds only, yield potential was the most important criterion on which farmers valued the seeds of any castor variety. Other qualities which had to be satisfied by a good castor variety were easiness of cultivation and resistance against drought, floods, pests and diseases.

6.4.2 RECOMMENDED

Both the Maheswaram project and AWARE, stressing the importance of higher yields and higher production, favoured improved varieties of all kinds of dryland crops. Although the propagated seeds were not indigenous to the area, both projects did not start from scratch: i.e. various government programmes had already begun distributing improved seeds to farmers in the area. However, these first generation seeds of dryland crops like *jowar* and *bajra* had mainly been bred for their high grain yield potential. These higher yields did not always coincide with criteria like resistance against drought, floods, pests and diseases, and palatability or yield of fodder. Knowing that most farmers in semi-arid India had rejected these first generation seeds, these were no longer propagated by both projects and the T&V extension system. Instead, recommending CSH-5 and CSH-6 for *jowar* and HY-2 for redgram first, in later periods the choice was enlarged to CSH-9, SPV-386 and SPV-475 for *jowar* and HY-4, BDN-2 and LRG-30 for redgram. It was claimed that the latest improved varieties of *jowar* combined relatively high yields of tasty food grain with relatively high yields of good quality fodder. Concerning castor, initially *Aruna* and *Gujarati* were advised. In later periods, *Gauch-1*, GCH-2 and GCH-4 were propagated as well. The cultivation of improved varieties of less important dryland crops, like groundnut, sunflower, maize, *bajra*, *ragi*, sesame and various vegetables was stimulated as well, in particular by the Maheswaram project. Since this project did not deal with wetland agriculture, it did not advocate the use of improved seeds for paddy. AWARE did and advised the cultivation of various improved seeds like *Hamsa* and *Tellahamsa*.

6.5 Crop Management Practices: Plant Population

6.5.1 INDIGENOUS

Farmers argued that in case of *jowar* and other food-cum-fodder crops, they had to sow thin for grain and thick for fodder. Except for crops having tiny seeds like horsegram and safflower, seeds of most dryland crops were not broadcasted but sown behind the indigenous country plough. The quantity of seed used depended on various criteria, of which yield of produce (in grain or fodder), soil characteristics, soil moisture and crop geometry/spacing were the most important ones.

While sowing behind the plough, farmers made allowance for interculturing with a blade-harrow, the *dante* or *guntaka*. Hence, inter-row spacing was more or less similar to the width of these two implements, which was roughly 40 cm for *dante* and 70 cm for *guntaka*. Keeping these criteria in mind, most farmers used 10 to 15 kg/ha for *jowar* and 12 to 17 kg/ha for castor. For intercrops they used 2 to 5 kg for *bajra* and 1 to 2 kg/ha for redgram and cowpea.

6.5.2 RECOMMENDED

Both projects advised farmers to limit the number of plants per hectare. This recommendation fell in two parts: i.e. a low seed rate and wide spacing of plants, already advised in the 1930s. Following their advice to intercrop *jowar* with redgram and to use improved seeds of both, both projects and the T&V extension system advised sowing 8 kg/ha to 10 kg/ha of *jowar* and redgram each. For improved castor it was 10 kg/ha to 12 kg/ha. The recommended inter-row spacing was some 60 cm for *jowar*/redgram and 90 cm for castor.

6.6 Crop Management Practices: Tillage Implements

6.6.1 INDIGENOUS

Indigenous implements like wooden country ploughs, blade-harrows (*guntaka* and *dante*) and a kind of leveller (*jambu*), hardly differed from those used hundreds of years ago. Made of wood to which iron parts were attached, they were made locally by carpenters, costing between Rs 200 and Rs 300. The wooden plough opened the soil without turning it. The *guntaka* was used for preparatory tillage operations (a kind of shallow ploughing), for levelling and for interculturing to loosen the soil and to remove weeds between a standing castor crop. The *dante* was used for interculturing as well but between a standing *jowar* crop. The *jambu* was predominantly used in wetland cultivation, mostly to prepare and level paddy plots. If used in dryland it was to remove rubbish and to loosen the surface of the soil. This latter activity was carried out occasionally, only if pre-monsoon showers, fallen a couple of days after the sowing of *jowar*, had resulted in crust formation.

6.6.2 RECOMMENDED

Both projects advised farmers to change their wooden country plough for the iron one. In addition, the Maheswaram project also recommended the seed drill and the seed-cum-fertilizer drill. The iron plough not only differed from the indigenous plough in shape and materials used, but in its effects as well, ploughing deeper and putting the soil to one side of the furrow. Since it ploughs deeper, it produces a better tilth, manure is better mixed with the soil and water can infiltrate more easy into the soil. Iron ploughs, however, are more expensive than indigenous ones and cannot be made by local artisans.

Because broadcasting of seeds would produce a less uniform plant population and would complicate or even prevent interculturing, the Maheswaram project advised farmers to use a seed drill or a seed-cum-fertilizer drill. The advantages of these drills were various: i.e. labour saving, proper seed rate and a more uniform plant stand, line-sowing and the proper placement of seeds. The seed-cum-fertilizer drill was considered somewhat more useful, because it offered the possibility of placing chemical fertilizers together with seeds.

6.7 Crop Management Practices: Organic Fertilizers

6.7.1 INDIGENOUS

Respondents in the four study villages appreciated FYM above inorganic fertilizers saying that while FYM improved the texture of the soil, chemical fertilizers did not. In addition, they claimed that while FYM gave 'strength' to plants for two to three years or two to three cropping seasons, plants benefited from inorganic fertilizers for a much shorter period. Nevertheless, many farmers were not able to apply what they considered to be the proper dose of FYM of 6,000 kg/ha for dryland and 9,000 kg/ha for wetland every two to three years (see also Jha and Sarin, 1984: 33).[1] On average, most dryland plots were manured in rotation and got this dose once in four years only (see also Walker and Ryan, 1990: 53; Sunitha Raju, 1990).

The limited availability of FYM for field application was mainly the result of insufficient production and, to a lesser extent, because it was put to alternative uses like for making dung cakes used as fuel and for 'cleaning' of walls and floors of houses with a mixture of dung and water. The desperate need for manure is shown by the fact that only four respondents sold FYM in the three year research period. Especially medium and large farmers supplemented their own FYM production by buying sheep and goats dung from shepherds or chicken manure from poultry farms. A few others bought dung from dairy farmers in Hyderabad, which they used for high value cash crops like mulberry, grapes and mangoes. Perhaps the best illustration of the desperate need for manure is the fact that some respondents 'manured' with solid waste from Hyderabad: the plastic bags, shoes and batteries were clearly visible in their plots.

Besides being limited in quantity, FYM was poor in quality. Most animal droppings together with all kinds of waste material from crops and the household were dumped in a pit or on a hill near the house, sometimes near the plot. Under normal circumstances, it remained there for a couple of months, meanwhile being affected by the weather. Some days before the first preparatory tillage operation, it was transported to the plots and spread.

6.7.2 RECOMMENDED

Both the Maheswaram project and AWARE promoted this indigenous practice. In fact, observing that limited financial resources prevented many farmers from applying the recommended doses of chemical fertilizers, the Maheswaram project put more emphasis on organic farming after the Mid Term Review (GoI, 1988: 16). However, neither project was very precise about the quantity of organic fertilizers to be used, let alone the timing of the operation and the methods of application. From discussions with project staff and T&V extension officers it is concluded that they recommended 5,000 kg/ha per year, to be applied some four to six weeks before sowing.

Since serious losses of nutrients take place if FYM is exposed to sun and rain, composting is a kind of dire necessity. Realizing that insufficient and low quality FYM was produced, in its final phase, the Maheswaram project advised farmers to dig compost pits. As an incentive, Rs 50 per compost pit of 10 cu. m was provided towards digging cost.

Nutrient management by way of chemical fertilizers was included in both projects from the start. However, since chemical fertilizers were used by farmers before the intervention of both projects, the projects did not initiate it. Both recommended relatively low levels of inorganic fertilizers since they considered that the risk of crop failure and hence the risk of not recovering an expensive input, was rather high. For *jowar*, the recommended quantity needed to be applied in two doses, a basal dressing of 10 kg N and 30 kg P per ha, and a top dressing of 30 kg N per ha about one month after sowing. They recommended not to apply the basal dressing in the furrow near the seeds if *jowar* was sown before the onset of the monsoon, since insufficient moisture in the soil profile would harm the seedlings. The top dressing needed to be applied some 5 to 10 cm from the stem of the plants. For castor, the recommended dose needed to be applied in two doses too, a basal dressing of 30 kg N and 30 kg P per ha, and a top dressing of 20 kg N per ha about 6 weeks after sowing.[2] The basal dose could safely be applied near the seeds in the furrow, whereas top dressing needed to be done as side placement some 5 cm to 10 cm from the stem of the plants.

6.8 Crop Management Practices: Time of Sowing

6.8.1 INDIGENOUS

The time of sowing of *jowar* and castor was heavily influenced by farmers' perception of soil moisture availability at the beginning and of possible pest attacks later that agricultural season. Based on their experience they knew that there was a high probability that pre-monsoon thunderstorms would occur in March, April and May. In addition, they knew that if sown early *jowar* would have ripened to such an extent that the September rains and the ensuing pests and diseases would hardly have a negative effect on yield. Their ideas were confirmed by astrology and the almanac that taught them that it was best to sow *jowar* in *Rohini karthe* (an auspicious fortnight), the last week of May and the first week of June.[3] However, astrology was just a kind of a guide, not a strict rule. Most respondents affirmed that pre-monsoon showers were risky (Binswanger et al., 1982) and that astrology would only be followed provided these showers had been good. In case they had been bad, most respondents would postpone sowing for a couple of days until the soil had become sufficiently moist.

The uncertainty of these pre-monsoon showers, together with the fact that farmers usually had only two to four days to sow *jowar* after they had fallen and before the upper soil layer had become too dry for germination, made the cultivation of this crop a risky undertaking. When these pre-monsoon rains were late (meaning overdue by more than a fortnight) farmers often concluded that the planned cropping strategy was no longer the most optimal one and often decided to adapt it by changing, for instance, to castor later that season, or to leave the land fallow. Hence, re-sowing of *jowar* hardly occurred. Most farmers were reluctant to re-sow because they often hoped for the best, thinking that if *jowar* failed as a food crop it might succeed as a fodder crop.

As in *jowar*, the best time to sow castor was influenced by a combination of rainfall and pest risks. Farmers knew that to avoid an attack by the Red Hairy Caterpillar (RHC), which would normally come by the end of June, they had to sow castor some four weeks after *jowar*. Astrology and the almanac predicted that the last week of June and the first week of July (*Arudra karthe*) was the ideal sowing time. As in *jowar*, respondents claimed that if rainfall had been insufficient, they would postpone sowing by a couple of weeks, preferably not more than four. Though re-sowing of castor was possible, most farmers preferred to play a waiting game, hoping that castor would recover the moment rain fell.

The time of sowing (or any other operation that needed draught-animals) was also influenced by the availability of bullocks or draught power in general, but this influence was far less compared to rainfall and pest attacks. Some 40 per cent of the respondents in Debbadaguda, Mansanpally and Ekvaipally, and some 30 per cent of the respondents in Ghatpally claimed to have no or insufficient draught power to cultivate all their land, whether dry or wet. These were predominantly the marginal and large farmers, the latter because they owned large areas of land. To alleviate this problem of insufficiency, farmers ploughed with dairy cattle and/or hired bullocks from other farmers. Using dry cows or she-buffaloes together with or instead of bullocks occurred rather regularly in Ekvaipally, where 34 per cent of the respondents said that they used this option, while it was done by 17 per cent in Mansanpally, 15 per cent in Ghatpally and by 5 per cent of the respondents in Debbadaguda. Hiring bullocks was used most by respondents in Mansanpally (39 per cent), followed by Ekvaipally (33 per cent), Debbadaguda (24 per cent) and Ghatpally (18 per cent). Being thrice the salary of a daily wage labourer, Rs 50 per day for a team of bullocks with driver was a price many marginal and small farmers could not afford.[4] As a result, some farmers sowed very early or delayed sowing. Other options were to leave a part or all the land lie fallow, to lease or sharecrop out land or to hire a tractor. The latter alternative has been a rather recent one and certainly not possible for the large majority of farmers.

The time of sowing of dryland crops was hardly influenced by labour availability. First of all, as the sowing period of dryland crops hardly competed with agricultural activities on wetland, agricultural labour was not in short supply. Most wetland plots were still lying fallow because tanks were not completely refilled. Moreover, many of those farmers who worked elsewhere in or outside agriculture, had returned to their village before the start of the agricultural season. Secondly, labour input for sowing was relatively low. Most farmers, used two to three labourers from their own household (one male worker for ploughing and one or two female workers for sowing and to apply chemical fertilizers) and had completed sowing within a couple of days. In fact, with three persons, a one hectare plot could be sown within three days. Only in a few cases household labour needed to be supplemented with paid labourers. For reasons of status, some (and most certainly not all) FC farmers were reluctant to till the soil and wanted to use only paid labour. In addition, since most farmers preferred to complete sowing within a couple of days, large dryland plots would be sown with labour from outside the household. Most of these plots were owned by large farmers.

6.8.2 RECOMMENDED

Both projects followed farmers' preferences. They advised early sowing, or dry sowing as it was sometimes called, for *jowar* and sowing of castor about one month later. What was added for *jowar* by the Maheswaram project was that if pre-monsoon showers had not been able to recharge the topsoil in *Rohini karthe*, sowing had to be postponed by a couple of weeks.

Agronomists claimed that early sowing of *jowar* had many advantages above sowing after the arrival of the monsoon: i.e. better moisture conservation possibilities, good seedling vitality, longer growing season and avoidance of weeds, pests and diseases. The most important pests and diseases like the earhead-bug, shoot fly and downy mildew are weather bound and attack *jowar* around September. They are avoided through early sowing. Another advantage of early sowing is that it enables the cultivation of a second, *rabi*, crop.

Castor is one of the exceptions to this rule of early sowing. If sown early it will be attacked by pests like the RHC and the semilooper, while it might grow tall and become too vegetative. Because it can stand the relatively cold night temperatures which set in after the *rabi* period and can survive on the moisture stemming from the overnight fog in that period, it was advised to sow this crop rather late, up to one month after the arrival of the monsoon. The end of June or the beginning of July was considered to be the most appropriate time in both projects.

6.9 Crop Management Practices: Weeding and Interculturing

6.9.1 INDIGENOUS

Aware that weeds were harmful for their crops farmers had two indigenous options: i.e. to remove them within plant rows by hand and/or between the rows through interculturing with the animal-drawn *dante* or *guntaka*. While almost all manual weeding was done by women, interculturing was carried out by men. However, unless weed infestation was high, most farmers considered manual weeding in dryland a waste of time and preferred interculturing. It was not only favoured because of its lower costs but more so because it was considered a multi-purpose activity. It not only removed weeds, it was an *in-situ* moisture conservation measure as well (see section 6.2.1).

6.9.2 RECOMMENDED

Since weeds compete with a crop for nutrients, water and light, and carry pests which may attack a crop as well, both projects not only advised off-season tillage and early sowing, operations that were discussed earlier, but also interculturing by blade harrows (the indigenous *dante* and *guntaka*) and manual weeding.[5] The period of severe weed competition in most dryland crops is up to 30 days after sowing, while it is up to 40 days in longer duration crops like castor. Both projects advised that for most crops the first weeding, preferably a manual weeding, had to be completed within 20 to 25 days after sowing. The recommendation differed slightly for castor: i.e. a first weeding should have been undertaken by blade harrow followed by a manual weeding within 40 days after sowing. Well aware that the indigenous practice was limited to one to two interculturings, the projects and the T&V extension system advised farmers to remove weeds in *jowar* through one intra-row manual weeding and many interculturings with the *dante*. Concerning castor they advised one hand weeding and frequent interculturings (up to seven times) with the *guntaka*.

6.10 Crop Management Practices: Pest and Disease Control

6.10.1 INDIGENOUS

In spite of the fact that local names for many pests and diseases existed (which shows farmers' awareness) and the fact that several indigenous practices to protect crops had been developed, farmers considered attacks by animals, pests and diseases less of a problem than drought. Thus farmers preferred multi-purpose activities like ploughing, removal of stubble, proper rotation of crops, spacing of plants, sowing at a specific time and interculturing, measures that also combated pests and diseases and did not put them to additional expenses. Consequently, farmers reacted only occasionally with specific measures to control pests and diseases. Similar to the eradication of weeds, it seemed as if a certain threshold had to be passed first before action was undertaken: farmers had to be convinced that the pest attack was serious, that control measures would be successful and that the incremental yield would cover at least the costs involved.

Early sowing of *jowar* to prevent an attack of the earhead-bug and downy mildew later in the agricultural season, was one of the few indigenous precautionary measures applied to this crop. Although risky, almost all farmers had adopted this measure because of the negligible costs involved. Other pests and diseases were considered to be either less fatal or less curable and, except for the removal of affected plants, other indigenous curative measures were not taken. Most probably one of the other reasons why farmers used only a limited number of indigenous control measures, had to do with the fact that *jowar* was neither considered to be a highly valued cash crop nor a highly valued food crop.

Although the pod borer had attacked redgram almost every year during the three years of the fieldwork, respondents were convinced that it was difficult to prevent or control it, or any other pest or disease in this crop. The only indigenous measures used were increasing the spacing between the rows (that is, sowing in a higher ratio than four rows of *jowar* against one row of redgram), or sowing *jowar* as a monocrop.

Being a highly valued cash crop, more than for *jowar* and redgram, indigenous precautionary and curative measures were available for castor. All were aimed at controlling the RHC and the semilooper and the negative effects of other pests and diseases were considered to be of less importance. However, because farmers considered the effects of many indigenous measures to be dubious or less effective than modern measures, most had become obsolete.[6] Besides the adjustment of the sowing date, the only other indigenous activities in vogue during the fieldwork period were the application of an indigenous pesticide based on the fruits of the neem tree and removing the caterpillars by hand and killing them with sticks or by putting them in a container or in a pit. The only measure widely practised during *kharif* 1992/3 and 1993/4 was the adjustment of the sowing date, a measure which had negligible additional costs. The two other control measures, involving more expenses, were hardly undertaken. Neem application was, for instance, not practised at all. Removing the caterpillars by hand was done on 5 per cent of the plots, mostly in Mansanpally (13 per cent), followed by Ghatpally (5 per cent), Ekvaipally (2 per cent) and Debbadaguda (2 per cent). In all villages, most of the treated plots belonged to small and small-medium farmers of the ST community.

6.10.2 RECOMMENDED

The final recommendations for *jowar* and castor by the projects concern integrated pest and disease management. While some elements were in line with indigenous practices and others had been introduced before the start of the project, its integration was completely new to the farmers. To prevent the *jowar* earhead-bug, farmers were advised to sow *jowar* early or to use improved, short duration seed varieties. For heavy invested plots, the projects recommends chemicals: one dusting with BHC 10 per cent or Ekalox 4 per cent. Shoot fly, a minor pest of *jowar* attacking in the early seedling stage of the plants, could be controlled by sowing the crop by the end of June or the first week of July. To control the stem-borer, the projects recommended spraying with Endosulphan once or twice.

The RHC attacks castor throughout the year with a peak about one month after the first monsoon showers. Another caterpillar attacking castor is the semilooper. Egg laying usually starts with the onset of the monsoon but a peak build-up of semiloopers occurs mostly during the first fortnight of August. Control measures recommended by the projects were, late (second half of July) sowing, removing and destroying eggs and caterpillars, removing and burning infested plants, and dusting or spraying with Endosulphan or Thiodan. In case of the semilooper, the Maheswaram project also advised the utilization of the presence of two important natural parasites which were generally in abundance in the second half of August as a biological control. To control attacks of the pod-borer on redgram, the projects recommended several kinds of pesticides.

6.11 Interventions by the Maheswaram Project in Mansanpally and Debbadaguda

As stated before, Mansanpally was short-listed by the Maheswaram project management as an exemplary village of its intervention strategy. Based on the outcome of a report covering the first fieldwork period the projet management objected against the inclusion of the village and requested to study another. Hence, Debbadaguda was selected.

While Mansanpally was among the first five villages to experience the activities of the Maheswaram project in 1985/6, operations only started in Debbadaguda in 1990/1, in the second phase, after the project was extended to the south. According to project documents, with 66 ha the village territory of Mansanpally was only partly incorporated in the project, meaning that only plots owned by the Lambadas were to be treated and that only these households were eligible for other kinds of project support. In practice, however, project staff did not make a clear distinction between Mansanpally villagers owning plots inside and outside the watershed. Since the two project officers did not live in Mansanpally's Lambada *tanda* but in the main village, they mainly contacted people living in the main settlement. Hence it is no surprise that several FC but also BC and SC respondents claimed to have received support from the project.[7] In contrast to Mansanpally, Debbadaguda's entire village territory was included in the project, meaning that, in principle, all farmers were eligible for participating in project activities.

Because of a change in project activities and a change in the way recommendations were communicated to farmers, changes that mainly came about after the Mid Term Review of 1988, some differences in activities were discernible between the two villages. The most striking difference concerned (*in-situ*) soil and water conservation measures. Dryland plots in Mansanpally were bunded with waste weirs first, while some plots were planted with *khus* lines later. In contrast, dryland plots in Debbadaguda were planted with *khus* only. Because *in-situ* soil and moisture conservation activities were emphasized after 1988, farmers in Debbadaguda were more exposed to these recommendations than farmers in Mansanpally. With the exception that in each village somewhat different improved seed varieties were distributed and that Debbadaguda farmers received more fruit trees than Mansanpally farmers, most activities undertaken by the project concerning improved crop management, were similar.

In both villages the Maheswaram project tried to induce farmers to adopt its recommendations by using a four-pronged strategy. Besides building soil and water conservation structures in farmers' plots itself and supplying subsidized inputs, the project informed farmers about the recommendations by exposing them to the mass media (for instance leaflets in Telugu, articles in regional newspapers and radio talks) and through personal contacts. Although the concepts of observability and trialability suggest that the project could benefit from demonstration plots located in the vicinity of the villages, plots which could accelerate the diffusion process, particularly among farmers who preferred to wait and see (Feder and O'Mara, 1981: 71), these were not established in either of the two villages.

While most elements of this strategy were similar in both villages, personal contacts differed somewhat. This does not apply to the Village Committees which were meant to facilitate communication between the project and the village and inform farmers about the recommendations and other project activities. These did not take off in either village. The way farmers reacted upon mass meetings, upon contacts with project staff and upon visits to agricultural research centres is described next.

While the project organized various large meetings for Mansanpally farmers in Maheswaram town, such meetings were not organized for Debbadaguda farmers. Mansanpally respondents claimed that all communities of the village were invited, but that most people did not attend, reasons being indifference and lack of time. Those who did attend stated that although the information given was useful, they were more interested in the promises made regarding subsidized inputs. Many of them admitted that initially these promises made them feel favourably inclined towards the project, but added that they felt more or less betrayed afterwards, when experiencing that a lot of these promises were not kept.[8]

Especially in the second phase of the Maheswaram project, farmers were invited to be trained at the Agricultural University of Andhra Pradesh, at CRIDA's research station in Hyatnagar and even at a horticultural research station in Tirupati. Although they received an attendance fee, on most occasions only a limited number of people participated in such training programmes. Sometimes members of the village elite in both villages opposed the participation of SC farmers, in Mansanpally they were also against ST farmers. More often, farmers were not interested in participating, were not confident about the practical use of the new technologies from the outset or were in need of daily wages and went to work elsewhere. Hence, only 7 respondents from Mansanpally against 33 respondents from Debbadaguda had attended such training. The fact that 4 out of 7 Mansanpally trainees were BC farmers shows that the project did not restrict itself to the *tanda*. With respectively 11 and 10 respondents, most trainees from Debbadaguda were FC and ST farmers, followed by 7 BC and 5 SC farmers.

Those who participated were neither impressed by the training, nor by the agricultural scientists, nor by the research centres. They said that although they did not always know the instructions in detail, they were familiar with most recommendations. Moreover, they felt that since the conditions prevalent at these research centres differed considerably from those at their own farms, the recommendations were either not suitable for their requirements at all or had to be adapted to make them suitable. This feeling was based on three main considerations: (1) the situation under which the dryland crops were cultivated and the agricultural innovations were tested was far superior to their own; (2) much more than they could ever afford, considerable amounts of money were invested in dryland agriculture at these centres; and (3) the agricultural scientists working at these stations were not considered 'real' farmers, people who were familiar with the cultivation of dryland crops under 'normal' conditions.

The most important way in which farmers in both villages were informed was through personal contacts, i.e. through project staff supplemented by the regular T&V staff of the Department of Agriculture. In Debbadaguda they were assisted by a village volunteer, a young BC farmer owning a small plot of dryland. After receiving a month's training at the Agricultural

University of Andhra Pradesh, he worked for the project for eight months in 1992 and in 1993. Being young, owning only a small dryland plot and belonging to the BC community, he was not regarded as a respected farmer by the majority of the farming community. Since he had not adopted all recommendations himself, he was neither the model farmer nor a farmer innovator for others to follow.[9] In addition, the low salary, the disrespect he received from some project staff and the unauthorized dealings between some project staff and some large FC farmers, prevented him from doing his utmost.

The lengthy stay of two project officers in Mansanpally who supervised project activities and who were the project's change agents in the village for eighteen months, was not repeated in Debbadaguda. Project officers were supposed to travel to this latter village only in case specific activities had to be carried out and were supposed to meet farmers in the field before the start of the growing season to give pre-monsoon training. This way of contacting farmers was adopted later in Mansanpally too. Whatever this strategy, except for a couple of large, mostly FC farmers, many respondents in both villages said that they had met these officers only a couple of times during the project period and that too very irregularly. Most respondents also claimed that if they were contacted at all, they met project officers more after than before they had sown their plot with *jowar*. This irregularity and the limited number of contacts pertained even more to the visits of the T&V staff, who were supposed to meet contact farmers every fortnight. A large majority of the respondents stated never to have met these extension officers and were not able to mention the name of any T&V officer or contact farmer. In fact, not a single farmer in the two villages was selected contact farmer. In this, the two Maheswaram project villages did not differ from Ekvaipally or Ghatpally.

Although a large majority of the respondents in Mansanpally and Debbadaguda had heard about the project and knew about the visible activities (in particular about the contour/graded bunds and *khus*), most respondents had only a very faint knowledge about most project activities. In particular the marginal and small farmers in both villages stated that they were never properly informed about these activities, about the benefits of the recommendations, let alone be asked to give consent for constructing bunds or to plant *khus* in their plots. These and other farmers felt that the project was not in their interest but in the interest of the government and large FC farmers. Most respondents felt outsiders, felt that they had only limited influence. Hence, for most of them participation was just restricted to receiving subsidized inputs.

Included in the small minority of mostly FC respondents who did not feel as though they were outsiders, was the tribal leader of Debbadaguda *tanda*, a man with a strong character. Although not educated he was clever and familiar with various government programmes targeted towards STs and with the procedures. He knew his rights and was not afraid to claim them, even if he had to contact Reddies, government officials or bankers or had to travel to Hyderabad to meet senior officials.[10] In fact, the village elite of Reddy farmers spoke with envy of the Lambadas since their leader had been able to get access to many government funds, much more than ST and SC communities in the other study villages. Being very well informed about the various components of the Maheswaram project, he insisted that plots of the tribals be treated with *khus* and other soil and water conservation structures and demanded seedlings of fruit trees, a multi-purpose water point, improved seeds, chemical fertilizers, compost pits and trees for farm forestry. Perhaps the fact that most of the plots owned by the tribals were located rather close to a dirt road was helpful; or that a supervision team under the aegis of the World Bank which had notified the project management that it wanted to see the progress made in *tandas*, was of influence. In any case, many ST farmers had received *khus*, fruit trees and the like from the project.

But these Lambada farmers were a minority. Perhaps because they received less from the project or for other reasons, most of the few respondents who had met the project officers were not impressed by their qualities and complained about their ill-mannered attitude. They expressed their doubts about the knowledge of agriculture of much of the staff, arguing that while promoting the agricultural innovations, they were not always able to explain their relative advantages. Respondents also argued that project officers were not able to convince them that these recommendations would not lead to lower yields compared to the indigenous practices. They were most certainly not reassured because they felt that these officers showed contempt for these indigenous technologies. Some farmers in Mansanpally reported that while sowing up and down the slope, sometimes field officers used brutal force to make them sow across the slope. Sometimes even fights broke out. As a result, they felt looked down upon, felt that they were not taken seriously. Smaller farmers, in particular those belonging to the SC community, felt humiliated because of the language used by the staff, of their rude behaviour towards them and disapproved of the fact that these people thought them to be 'innocent'.

In spite of this criticism of project officers' behaviour towards them, smaller farmers were not pleased that project officers focused on larger

FC farmers. In Mansanpally, for instance, influential FC and BC farmers, whose dryland plots were located outside the project area, said that it was because of their friendship with the field staff who had lived in the village, that these plots had been bunded and that they had received fruit trees, inorganic fertilizers, improved seeds, iron ploughs and smokeless *chullas*, either free or for a low price. The first evaluation study carried out for the project, arguing that it was difficult to identify the soil and water conservation structures in the field as there were deviations between reality and the plans (Sitapathi Rao et al., 1990: 14), seems to confirm this. Some large FC farmers living in Debbadaguda very openly admitted that although they were not always eligible to receive agricultural inputs, fruit trees and implements, these had been given by project officers as a 'compensation' for allowing them to plant *khus* in their plots.[11]

Project staff argued that if the project was to make any speedy success, it was necessary to satisfy large FC farmers first. They added that if these farmers were not allowed to benefit more than others, it would be quite probable that these farmers would obstruct project activities. Consequently, they would not reach their targets, which would infuriate their superiors, who needed to satisfy the executives, who in turn needed to satisfy the funding agency. Hence, pressed by their superiors who knew that the project was behind schedule, it was for efficiency reasons that they did not select beneficiaries on the basis of poverty, neither selected plots on the level of degradation, but concentrated on large plots, which were generally owned by BC and especially FC farmers.

The fact that agricultural officers often lacked sufficient knowledge of agriculture, that these officers often treated farmers with disrespect and that they were biased in favour of large FC farmers who consequently benefited more, resulted in distrust on the part of many respondents who had met these officers. In fact, many respondents felt that field staff did not always tell the truth. This feeling was based on the fact that on a number of occasions project staff had not kept their promises, like giving all kinds of inputs for free, if they were allowed to construct bunds or plant *khus* on their plots. They also felt that they were not always well informed about the planning of activities and about which farmers were entitled to benefit from a programme and which were not. Even FC respondents were convinced that by withholding information project officers were able to cheat. They explained, for instance, that by not informing farmers about the availability of subsidized inputs, or misinforming them about the conditions of the scheme, or by just telling them that the inputs were out of stock, officers were able to lay their hands on these inputs and were able to sell them to shopkeepers. Some FC informants said that this was the way by which they received inputs for free or for low prices. Distrust was also aroused by stories of some Mansanpally farmers on whose land bunds had been constructed without their oral or written permission. In addition, illiterate farmers in Mansanpally and Debbadaguda stated that since they had not been aware of the exact contents of the 'agreement bond' they had signed, the construction of bunds and the planting of *khus* in their fields had taken them by surprise.

6.12 Interventions by AWARE in Ekvaipally

As stated before, Ekvaipally was selected after consultation with AWARE's chairman as an exemplary village of AWARE's intervention strategy in its early years.

Most probably, AWARE's intervention in Ekvaipally took from 1977 to 1988. It was AWARE's chairman who informed inhabitants of the main village and the Lambada *tanda* personally about AWARE and who motivated them to form *sanghams* in their village and *tanda*. A well educated large farmer from the toddy-tapper caste (BC community) was assigned by him to be the local organizer. Together with AWARE's field staff, it was this organizer's duty to motivate the rural poor in the vicinity of Ekvaipally, to establish *sanghams* and to assist the rural poor in the process of social, political and economic development.

In addition to the two *sanghams* created in 1977, various other *sanghams* were set up over the years. In 1982, AWARE's chairman personally gave permission to a Reddy farmer to start a *sangham* in Maripally, one of Ekvaipally's hamlets. In 1986 two Women's Associations and one Youth Association were established. In 1993, on paper, the membership of all five *sanghams* totalled some 150 men, 50 women and 25 youngsters, who had paid an entry fee of Rs 11 each. Members belonged to all communities, including a small number of FCs and a larger number of BCs. From the respondents selected for this study, 36 males were members of the *sangham*, 11 females were members of the Women's Associations and 2 youngsters of the Youth Associations. On the other hand, over the years 35 respondents had resigned their membership.

Since *sanghams* play such a pivotal role in AWARE's development and empowerment strategy of its target groups, it is necessary to deal with them at some greater length. Without doubt, AWARE succeeded in establishing five local organizations and managed to induce 225 people

(10 per cent of the population) to join these. However, all kinds of (fundamental) problems which often arise while organizing people occurred over the years and neither AWARE nor its local staff were able or willing to solve them. These problems come under the headings of leadership, internal decision-making and participation of the members.

According to AWARE's rules, the board of each *sangham* had to be composed of a president, a vice-president, a secretary and some additional members. However, most boards in Ekvaipally had a president and a secretary only. According to AWARE's rules, each board had to be democratically elected by the members of that particular *sangham*. Nevertheless, the boards of the first two *sanghams* were appointed by AWARE's chairman. His reasons for doing so are not known. One might argue that given the situation that the target groups were unfamiliar with the idea of having their 'own' local organizations, given the fact that AWARE was still a young organization in which procedures were not very clear yet, and given the fact that for efficiency reasons it was better to appoint more educated people, it was the best possible option at that time. AWARE's chairman could not have expected that the man he had assigned as local organizer turned out to be someone who used the *sanghams* not only as vehicles to promote the economic and political development of the target group, but more to further his own development and that of a few allies.

Within 15 years, this local organizer had used his position in such a way that he had become a very powerful person, not only in the village but also in the *mandal*. Very much in line with AWARE's ideas of empowerment, he had succeeded in getting political leverage in the village and the *mandal* and was regularly seen together with Congress politicians in Hyderabad. It was a great success for him and AWARE that he managed to beat the Reddy *sarpanch* and himself became *sarpanch* of Ekvaipally. He had strengthened his position to such an extent that when *sangham* elections were held in 1988 he managed to nominate his BC, SC and ST supporters in the boards of all *sanghams*, except the one in Maripally. In this latter *sangham*, elections were simply never held, meaning that since its establishment it remained under the control of the same Reddy farmer. Although this Reddy president had more room for manoeuvre (he was of higher social status than the BC organizer, was a fairly well-to-do farmer and a respected person in his own village/hamlet), he was dependent on this BC organizer for AWARE and government funds.

Except for the fact that they were not controlled by the village elite, the boards of the *sanghams* very much resembled the 'democratically' elected village *panchayats*. Like so many of these traditional local organizations, Ekvaipally's *sanghams* lacked accountable leadership, internal democracy and were not able to promote more than superficial participation of its members. Being appointed by AWARE's chairman and being one of the few educated people among the lower castes, the local organizer was able to manage the *sanghams* more or less as his own kingdoms, especially after he had been elected *sarpanch*. Various board members said that he more or less controlled most *sanghams* in Ekvaipally and that final decisions were made by him. As a consequence of his authority and patronizing behaviour, of weak leadership of the presidents and the other members of the board and also the complying nature of most *sangham* members, internal democracy was lacking in the *sanghams*. In fact, because the boards more or less lacked any decision-making capacity, they hardly met. Although AWARE's field staff were very much aware of this, they did not intervene. Key persons informed me that this was related to the good contacts between the local organizer and AWARE's headquarters and with the fact that he let the field staff share in the benefits.

AWARE and this local organizer managed to make many SCs, STs, BCs and some FCs join one of the *sanghams* in the village. However, many respondents said that they had not joined the *sangham* because of empowerment reasons (because, as a group, they could oppose the landlords or the civil servants), but more for the economic benefits and also because they were forced to do so by respected people in the village (caste or tribal leaders) or the local organizer. Consequently, for many respondents participation in the *sangham* remained confined to paying the membership fee and attending meetings now and then. It certainly did not induce their participation in group activities on a large scale.

During the first couple of years of AWARE's involvement, the *sanghams* met rather regularly. Encouraged by the prospect that schemes would be distributed by AWARE, *sanghams* must have met some four times per year. But these were fewer than AWARE's recommendation to meet at least twice a month (Sohoni and Singh, 1990: 86). In later years, the number of meetings decreased and after AWARE's withdrawal in 1988, *sanghams* met perhaps once a year. Besides the reduction of direct AWARE involvement, growing disinterest and even dissatisfaction among the members were the main reasons for this drop. AWARE's field staff, who were supposed to visit the village regularly, but who were only marginally involved in local affairs, had not been able to create a feeling that these *sanghams* had been established for the sake of its members. It turned out, for instance, that the members of the *sanghams* did not participate in project design. Thus, not a single respondent had ever seen, heard or discussed a village action plan. Many of them were convinced that the activities undertaken by AWARE and the *sanghams* were designed by the local organizer and AWARE staff, which gave them a feeling that the *sanghams* were more a vehicle to satisfy the needs of these people than their own.

It seems as if those who joined the *sanghams* for quick economic benefits were the first to become disinterested. Feeling that most meetings were a waste of time, many respondents stopped attending regularly. Many said frankly that they only went to meetings when the 'discussion' was about loans or schemes to be distributed by AWARE or the government. Feeling that the BC organizer favoured relatives, households belonging to his own caste and supporters when distributing schemes, a considerable number of people no longer felt the need to participate during these meetings, while others resigned their membership. Consequently, during the fieldwork period the *sanghams* were more or less defunct. Their main activity was to distribute schemes supplied by government programmes like the IRDP once a year. However, convinced that they did not belong to the inner circle of the local organizer, many members did not take the trouble of attending such meetings.

Consisting of a social (organization and awareness building), a basic needs and an economic programme, AWARE's activities in Ekvaipally resembled those carried out in other 'AWARE villages'. Besides the establishment of the *sanghams*, AWARE started almost immediately with motivational and other kinds of training courses in the village itself. Of these small training courses 46 were imparted during the 10 years of its direct involvement (AWARE, 1994b: 4). In later years, members of the various *sanghams* participated in mass rallies (*mahasabhas*) held outside the village, often in other districts.

AWARE started its basic needs and income and employment generation programmes in Ekvaipally after 1980. Besides training on subjects like health and hygiene, necessity of education, AWARE staff advised the target group on topics like land reforms, minimum wages, child labour, dowry system, gender equality, abolition of bonded labour and human rights. During AWARE's active involvement, two village hand-pumps were installed and some government financed houses were built for SCs. During the fieldwork period, after AWARE had withdrawn, and much later than in the other study villages, government financed houses were constructed in the two *tandas*. Under AWARE's community programme for common welfare, mulberry plants for sericulture were distributed to the Reddy president of the Maripally *sangham*.

Besides the usual training courses, AWARE's programmes on animal husbandry and cottage industries (coming under its economic programme) consisted mainly of distributing different kinds of assets. According to AWARE's own figures, it distributed bullock carts to 17 people, sheep and goats to 18 people, while 12 people received assistance in chappal making and 15 in vegetable vending (AWARE, 1994b: 8).[12] While most of the respondents who received money for bullock carts and sheep and goats were satisfied, most of those who had been given loans for chappal making and vegetable vending were somewhat displeased. The women who were supposed to start a vegetable vending business complained that it was not only pointless to give money for the same enterprise to so many people, but it was not very fruitful either, given the fact that vegetables were hardly cultivated in the village. One of its last activities under this programme was undertaken in 1993, at a moment when AWARE had withdrawn from the village. During a large meeting with civil servants and bank officials, AWARE's chairman put money in the bank that functioned as an insurance in case AWARE beneficiaries were to default. Except for this deposit, board members of the *sanghams* assumed that the total loans supplied by AWARE under its basic needs and economic programme must have been some Rs 5,00,000. Five years after AWARE had left the village, in 1994, less than Rs 1,00,000 had been repaid.

As indicated in Chapter 9, AWARE had developed a five tier approach for its agricultural programme: (1) improving the availability of irrigation water; (2) soil and water conservation on dryland; (3) training of farmers; (4) supply of agricultural inputs and implements; and (5) marketing of agricultural surplus. In Ekvaipally its programme was limited to the first, the third and the fourth tier and, in spite of AWARE's claims, the soil and water conservation programme was never implemented. Respondents stated that the few graded bunds constructed on some dryland plots had not been built by AWARE but by the government, before the arrival of the organization, most probably in the early 1970s. Later, staff acknowledged that AWARE had not built any bunds as farmers did not favour these kinds of structures. Nevertheless, one might expect some impact of their training courses concerning soil and water conservation and of the information passed on to the villagers by the T&V extension agents, which were supposed to visit Ekvaipally on the request of AWARE. The least one might have expected is that AWARE would have induced its target groups to construct indigenous border bunds on drylands.

AWARE also had problems implementing its last tier, the marketing of crops by cooperatives. It turned out later that AWARE had failed to stimulate the marketing of agricultural products by cooperatives or by the higher level target group organizations (the District or Area Development Service Societies) in any of its villages. In fact, AWARE had major problems filling the position of marketing officer at its headquarters in Hyderabad.

AWARE's first and fourth tier were mainly meant to improve the availability of agricultural inputs: i.e. irrigation water in the first and improved seeds, chemical fertilizers, pesticides, implements, plough bullocks and dairy cattle in the fourth tier. To improve this availability AWARE used both its access and parallel strategy: i.e. it availed itself of government programmes and used its own funds. The distribution of these inputs was in the hands of the *sanghams*, meaning the local organizer/ *sarpanch*. He used more or less the same selection procedure for the distribution of AWARE and government schemes as *sarpanches* did in the other three study villages. Although beneficiaries had to be selected on the basis of poverty by the entire board of each *sangham*, he took care that non-poor and non-target group members also received these subsidized schemes.

AWARE reported that 19 irrigation wells and 16 pump sets had been distributed during its direct intervention and that 100 ha of dryland was converted into wetland (AWARE, 1994b: 6, 8).[13] In addition, another 60 households had been assisted with chemical fertilizers. These are impressive figures. But it was not mentioned who the beneficiaries were. However, a list pertaining to schemes distributed mainly in 1985/6 under AWARE's basic needs and economic development programmes made by the president of one of the *sanghams* indicated that non-target group BCs benefited most. Of the 58 households which were assisted, 32 households belonged to the BC community, receiving schemes worth Rs 81,500, 16 households were of SCs and the value of their schemes was Rs 41,500, while another 10 ST households were assisted through schemes worth Rs 25,500. It is a clear example of patronage by the local organizer: he used his position to get access to AWARE and government schemes which he, after having received a commission, distributed to friends and supporters who in turn supported him during elections. It clearly led to resentment and to faction fighting in the various communities in Ekvaipally between those who benefited from his kindness and those who did not.

Respondents stated that the ways 'to cheat and eat' used by this BC organizer/*sarpanch* in cooperation with board members of *sanghams*,

CDOs, other organizers and board members of the supra-local organizations of the target groups (the District or Area Development Service Societies) resembled the ways the landed elite did in cooperation with civil servants.[14] Key informants stated that on many occasions these intermediaries or brokers between the villagers and AWARE and government agencies gave less money to beneficiaries than the villagers had signed for; sometimes they gave 'strange' money to beneficiaries;[15] sometimes kept back part of the schemes for themselves; and sometimes put fake names (*binami* names) on lists which allowed non-target group people to get loans. In fact, it seemed quite a normal procedure to ask for 10 to 20 per cent of the loan 'to cover expenses'.

The practices of the BC organizer/*sarpanch* had grown to such an extent that during the study period Naxalites threatened to make an assault on his family and property. This more or less worked as an 'outside' check on extreme corruption. Except for a small faction of antagonists who said that they would fight him at the next *panchayat* elections, there was much less effective resistance to his practices in Ekvaipally.[16] Many respondents argued that they were more or less accustomed to such dealings saying that 'everybody is doing it'. Moreover, they felt that in all likelihood another organizer or *sangham* president would act in a similar manner. In addition, a considerable number of villagers benefited in one way or another from AWARE's activities, either because they benefited from overall village improvement or because they had received loans. Although most beneficiaries received less money than they had signed for, they were convinced that neither the BC organizer/*sarpanch* nor AWARE field staff would press them hard to repay it. They regarded the schemes as a gift from heaven, a gift from AWARE's chairman, who was considered a god, and not a loan. For that they were thankful.

AWARE's third tier consisted of agricultural extension, i.e. informing farmers about soil and water conservation and improved crop management. Most information given by AWARE was through personal contacts. Having an access strategy, AWARE did not care much about employing people with an agricultural background, be it farmers, agronomists or extension officers. Thus, agricultural extension was supposed to take place by inviting T&V staff who had to give special technical training courses in Ekvaipally. According to respondents four to six of these courses were held during the years, or about one each year during the later phases of AWARE's direct intervention. Information about the number of people attending is lacking, though. Since these courses were accessible to all villagers many BC and also some FC farmers participated. On the other hand, quite a few SC and ST respondents claimed not to have ever heard about these courses or to have participated.

Subjects discussed during the courses were in line with AWARE's modernization ideology: i.e. the use of new agricultural technologies for wetland and dryland, information on soil and water conservation practices and the way people could get access to government schemes for minor irrigation, agricultural inputs (subsidized improved seeds, chemical fertilizers and pesticides) and animal husbandry (plough bullocks, dairy cattle, goats and sheep). In spite of its access and empowerment strategy, neither AWARE staff nor the *sanghams* after AWARE had withdrawn, succeeded in persuading T&V extension officers to visit the village regularly. In addition to these courses, for a small section of people, mostly those in favour of the BC organizer, some agricultural training programmes were given in AWARE's Human Resource Development Centre. These programmes focused mainly on wetland agriculture.

Respondents in Ekvaipally were quite aware that most of AWARE's field staff were raised in an urban environment and lacked agricultural knowledge. Hence, they did not turn to them for information about agricultural innovations. Most villagers did not consider the T&V extension officers, who visited the village after a request from AWARE and the *sanghams*, as a good source of information either. The complaints voiced by the villagers were very much in line with those expressed by farmers in Mansanpally and Debbadaguda: i.e. except for the few times they were specifically invited, they hardly ever visited Ekvaipally; during the few visits they made they preferred to talk with large FC farmers; and they also showed contempt for the indigenous agricultural technologies. It made many respondents doubt their capacities, interest and intentions.

The final remark in this section relates to an assessment of the overall effect of AWARE's intervention on village development in Ekvaipally. As it was not the purpose of this study to scrutinize social, political and economic development in depth but to concentrate on the effect of AWARE's dryland activities, the next few lines, although based on detailed fieldwork, are just a quick assessment. It was found that the four study villages were largely similar, i.e. AWARE's influence was certainly not manifest. But then, the villages were purposely selected and had to be more or less similar on various social and economic criteria. Comparing the past with the present situation in the village, respondents were convinced that although Indian society had improved over the years, social and economic development in Ekvaipally was partly a result of AWARE's activities. They argued that AWARE's motivational training courses had most effect, indicating that villagers feared the landlords less than they did in the past and that untouchability in the village had decreased. But they were convinced that the proximity to Hyderabad and improved communications had also considerable effects. Economic development, they claimed, was less a consequence of AWARE's activities than of overall economic growth. It led, they argued, to an increased demand for toddy and increased income for an increasing number of toddy-tappers. Despite AWARE's strong anti-alcohol campaigns, AWARE never asked the toddy-tappers to stop tapping and people continued drinking, also during AWARE's intervention period.

6.13 Sources of Agricultural Information

What became clear from the last two sections is shown in Table 6.1. The T&V extension system failed to function as an important source of information and advice for dryland agriculture. Whatever the project activities and irrespective of village, in 1991/2-1992/3 most respondents depended more on fellow farmers, relatives and friends and, in addition, on shopkeepers and merchants. The difference between the somewhat higher figures for 'projects' and 'T&V' in Debbadaguda compared to Mansanpally and Ekvaipally was due to the fact that while the Maheswaram project was still operative in the former village, project activities had ceased to exist in the latter two villages. In all villages, it was predominantly FC respondents who met T&V extension officers. In Debbadaguda, it was primarily ST farmers who met extension officers of the Maheswaram project. Nevertheless, the number of contacts between respondents and project officers or T&V extension agents was very low during the study period. Not surprisingly, arguing that they had never met any such agent, 43 per cent of all respondents said that they had no opinion on the functioning of extension agents working on behalf of both project and on the T&V system. An additional 54 per cent of all respondents stated that the service provided was bad to very bad.

Although Long and Villareal (1994: 46) indicate that opinion leaders, whom they call 'gatekeepers' or 'brokers', may not only facilitate but also block change, social learning theory suggests that obstruction of agricultural change can (partly) be circumvented by observing how these opinion leaders put agricultural innovations into practice in their own fields; that this suggestion may be valid is evident from the fact that many BC, SC and ST respondents claimed that experienced and large Reddy farmers were the most important source of agricultural information. In

TABLE 6.1. MAIN SOURCE OF INFORMATION AND ADVICE FOR DRYLAND AGRICULTURE IN 1991/2 AND 1992/3, BY VILLAGE (*percentages*)

Sources	Debbadaguda	Mansanpally	Ekvaipally	Ghatpally	Total
Farmers, relatives, friends	80	66	70	89	74
Shopkeepers, merchants	7	31	25	5	19
Projects	7	1	1	0	2
T&V	4	1	2	4	2
Others	2	1	2	2	2
Total	100	100	100	100	100

Source: Own survey.

fact, many of them said that they imitated large FC farmers after having seen the results of the recommendations in their plots. Observability, however, is more than that. Many smaller farmers reported that they had learnt about an agricultural innovation while working as a daily or permanent labourer for a large FC farmer who had already adopted it. By working with an innovation a farmer gets experience, which reduces uncertainty and may ultimately lead to adoption. During the study period there was only one farmer (a large FC farmer from Ghatpally) who tried 'to block agricultural change' by preventing other farmers from observing his wetland plot cultivated with a new variety of chillies.[17]

Some farmers had the advantage that their information network was larger and more diverse than that of other farmers, which allowed them to turn to more and to different sources of agricultural information, especially those located outside the village. Having better access to mass media, some FC farmers had introduced an agricultural innovation in their village. Thus, after having read some articles in a newspaper a Reddy farmer from Ghatpally imported fruit trees from Rajasthan and another Reddy farmer from Ekvaipally began to experiment with vegetable seeds. As was also demonstrated in other situations by Tiffin et al. (1994: 265) and Sandee (1995: 110, 138-42), it shows that personal contacts were not limited to their own village or immediate surroundings. Although personal information networks of FC farmers were larger than that of other farmers, smaller farmers also carried information about agricultural innovations from quite a distance back to their home village. While most FC farmers heard about these innovations during visits to relatives, civil servants and sometimes agricultural universities, smaller farmers heard, observed and gained experience with innovations working as seasonal migrants. Nevertheless, it was still predominantly the larger FC farmers who imported information about agricultural innovations into the villages.

6.14 Conclusion

The section on indigenous technologies for dryland agriculture has indicated that the crops and the cropping pattern were well-suited to the agro-ecological situation in the area and that farmers had multiple objectives with many activities. Border bunds were not only constructed to prevent soil and water loss from the plot but also to demarcate ownership boundaries. Interculturing was not only an *in-situ* soil and moisture conservation activity but was also meant to remove weeds and to stimulate plant growth. Sowing up and down and across the slope in alternate years was not only meant to prevent soil and water loss from the plot but also to prevent the growth of plants in the same spot as the previous ones. *Jowar* was not only cultivated for food but also for fodder. Redgram was not only intercropped with *jowar* for reasons of soil fertility but also for its protein. Consequently, agricultural innovations which aim to improve only one element of these multi-purpose activities (e.g. concentrate on improving grain yields of *jowar* meanwhile neglecting fodder) have a disadvantage.

Farmers in Debbadaguda and Mansanpally participated only during the implementation phase of the Maheswaram project. But participation was more or less limited to adoption of the recommended technologies. The project tried to induce farmers to do so by building soil and water conservation structures in farmers' own plots, by supplying subsidized inputs and by informing farmers about the recommendations through mass media and more particularly through personal contacts. In both villages, the proposed Village Committees meant to improve communication between villagers and project staff did not take off. In addition, many farmers were not interested in attending mass meetings, whilst those who visited agricultural research centres were not convinced that the recommendations were better than the indigenous practices. Consequently, much weight was given to the contacts between project staff, T&V extension officers and villagers. However, the number of contacts was limited, particularly those between villagers and T&V extension agents. Most respondents, especially small and marginal SC and ST farmers, who met project staff felt uneasy. These contacts were strained because many respondents doubted the agricultural knowledge of the staff and felt unhappy about their behaviour. Contacts between large FC farmers and staff were better. Both sides seemed to realize that they more or less depended on each other: i.e. staff needed large farmers to speed up project implementation and large farmers needed project staff to get cheap inputs.

ST farmers in Debbadaguda also managed to benefit from certain schemes distributed by the project, mainly due to the tribal leader who claimed his rights. Most probably because the project was dissatisfied with the activities of the T&V extension service and the reaction of many farmers to their actions, it appointed a village volunteer in Debbadaguda. However, he was not a highly respected model farmer and found it difficult to persuade farmers to adopt the recommendations.

Within a couple of years after its first involvement in Ekvaipally, AWARE managed to establish a couple of local organizations in the main village and its hamlets. During its direct intervention which took ten years, many activities under its social (organization and awareness building), basic needs and economic programme were carried out. Especially the social programme had an effect on the village life. Although participation of the members of the *sanghams* included decision-making on project design (village action plan), this failed to take shape in Ekvaipally. In fact, decision-making by the members was restricted as it was not in the interest of the boards. More than being vehicles of empowerment of the target groups, the *sanghams* were vehicles of improving the economic position of its leadership, especially of the BC organizer, later *sarpanch*. Lacking accountable leadership, internal democracy and not being able to promote more than superficial participation of its members, they resembled in more than one way the 'normal' village *panchayats*. While on paper AWARE followed a five-way strategy to induce farmers to improve agriculture, it neither implemented its soil and water conservation programme nor its marketing programme in the village. It implemented its programme on minor irrigation, input supply and extension by using both its own resources and those provided by the government. The BC organizer/*sarpanch* in co-operation with others used his position to get access to AWARE and government schemes which he mainly distributed to friends and supporters. Using its access strategy and lacking experienced staff, extension was mainly given by T&V staff. However, in spite of AWARE's empowerment strategy, the *sanghams* did not succeed in inducing T&V extension officers to visit Ekvaipally regularly.

Whatever the activities undertaken by both projects to inform farmers about the recommended technologies, farmers still relied on their indigenous sources of information, i.e. fellow farmers, relatives and friends. Many of the respondents who met extension officers criticized these change agents for their limited agricultural knowledge; for the way they behaved towards farmers; for the way they tried to induce farmers to stop using their indigenous agricultural practices; for the fact that they gave preferential treatment to certain farmers; and for the fact that some of them 'ate' money. Hence, trust was often lacking and T&V extension officers were regarded less as opinion leaders than their own villagers. What it meant for the adoption of the recommendations will be shown in the next chapter.

NOTES

1. Farmers could choose between FYM from cattle, goats, sheep, and poultry. Although farmers favoured chicken manure, at least if it was pure, most used manure from cattle because it originated from their own farm and, if bought, was cheaper than other types of FYM. Nightpenning of a shepherd's flock of sheep or goats in a plot hardly occurred. Manure from poultry was mainly used by large farmers to manure commercial crops on wetland. They bought it from owners of large poultry farms which had mushroomed around Hyderabad during the last decade or so. However, chicken dung was more and more polluted with seeds of weeds. Consequently, many wetland plots had become overgrown with weeds which were rather difficult to remove.
2. Initially, the recommended dose for castor had to be applied in three doses: a basal dressing of 10 kg N and 30 kg P per ha, and two top dressings of 20 kg N per ha each about 6 and 12 weeks after sowing (GoAP, 1981: 59).
3. Based on astrology, Telangana farmers divided the agricultural year into 26 fortnights or *karthes*. Since each fortnight was considered to have certain (dis)advantages, farmers performed certain operations only in certain 'auspicious' *karthes*. Every year, the Department of Agriculture of Andhra Pradesh published an agricultural almanac in Telugu which used these *karthes*.
4. Once, an Ekvaipally farmer had no other option but to supplement the traction power of one cow with his own feeble force.
5. Initially, several kinds of herbicides were included in the plans of the Maheswaram project too (GoAP, 1981: 52). However, because of the high costs involved and their labour replacing nature, they were not recommended.
6. Most respondents seemed to feel some embarrassment mentioning the following outdated indigenous measures for castor: praying to the gods and sacrificing the first five rows in the plot, making a small trench around the plot to prevent the caterpillars from entering it and spreading cooked rice mixed with curd in or around the plot to attract birds which would eat both rice and caterpillars.
7. Thus, a statement showing the activities undertaken in Mansanpally reported that 118 ha were treated with bunds and 172 ha with *khus* (GoAP, 1992b).
8. Since we do not know what promises were made and if these were kept or not, we do not know whether this feeling is justified or not.
9. The *khus* planted in his plot died; he did not use improved *jowar* and redgram

seed; intercropped *jowar* with redgram in a 4:1 ratio; used less than the recommended quantity of chemical fertilizers and did not give a topdressing; did not use a seed-cum-fertilizer drill; and did not use pesticides.

10. In fact, his experience with senior officials had taught him to meet them on Sunday in the quiet atmosphere of their homes.
11. Large FC farmers explained that although the planting should have been paid for by the project, it was paid for by them using their own permanent labourers. Project staff more than compensated them by giving agricultural inputs and the like, meanwhile showing fake bills stating expenses on labour to their superiors and being refunded for money they never spent.
12. The fact that these schemes were identical to the schemes distributed by the government under the IRDP shows less of AWARE's innovativeness and more of its access strategy.
13. Especially the area converted into wetland seems somewhat out of proportion.
14. Various respondents mentioned a couple of names of AWARE staff and board members of *sanghams* and higher level target group organizations in the Amangal area who had 'eaten' considerable amounts of money, even up to Rs 1,00,000.
15. A couple of respondents stated to have received white pieces of paper that were wrapped up in normal bills and one respondent stated that a part of the loan consisted of 'Rs 5 bills' which had been printed by AWARE for its boat hospital.
16. Some respondents, expecting repercussions from AWARE staff and the friends of the BC organizer, admitted that they were not eager to inform us about what had happened in the village.
17. He had a couple of defence lines. First of all, his plot was located at quite a distance from the village. Secondly, part of his wetland plot was planted with fruit trees. Thirdly, on a small nursery, screened off from this orchard, he had sown his new crop. Of course, trespassers were not allowed.

7. Adoption of Recommended (*in-situ*) Soil and Water Conservation Technologies

M. Put

Irrespective of their ideological differences and dissimilar approaches, both the Maheswaram project and AWARE tried to disseminate almost similar recommendations for dryland agriculture for which they relied heavily on the T&V system of agricultural extension. For analytical purposes these recommendations are grouped under two headings, to be discussed in separate chapters. This chapter is structured along the lines of the recommendations for soil and water conservation and the next along the lines of the recommendations for improved crop management. For a description of both indigenous and recommended technologies, I refer to Chapter 6. For each recommendation separately, the following questions will be answered. Did respondents know about (the exact contents of) the recommendation and what was their main source of information? How many respondents actually adopted it? Was it full or partial adoption or did respondents adapt it? Was (partial, adapted) adoption the result of project intervention? What were the reasons for adopting, adapting or rejecting it? This last question deals with how respondents have assessed each recommendation on characteristics like relative advantage, compatibility, complexity, trialability and observability. Observed adoption rates per recommendation will be compared with the hypotheses as formulated in Chapter 4 and a summary of the outcome is presented in a separate table.

This analysis is made at plot and crop level. Where necessary, the information is consolidated for the entire three-year research period, using holding-size and community as intervening variables to differentiate between the respondents in the study villages. In addition to this

[*] This Chapter is included in Put (1998). Courtesy of Thesis/Thala Publishers, Amsterdam.

quantitative plot and crop level data, qualitative information from 88 open interviews is used. The same procedure is followed in the next chapter on improved crop management. To begin with, two representatives of groups which are often labelled 'progressive' and 'laggard', a large FC farmer and a marginal SC farmer, are given the floor. Their names are fictitious.

Mr X.Y. Reddy, a large and influential FC farmer in Debbadaguda (one of the study villages incorporated in the Maheswaram project), on improved soil and water conservation measures for dryland agriculture:

I own seven dryland plots, totalling 32 ha, and three wetland plots, totalling some 13 ha. There are two main reasons why I concentrate on arable farming on wetland. First of all, for all kinds of reasons, yields of dryland crops are much lower and more uncertain than yields of wetland crops and it is not possible to predict whether you will get a good return on your investments or not. Another reason is that I own many animals and hence need a lot of dryland as pastures. Consequently, I have no other option than to lie a large part of my dryland fallow.

Like so many farmers, I am quite aware of soil and water loss on my dryland plots, but I am not very interested in preventing it. Of the seven dryland plots, six have only a minor gradient. Of these six plots, four have no bunds at all, while the other two have one bund at the top of the plot and one bund at the lower end. These bunds were made by my father after a land dispute many decades ago. They are meant for demarcation and to prevent water from flowing downwards, to keep it inside the plot. Although the gradient of the seventh plot is larger than that of the other six, our family was not interested to construct bunds on this plot. However, some years ago, officers from the Maheswaram project planted five lines of *khus* (hedge of Vetiver grass) along the contour on part of this plot. This *khus* was planted very late, somewhere in September or October, and since rainfall was scanty after planting, some 80 per cent of the *khus* died within a couple of months. I preferred to have contour bunds, like the ones the Maheswaram project constructed near Maheswaram village, but because the project did not have a budget for contour bunds in this part of the project area, I had no other choice than accepting the *khus*. Anyway, together with the *khus* they gave improved seeds and chemical fertilizers for free.

The timings of all operations on dryland depend on the rains, but *karthes* (astrological calendar) are important guidelines. Land preparations for *jowar* normally start in the first week of May. For castor this date is less fixed and preparatory tillage may start somewhere in June or July, or even later. Preparatory tillage means that my permanent labourers plough or harrow the land at least twice, the first time against last year's sowing direction and the second time in the same direction as last year's sowing direction.

Regarding the direction of sowing, I weigh the pros and cons of soil erosion and plant performance or yield. In the plot with the steep gradient, I consider soil erosion more important than yield, hence I sow across the slope. In the more or less level plots, where erosion is less, I sow in alternating years once up and down the slope and once across the slope. This is the best way to get higher yields. These practices I learnt from my forefathers. On advice of the Maheswaram project officers I started contour cultivation in *kharif* 1993/94, but it was very difficult. I managed to do it because I used a tractor for land preparation. I used bullocks for sowing, but since they are not used to turn after every furrow, it was complicated. Besides, because you do not get straight lines, people ridiculed me, stating that the labourers who sowed my crops must have been drunk at the time. These are the main reasons why I have decided not to cultivate along the contour during the next season.

Mr Y.Z. Jangaiah, a marginal SC farmer of Ghatpally (the non-intervention village) said:

I own and cultivate only one dryland plot measuring about one hectare. Since this plot is almost level I am not really concerned about soil and water loss. The only measure I have taken is the construction of bunds along its borders. These bunds serve two purposes. Next to soil and water conservation they function as demarcation, which is in fact more important. I have heard about and have seen both contour bunds and *khus* near Maheswaram village. I will not construct contour bunds myself because labour costs are high, because water will stagnate in front of the bund and because I have doubts about the positive effects on yields. If the government decides to construct these type of bunds in my plot I will allow them only if all farmers in our village are in favour of construction and if some changes are made. I do not favour *khus* because I have seen that much of it has died.

Normally, preparatory tillage for *jowar* starts somewhere in April. Although it is less fixed for castor, I prefer to have this kind of tillage completed by the end of June because an early harvest of castor gives a higher price. I prepare the land by ploughing once up and down the small slope and once across, or the other way around. This criss-cross ploughing is the only proper way to open the soil and to make it soft. The first turn of the plough or harrow is always against last year's sowing direction. It is better for the plants not to grow in the same furrow as last year's plants and by sowing up and down the slope in one year and across the slope in the next, this is avoided and all the fertility present in the soil is used by the new plants. In our village it is mutually understood that everybody sows in the same direction, at least for the more or less level plots. It is not a kind of strict law because there is no punishment for those who do not adhere to this rule. Only a few people, particularly those who have plots on a steeper gradient, sow across every year. They are more interested in conserving soil and water and less interested in getting good yields. I have never seen or heard about contour cultivation and dead furrows.

7.1 An Overview

In a nutshell Table 7.1 demonstrates that the soil and water conservation programme of both projects were not very successful. The highest adoption rates of respondents was for indigenous technologies promoted by both projects. Except for the construction of graded/contour bunds and the planting of *khus*, the adoption of 'real' innovations introduced by the Maheswaram project and AWARE, was negligible. But with 10 per cent or less, the percentage of respondents who had adopted these bunds and/ or *khus* was not very spectacular either. The next sections take the analysis further, concentrating on the characteristics of the recommendations, who adopted them and who did not and the reasons for the same.

7.2 Graded/Contour Bunds and *Khus* Lines

The Department of Agriculture had already built some graded bunds in the Maheswaram project area before the commencement of the project. However, these were neither constructed in Debbadaguda and Mansanpally (the 'Maheswaram villages') nor in Ghatpally (the 'non-intervention village'). The same Department carried out a graded bunding programme in Amangal *mandal*, also in Ekvaipally, before AWARE started its intervention in the area. Besides promoting graded/contour bunds and *khus* by informing farmers about their existence and advantages, the Maheswaram project also constructed graded/contour bunds and planted *khus*. In Mansanpally it predominantly built graded/contour bunds and to a lesser extent planted *khus*. In Debbadaguda it planted *khus* only. In contrast to what was said by the AWARE management, AWARE had not

TABLE 7.1. RATE OF ADOPTION OF RECOMMENDED SOIL AND WATER CONSERVATION TECHNOLOGIES, BY VILLAGE

Recommendation	Debbadaguda	Mansanpally	Ekvaipally	Ghatpally	Total
Contour bunds and *khus*[3]	10	3	7	0	5
Post-harvest tillage[3]	0	0	0	0	0
Summer harrowing (*jowar*)[1]	33	88	18	44	48
Contour/across sowing[1]	63	0	54	16	34
Dead furrows[3]	0	0	0	0	0
Average	18	15	13	10	15

Note: Since not all recommendations were innovations, I distinguish: 1 = indigenous technology; 2 = technology introduced before the start of the projects; and 3 = (elements of a) technology introduced by the projects.

Source: Own survey.

built any of these structures or planted *khus* in Ekvaipally. AWARE informed farmers on graded/contour bunds, not on *khus*, during their training courses. Hence, one would expect some effect of their social motivation and agricultural training programmes and of the T&V extension officers, which were supposed to visit Ekvaipally on request of the *sanghams* and inform their members about soil and water conservation.

The open interviews suggested that neither the Maheswaram project nor AWARE were very successful in informing farmers orally or in writing about graded/contour bunds. Only three respondents in Debbadaguda and six respondents in Mansanpally said that they had been informed by project officers of the Maheswaram project. In Ekvaipally, 6 persons reported receiving oral or written information. However, as sources they mentioned fellow farmers and the T&V extension service, not AWARE. Nevertheless, for many respondents the Maheswaram project was a better source of information than AWARE. This was the result of the graded/contour bunds which had been constructed in the project area covered by the first phase. They were clearly visible, which allowed observation. Moreover, the mighty bulldozers which had been used to construct these bunds had impressed many respondents. Thus, during the open interviews most respondents living in Debbadaguda and Mansanpally said that they knew about graded/contour bunds because they had seen these in the project area. The fact that almost all respondents from Ghatpally (the non-intervention village) knew about these bunds, shows the importance of observation. All twenty-five interviewed in Ekvaipally reported to know about contour/graded bunds too. Most of them had seen the few graded bunds present in the village itself, on plots of Reddy farmers, and a minority said that they had seen these bunds in the Maheswaram project area.

The same open interviews indicated that the Maheswaram project was more successful in informing farmers orally or in writing about *khus* than about bunds. Although somewhat less important than with graded/contour bunds, observation was also important. All the respondents who I met for the open interviews in Debbadaguda knew about *khus*. Of those, fifteen respondents were informed by the Maheswaram project and the remaining six had seen these vegetative hedges in the field. Of the nineteen interviewed in Mansanpally, fifteen knew about *khus*, and ten persons had seen these hedges in the field. In contrast to the graded/contour bunds which were clearly visible, for respondents in Ekvaipally and Ghatpally it was not so easy to observe *khus* in the Maheswaram project area because much of it had died. In fact, many of the more informed Reddy farmers in these two villages also often confused *khus* with the tall indigenous grasses which grew everywhere in the villages. Nevertheless, still six persons in Ekvaipally and thirteen interviewed in Ghatpally knew about these hedges. Almost all of them knew about the *khus* lines because they had seen them in the area, and not because they had been informed by extension agents. In fact, only one Ekvaipally farmer had been informed by AWARE. This might seem strange since AWARE did not promote *khus* in the area. However, he had seen khus lines in AWARE's farm-cum-training centre where they had been planted by the Maheswaram project.

Before analysing the adoption rates of farmers, one remark needs to be made about the sincerity of AWARE's soil and water conservation programme. As AWARE's farm-cum-training centre was located in the Maheswaram project area (see Map 5.1), this latter project had constructed contour bunds and had planted *khus* lines. However, AWARE preferred irrigated agriculture and horticulture (with drip irrigation) above dryland agriculture and was not very interested in maintaining these bunds and these *khus* lines. In fact, some bunds had partly been demolished by ploughing and sowing up and down the slope, over the bunds. Other bunds had been demolished by deep pits which had been dug to plant fruit trees. As a kind of self-fulfilling prophecy (because many *khus* plants had died, *khus* lines had been interrupted in many places and plants had come up in the plot itself) AWARE staff did not maintain the *khus* properly.

Since the adoption rates for graded/contour bunds and *khus* hardly differed in the study period, only the results for *kharif* 1992/3 are presented in Figures 7.1 to 7.3.[1] At the outset it has to be stressed that the figures only relate to those dryland plots in which the recommended technologies were still found. The comparison of the observed average adoption rates at village level (Figure 7.1) with the hypothesis on expected adoption rates (see Figure 4.1), gives a clear indication that project interventions did generate the expected effect, meaning that the hypothesis was confirmed. But there is no conclusive evidence, though. First of all, the averages indicate that even in the best performing village, Debbadaguda, only 10 per cent of the respondents had adopted these recommended technologies on dryland. That *khus* was an outright failure in the Maheswaram project was reported by Sivamohan et al. (1993) too. This low figure must be very disappointing from the project's point of view. In fact, the dryland protected by the recommended technologies in all four study villages was limited to a mere 73 ha of the total 971 ha owned by the respondents (8 per cent): in Debbadaguda to 20 ha, in Mansanpally to

4 ha and in Ekvaipally to 48 ha. As expected, no dryland plot in Ghatpally had contour bunds or *khus*. Second, the observed average adoption rates in Ekvaipally were the result of government activities prior to AWARE's arrival in the village and not of AWARE's intervention. Finally, most of the respondents were partial adopters. The graded/contour bunds were constructed or the *khus* planted by the projects in only a part of their dryland holding.

The comparison of the observed average adoption rate by community in the four study villages (Figure 7.2) with the hypothesis on expected adoption rates (see Figure 4.2), gives the following picture. In Debbadaguda, relatively more ST respondents than FC and SC respondents (having the same average adoption rate) had *khus* in their dryland plots. The fact that only a few FC respondents had *khus*, can be explained by the fact that much of the *khus* in their plots had died, that many had destroyed their *khus* lines by harrowing/ploughing their plots up and down the slope and because quite a few had converted their treated dryland plots into wetland. Since it was not expected to find a relatively higher number of treated plots of ST and SC respondents than of FC and BC respondents, the hypothesis for Debbadaguda was rejected.

Although the Maheswaram project was supposed to cover only the ST territory in Mansanpally, graded/contour bunds were constructed and *khus* was also planted elsewhere. Some Reddy farmers said that they, because of their contacts with project officers and with a Member of the Legislative Assembly who owned a grape garden in the village, received these technologies too. Thus, Figure 7.2 shows that quite a number of the ST and FC respondents had adopted these recommendations, in the order as expected. However, since none of the BC and SC respondents had adopted these recommendations, the hypothesis was neither confirmed nor rejected.

Relatively speaking, most of the treated plots in Ekvaipally were owned by FC respondents, only a few by BC and ST respondents (having the same average adoption rate). That the average adoption rate for ST and especially SC respondents was low to negligible can be explained by the fact that it was not AWARE but the government that had constructed these bunds. It meant that the hypothesis was rejected. This also applies to Ghatpally, where adoption of these recommendations was totally lacking.

As regards the expected and observed average adoption rates of the communities between the four study villages, these differed considerably. The average adoption rate of FC and BC respondents was highest in Ekvaipally while it was expected to be highest in Debbadaguda. The averages for SC and ST respondents were highest in Debbadaguda while

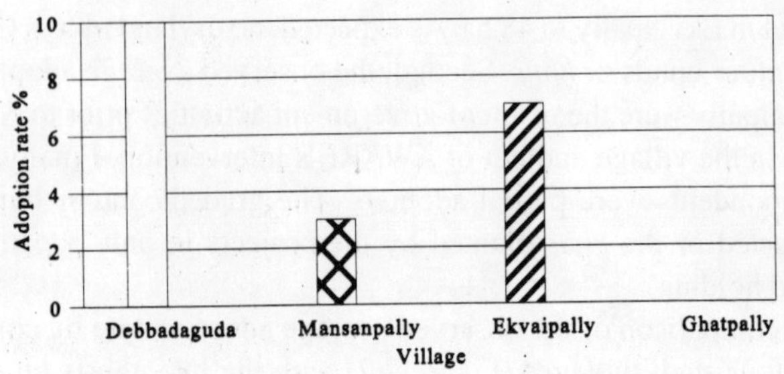

Source: Own survey.

FIGURE 7.1. OBSERVED ADOPTION RATE (%) OF RECOMMENDED SOIL AND WATER CONSERVATION STRUCTURES ON DRYLAND PLOTS, BY VILLAGE (*kharif* 1993/4).

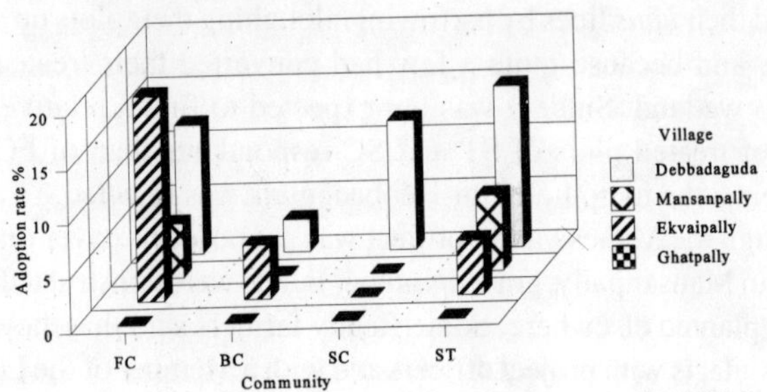

Source: Own survey.

FIGURE 7.2. OBSERVED ADOPTION RATE (%) OF RECOMMENDED SOIL AND WATER CONSERVATION STRUCTURES ON DRYLAND PLOTS, BY COMMUNITY, BY VILLAGE (*kharif* 1993/4).

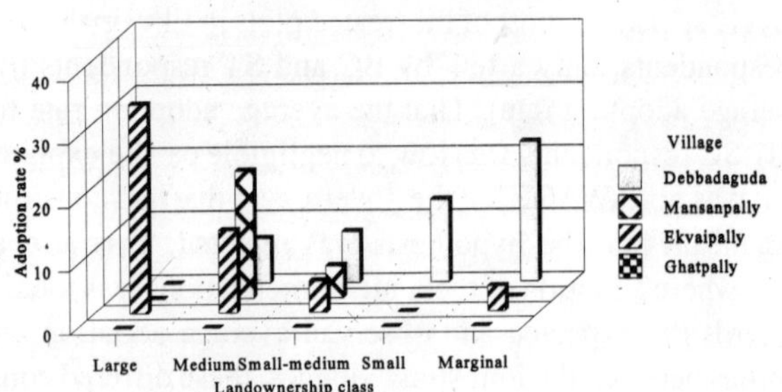

Source: Own survey.

FIGURE 7.3 OBSERVED ADOPTION RATE (%) OF RECOMMENDED SOIL AND WATER CONSERVATION STRUCTURES ON DRYLAND PLOTS, BY LANDOWNERSHIP CLASS, BY VILLAGE (*kharif* 1993/4).

they were expected to be highest in Ekvaipally. Since these observed rankings differed from what was expected, the hypothesis was rejected. It clearly meant that for the recommended soil and water conservation measures at least, the activities of the Maheswaram project had more effect on SC and ST respondents than the activities of AWARE. This was largely a consequence of the planting of *khus* in Debbadaguda, which was said to be an affordable, simple and replicable technology (GoI, 1988: I). But since most *khus* had died, it is highly doubtful whether this is a sustainable technology.

The comparison of the observed average adoption rates by holding-size in the four study villages (Figure 7.3) with the hypothesis on expected adoption rates (see Figure 4.3), gives the following picture. Relatively speaking, most of the surviving *khus* in Debbadaguda were found in dryland plots belonging to marginal and small farmers. Since this was in complete contrast to what was expected, the hypothesis was rejected. In Mansanpally, relatively most of the treated plots belonged to medium, followed by small-medium farmers. Since it was expected that small farmers would perform as well as medium farmers and marginal farmers as well as small-medium farmers, the hypothesis was rejected.[2] In Ekvaipally relatively more large and medium farmers had these new structures in their dryland plots. Since this finding was not expected, the hypothesis was also rejected. The same conclusion applies to Ghatpally, where adoption of the recommended structures was completely lacking.

As with community, the average adoption rates of the various land-ownership classes between the four study villages, differed considerably from what was expected. Whilst it was expected that the average adoption rate of large and medium farmers would be highest in Debbadaguda, in reality such was not the case. For large farmers the average was highest in Ekvaipally; for medium farmers it was highest in Mansanpally followed by Ekvaipally and Debbadaguda. Whilst it was expected that the average adoption rates of small and marginal farmers would be highest in Ekvaipally, they were highest in Debbadaguda. Only for small-medium farmers the observed ranking was similar to the one expected. Since these observed rankings differed from what was expected, the hypothesis was rejected. Again, it clearly meant that for the recommended soil and water conservation technologies, at least, the activities of the Maheswaram project had more effect on small and marginal farmers than the activities of AWARE. Whatever this outcome, it must be kept in mind that adoption rates were very low.

TABLE 7.2. COMPARISON BETWEEN OBSERVED ADOPTION RATES AND HYPOTHESES FOR RECOMMENDED SOIL AND WATER CONSERVATION TECHNOLOGIES

Intervention level	Confirmed	Neither confirmed nor rejected	Rejected
Village	Yes but low		
Community in village		M	D, E, G
Community between villages			Yes
Landownership class in village			D, M, E, G
Landownership class between villages			Yes

Note: D = Debbadaguda; M = Mansanpally; E = Ekvaipally; G = Ghatpally.
Source: Own survey.

The question arises why adoption rates were so very low and why respondents did not built the new structures themselves? To answer the first question partly, they were low because of project activities. Thus, adoption rates for graded/contour bunds and *khus* in Mansanpally were low because the village territory was only partly included in the project. The construction of these bunds and the planting of *khus* on dryland plots located outside the project area, the result of the wheeling and dealing between the landed elite in the village and project officers, had hardly improved the situation. The rates were low in Ekvaipally because AWARE had not implemented its soil and water conservation programme in this village.

They were also low because lack of local participation had created resentment among many respondents towards the Maheswaram project. Resentment was also present in Ekvaipally, not towards AWARE but towards the government which had executed the work. For not being consulted during the identification and formulation phases, most respondents in all three project villages felt that they had not been able to influence its contents. During the implementation phase, if they participated at all, many found that they had no other option but to allow construction or planting activities in their plots. Dissatisfaction was, for instance, present in Debbadaguda where the project planted *khus* while many farmers clearly preferred contour bunds. That many farmers none the less allowed these planting activities in their dryland plots was because they expected and also received other benefits from the project like free or subsidized improved seeds, inorganic fertilizers and fruit trees. Preferring these inputs above *khus* meant that when much of the *khus* died, most farmers just let it pass. This attitude had everything to do with the fact that they felt that these plants were planted for the sake of the project and project officials, but not for their sake or benefit. Overall, the objectives of the project were not in line with those of most farmers.

Many respondents also felt dissatisfied because they had not been properly informed about the contents of the soil and water conservation programme, the timing of the operations, the criteria for selection of beneficiaries and of plots, the financing and subsidization of the works and they criticized the high-tech operations which they were unable to grasp. In Mansanpally, but even more in other villages included in the first phase of the project (see for details Sitapathi Rao et al., 1990), misinformation and lack of willingness of farmers to participate led to confrontations between farmers and project officers. On a number of occasions it led to fights or near fights when farmers found that project staff had started constructing contour bunds without their consent.

But low adoption rates were a consequence of the technology itself too. This is evident from the fact that even though many respondents had seen the graded/contour bunds and, to a lesser extent the *khus* lines, not a single respondent had felt the need to construct or plant these in his own plots. Thus, observation does not necessarily lead to adoption. While almost all respondents were convinced that graded/contour bunds prevented soil and water loss, those who knew about *khus* were not convinced about its advantages. They knew that it filtered the water and that soil particles were prevented from running down, but the fact that so much *khus* had died, did not make them believers. In fact, an important explanation for the limited success of the Maheswaram project in Debbadaguda was the demise and deliberate destruction of much of the *khus*. Limited compatibility of this 'imported' technology to the agro-ecological condition in the area is evident from the fact that, according to respondents, on average less than 20 per cent of the *khus* lines in Debbadaguda had survived. Many plants, especially those planted too late in the season, died because of water-stress, or because they were attacked by termites. Limited compatibility of this innovation to the existing agricultural preferences is evident from the fact that many farmers continued ploughing and harrowing in a criss-cross pattern through these *khus* lines. In addition, some farmers, particularly the larger ones, converted their dryland plots planted with *khus*, into wetland. If allowed to choose between the two alternatives, most respondents preferred contour bunds. That this preference did not lead to construction was based on its properties, mainly relative disadvantage, incompatibility and limited trialability.

Contour bunds are a clear example of what Rogers (1995: 70, 217) calls preventive innovations. In answer to the second question asked earlier, lack of interest for constructing these bunds themselves was grounded on a firm belief that construction would cost too much labour and capital, while the immediate benefits were not clearly discernible. As will be demonstrated in the next chapter, it confirms the hypothesis which was formulated in the fourth chapter, that farmers are more reluctant to adopt soil and water conservation innovations than crop management innovations. It also means that informing farmers about the benefits of these structures is not sufficient and that change agencies have to use incentives like subsidized or free inputs to stimulate the adoption of these preventive innovations or even have to finance the construction themselves. Indeed, many respondents would accept these contour bunds if the government was to build them free of cost or if the government offered a grant or paid reasonable wages to the beneficiaries to construct the bunds themselves.

Another reason why it is extremely doubtful that farmers themselves will construct these new structures on dryland, is that a large majority of the interviewed farmers valued their own indigenous structures above contour bunds. That the relative advantage of these recommended structures was not sufficiently high is also shown by the fact that not a single respondent had demolished indigenous bunds and built the recommended ones. Moreover, they were not interested in adapting the recommended structures to their wishes. Insufficient relative advantage together with the fact that it was very expensive to build these kinds of bunds, prevented all respondents from giving it a try.

Moreover, respondents as well as other farmers who were interviewed in Mansanpally were not always pleased with the graded/contour bunds constructed by the Maheswaram project. Their dissatisfaction was primarily aimed at those contour bunds which were constructed by bulldozers, which they thought were very small and fragile. In fact, quite a few had been broken and had been repaired by the Maheswaram project. Other farmers complained that a gully had formed in their plot because excess water was not properly drained from a contour bund constructed in their neighbour's plot, which stopped at the border of their own.

The improved technologies have not been able to render the indigenous soil and water conservation technologies superfluous. A large majority of 63 per cent of the dryland area was protected by border bunds or other forms of indigenous technology. These indigenous structures were used on a relatively large scale in Ghatpally and Mansanpally (by some 90 per

TABLE 7.3. RATE OF ADOPTION OF INDIGENOUS AND RECOMMENDED SOIL AND WATER CONSERVATION TECHNOLOGIES, BY TYPE OF FARMER, BY VILLAGE

Type of farmer	Debbadaguda		Mansanpally		Ekvaipally		Ghatpally		Total		N plots
	Ind.	Rec.	Ind.	Rec.	Ind.	Rec.	Ind.	Rec.	Ind.	Rec.	
Pure dryland farmers	59	7	84	5	60	3	98	0	78	3	188
Farmers with < 1 ha wetland	49	14	91	0	40	3	96	0	67	4	189
Farmers with ≥ 1 ha wetland	22	9	82	9	55	19	87	0	55	11	80
Average	47	10	87	3	49	7	96	0	69	5	457

Note: Ind. = indigenous technology; Rec. = recommended technology by the projects.
Source: Own survey.

cent of the respondents) and on a much smaller scale in Ekvaipally and Debbadaguda (less than 50 per cent of the respondents). Hence, and in spite of project activities, most of the unprotected area was located in Debbadaguda (164 ha or 53 per cent of the dryland area owned by respondents) and in Ekvaipally (81 ha or 30 per cent), against 25 ha (13 per cent) in Mansanpally and 19 ha (9 per cent in Ghatpally). In Debbadaguda, in particular large FC farmers contributed to the unprotected area. In Ekvaipally, a relatively large part of the unprotected area was owned by marginal, small and small-medium farmers, mostly of the SC and BC community. If not for the new structures built in their plots by the Maheswaram project and by the government, FC farmers in both Debbadaguda and Ekvaipally would have had much more unprotected dryland plots.

Pure dryland farmers were more interested in taking protective measures against soil and water loss than those also owning wetland. Moreover, pure dryland farmers were more inclined to protect all their dryland plots while most of the other farmers were less interested in doing so. Table 7.3 shows that, except for Mansanpally and Ekvaipally, a negative relationship existed between the adoption of indigenous and recommended structures and the wetland area owned by a farmer. Why such a relationship was less evident in Mansanpally and Ekvaipally is not known. What is apparent in these two villages, though, is that a large number of those farmers who owned more than 1 ha of wetland, benefited from the Maheswaram project and the government bunding programme. In fact, what is obvious from the table is that most of the recommended structures have been built in plots of respondents who also owned wetland, farmers who were less interested in protecting their dryland plots from soil and water loss. This agrees with the earlier finding that most of these treated plots belonged to medium and large farmers, often of the FC community, a community which owned large tracts of wetland.

A clear relationship between the presence of soil and water conservation structures and the gradient of the plot was not discernible. The assumption that plots located on steeper slopes would be better protected than more or less level plots, proved wrong. Some 76 per cent of the more or less level plots were protected, while the same held true for 70 per cent of the plots located on medium slopes and 74 per cent of the plots located on slopes having higher gradients. This might be another indication that border bunds were not so much constructed to prevent soil and water loss but more to demarcate plots. It is also surprising to find that about half of the plots having graded/contour bunds or *khus* were located on what farmers called more or less level plots. But then, by writing that the slope in a major part of the project area was under 2 to 3 per cent (GoAP, 1985c: 15), the Maheswaram project implicitly admitted that soil and water conservation was not very relevant, at least for many farmers.

7.3 *In-situ* Soil and Water Conservation

7.3.1 OFF-SEASON TILLAGE

Contour/graded bunds and *khus* decrease the length of a dryland plot, thereby decreasing the velocity of the run-off, hence minimizing soil and water loss in the plot. However, to optimize soil and water conservation additional practices were considered necessary by both projects. These '*in-situ* soil and water conservation practices' or 'improved interbund treatments' were meant to increase the infiltration of rainwater into the root profiles, thereby improving crop performance, while on the other hand limiting soil erosion. One of the three elements promoted by the Maheswaram project and AWARE was off-season tillage, consisting of post-harvest and summer tillage. As indicated in Chapter 4, while post-harvest tillage was a 'real' innovation, summer tillage for *jowar* was an indigenous practice.

The only way farmers were induced to adopt off-season tillage operations was by informing them about the contents and benefits. Neither project gave financial incentives. Except four respondents who stated during the open interviews that they had heard about post- harvest tillage, all others were unaware of this recommendation. Since not a single dryland plot was ploughed after harvest (Table 7.1), it may be concluded that both the Maheswaram project and AWARE were not successful at all in stimulating farmers to adopt this recommendation. Consequently, the hypothesis concerning the effect of both projects on post-harvest tillage was rejected. This conclusion more or less proves what was stated in Chapter 6 about the irregular and insufficient supply of information by the extension service, at least for this recommendation. The negligible adoption rate makes a comparison between communities and landholding-size classes by village and between the expected and observed adoption rates not very relevant.

Above-mentioned finding was confirmed by the open interviews. Tillage after a *kharif* crop will only be carried out as preparatory tillage for a *rabi* crop like horsegram or safflower, not as a post-harvest tillage for a next *kharif* crop for the sake of soil and water conservation. After I had informed farmers during the open interviews about this particular recommendation, all argued that it was unacceptable for reasons of limited relative advantage and incompatibility. While they perceived an extra ploughing without sowing as a waste of time and money [most farmers preferred to get additional cash income from (agricultural) labour], they were not convinced that it would enhance the moisture retention capacity of the soil. Some respondents even maintained that opening of the soil by ploughing after *kharif* harvest would not be beneficial at all since the hot summer sun would evaporate the little moisture available in the top layer. They asserted that the crust formed after *kharif* was in fact beneficial since it reduced evaporation. Most farmers intercropping *jowar* with redgram opposed post-harvest tillage because its incompatibility with the cropping pattern. Tillage was difficult, they claimed, since the indigenous redgram variety, which ripened only in December or January, was still in the field.

In contrast to post-harvest tillage, many respondents (but certainly not all) harrowed their *jowar* plots in summer. In the absence of a clear definition given by both projects, summer is rather loosely defined as the period between March and the first fortnight of May. The second fortnight of May is excluded since it coincides with the period in which most farmers preferred to sow their *jowar* crop. Summer tillage for *jowar*, an indigenous practice, was undertaken on average on 48 per cent of the *jowar* plots in summer 1992/3 and 1993/4 (Table 7.1). Large differences existed between the two years. While in 1992/3, on average 63 per cent of the *jowar* plots were harrowed in summer, it decreased to 33 per cent in 1993/4. This difference is largely a consequence of the differing rainfall pattern between the years. Compared to 1992/3, the number and intensity of pre-monsoon showers was worse in 1993/4. Lack of good rains before the start of the agricultural season made most farmers decide to skip summer tillage and to harrow only a couple of days before sowing.

The comparison of the average observed adoption rates at village level (Figure 7.4) with the hypothesis on expected adoption rates (Figure 4.1), disclosed that project interventions were not able to generate the expected effect. This conclusion applies more to AWARE than to the Maheswaram project. A rather large difference existed between the four villages where Mansanpally scored best (on average 88 per cent of the *jowar* plots were harrowed in summer), followed by Ghatpally (44 per cent) and Debbadaguda (33 per cent), while Ekvaipally (18 per cent) came last. Except for Debbadaguda and Ekvaipally, which swapped places in 1993/4, the ranking of the other two villages remained the same in the two years. However, since another ranking was expected, the hypothesis was rejected.

On the basis of the observed average adoption rate by community in the four study villages (Figure 7.5) and the hypothesis on the expected adoption rate (Figure 4.2), the following conclusions can be drawn. In Debbadaguda, highest average adoption rates were for ST respondents, followed by SC respondents, FC respondents and BC respondents.[3] Since this ranking was not in line with what was expected, the hypothesis was rejected. In Mansanpally, the difference between the various communities was very small, which was not expected. Moreover, while the observed ranking for ST and FC respondents was as expected, the ranking of the other two communities differed from what was anticipated. Hence, the hypothesis was also rejected. As expected, the highest average adoption rate in Ekvaipally was for ST respondents. However, since the ranking of the other communities (BC respondents were next in line, followed by SC and FC respondents) was not as expected, the hypothesis was rejected. In Ghatpally, ST respondents had the highest averages, which was not expected. They were followed by FC, BC and SC respondents, a ranking that was expected. Hence, the hypothesis was neither confirmed nor rejected.

The observed average adoption rates of the communities between the four study villages differed considerably from what was expected. For all communities, the highest average adoption rates were highest in

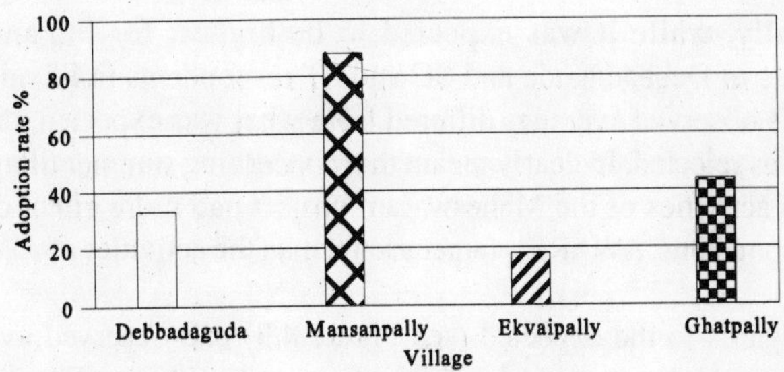

Source: Own survey.

FIGURE 7.4 OBSERVED ADOPTION RATE (%) OF SUMMER TILLAGE OF *JOWAR*, BY VILLAGE (*kharif* 1992/3-1993/4).

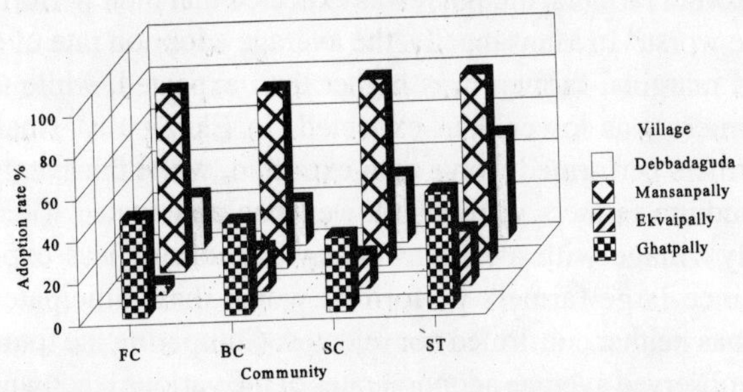

Source: Own survey.

FIGURE 7.5. OBSERVED ADOPTION RATE (%) OF SUMMER TILLAGE OF *JOWAR*, BY COMMUNITY, BY VILLAGE (*kharif* 1992/3-1993/4).

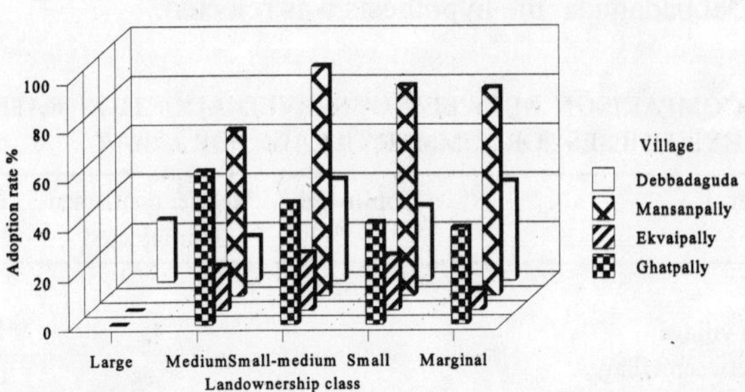

Source: Own survey.

FIGURE 7.6. OBSERVED ADOPTION RATE (%) OF SUMMER TILLAGE OF *JOWAR*, BY LANDOWNERSHIP CLASS, BY VILLAGE (*kharif* 1992/3-1993/4).

Mansanpally, while it was expected to be highest for FC and BC respondents in Debbadaguda and SC and ST respondents in Ekvaipally. Since these observed averages differed from what was expected, the hypothesis was rejected. It clearly meant that concerning summer tillage for *jowar*, the activities of the Maheswaram project had more effect on SC and ST respondents, AWARE's target group, than the activities of AWARE itself.

With regards to the expected (see Figure 4.3) and observed average adoption rates by landowner class in the four study villages (Figure 7.6), the hypotheses were rejected for all three project villages. In Debbadaguda, the average adoption rate was highest for small-medium farmers, followed by marginal and small farmers. All three classes had higher averages than large and medium farmers, though it was expected that their performance would be the worst.[4] In Mansanpally, the average adoption rate of small-medium and marginal farmers was higher than expected, while that of medium farmers was lower than expected. In Ekvaipally, small and marginal farmers performed worse than expected, while the averages of large and medium farmers were also lower than anticipated. Ghatpally was the only village with a ranking that was more or less expected. However, since large farmers performed worse than anticipated, the hypothesis was neither confirmed nor rejected. Comparing the four study villages, the observed average adoption rates of the various landownership classes differed considerably from what was expected. The averages for medium, small-medium, small and marginal farmers were highest in Mansanpally, followed by Ghatpally, Debbadaguda and Ekvaipally. Since it was expected that small and medium farmers living in Ekvaipally would have the highest average adoption rates, followed by large and medium farmers in Debbadaguda, the hypothesis was rejected.

TABLE 7.4. COMPARISON BETWEEN OBSERVED ADOPTION RATES AND HYPOTHESES FOR SUMMER TILLAGE FOR *JOWAR*

Intervention level	Confirmed	Neither confirmed nor rejected	Rejected
Village			Yes
Community in village		G	D, M, E
Community between villages			Yes
Landownership class in village		G	D, M, E
Landownership class between villages		Yes	

Note: D = Debbadaguda; M = Mansanpally; E = Ekvaipally; G = Ghatpally.
Source: Own survey.

Low adoption rates for summer tillage for *jowar* were only partly related to project performance. Thus, all except nine respondents (two in Debbadaguda, three in Mansanpally, one in Ekvaipally and three in Ghatpally) stated during the open interviews that they knew about this practice. Those who knew had been informed by their fathers, not by the Maheswaram project, AWARE or the T&V system. But the fact that farmers who knew this indigenous technology did not necessarily use it each and every year, indicates that a combination of 'non-project reasons' were more important. Most reasons related to the environment: to the weather and to soil characteristics. As indicated earlier, many respondents postponed preparatory tillage because pre-monsoon showers were insufficient to soften the soil and to facilitate harrowing. In fact, insufficient rain in summer made many respondents harrow only a couple of days before sowing. Second, farmers owning dryland plots with a specific soil type sometimes considered the topsoil already sufficiently soft to permit water to infiltrate with ease. Third, as with post-harvest tillage, some farmers were afraid that harrowing in summer would open and soften the soil too much, only stimulating evaporation. Summer tillage was less inhibited because of resource endowments, more specifically the lack of animal traction. Although bullocks were not always in a perfect shape on the eve of another agricultural season, the timing of the operation was such that sufficient bullocks were available.

One might assume a negative relationship between the area of wetland owned and adoption of summer tillage for *jowar*, arguing that farmers who depended entirely on dryland for their agricultural production would be more inclined to harrow their plots in summer than those farmers who owned wetland as well. Table 7.5 shows that such is not always the case. Only in Mansanpally and to a lesser extent in Ghatpally, pure dryland

TABLE 7.5. RATE OF ADOPTION OF SUMMER TILLAGE FOR *JOWAR*, BY TYPE OF FARMER, BY VILLAGE

Type of farmer	Debbadaguda	Mansanpally	Ekvaipally	Ghatpally	Total	N plots
Pure dryland farmers	28	93	10	41	48	115
Farmers with <1 ha wetland	40	85	24	49	51	127
Farmers with ≥ 1 ha wetland	28	77	18	37	35	28
Average	33	88	18	44	48	270

Source: Own survey

farmers harrowed more in summer than farmers owning more than 1 ha of wetland. In fact, in all villages except Mansanpally, summer tillage for *jowar* was practised more by farmers owning less than 1 ha of wetland. A reason for the lower than expected adoption rates of pure dryland farmers might be the fact that more than other farmers, they worked as daily labourer outside agriculture and often outside the village in summer. It is not known why Mansanpally is an exception. It might be that good rainfall during subsequent summers induced many farmers to harrow their *jowar* plots in this season. On the other hand, it might be that village customs (see the following section) prescribe farmers to harrow in summer. Also, farmers working as daily wage labourer in Hyderabad in summer often returned to the village for a couple of days to prepare their *jowar* plots.

7.3.2 CONTOUR CULTIVATION

Contour cultivation was another element of the *in-situ* soil and water conservation practices in dryland. It implied that tillage operations like preparatory tillage, sowing and interculturing, invariably had to follow the contours. However, both the Maheswaram project and AWARE were not very specific about the concept. Their documents show that it was watered down to across cultivation and reduced to sowing only. Both organizations as well as the T&V extension system argued that since many plots had multiple slopes (which was rather common in this undulating terrain) farmers would find it difficult to cultivate along the contour. Hence, increasingly they informed farmers about the contents and benefits of sowing across the major slope and not on all aspects of contour cultivation. Financial incentives were not given.

Most respondents were unaware about cultivation of dryland crops along the contour. Of the eighty-eight farmers questioned during the open interviews, sixteen stated that they had heard of sowing along the contour: eight in Debbadaguda, four in Mansanpally, one in Ekvaipally and three in Ghatpally. Most of them (eight farmers) had been informed by the Maheswaram project, five respondents had seen other farmers in the project area sowing along the contour and the other three farmers had heard of this recommendation from other farmers living in non-project villages.

A large majority of sixty-six respondents who participated in the open interview sessions were familiar with the advice to sow across the slope. Of them, nineteen lived in Debbadaguda, eight in Mansanpally, twenty-two in Ekvaipally and seventeen in Ghatpally. Quite a number of respondents living in the Maheswaram project villages said that project officers had informed them about sowing across the major slope. But most said that it was an indigenous practice and that their (fore)fathers already sowed across the slope every year or sowed up and down the slope and across the slope in alternate years. Only five respondents said that it was the Maheswaram project that had informed them about this recommendation. Nevertheless, a considerable number of respondents in Debbadaguda confirmed that on the basis of information given by the Maheswaram project, they had decided to sow across the slope. But while the Maheswaram project had some influence in Debbadaguda, it had none in Mansanpally. Only a minority of the Mansanpally respondents had met project officers who advised them to sow across every year. Some of them were very displeased with the way these officers tried to convince them. On a number of occasions officers had prevented farmers from sowing up and down the slope by taking hold of the bullock team and by ploughing in the recommended direction themselves. In Ekvaipally too, quite a number of respondents said that they had been informed about sowing across the slope by AWARE or by T&V extension officers when they paid a visit on behalf of AWARE. But they also stated that they were already familiar with this practice and that the information had not made them alter their sowing direction.

The next sections are based on the three practices which can be undertaken along the contour or across the slope: i.e. land preparation, sowing and interculturing. The section on land preparation can be a limited one since not a single respondent in the four study villages prepared his dryland by ploughing or harrowing along the contour or by doing it only across. It meant that the hypothesis on the outcome of project interventions on this improved way of preparatory tillage, was rejected. The negligible adoption rate makes a comparison between communities and landholding-size classes by village and between the expected and observed adoption rates not very relevant.

Concerning contour/across sowing, Table 7.1 shows that averaged for *kharif* 1992/3 and *kharif* 1993/4 rather large difference existed between the four villages. While adoption rates were high in Debbadaguda and Ekvaipally (where, respectively, on average 63 and 54 per cent of the dryland plots were sown in one of the recommended ways), they were low to very low in Ghatpally and Mansanpally (with, respectively, on average 16 and 0 per cent). Except for some minor differences in the rates of the two years, the ranking of the study villages remained the same.

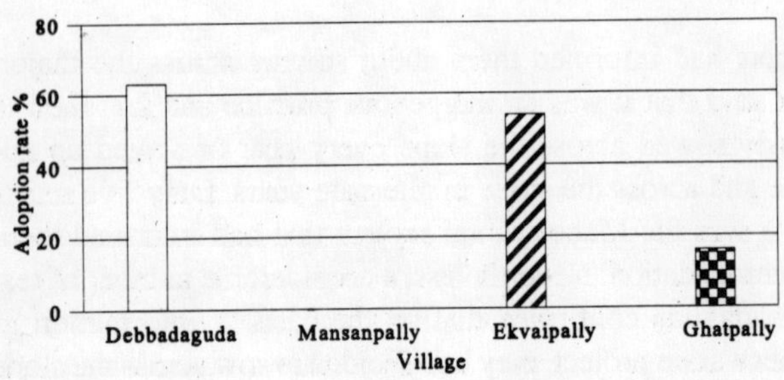

Source: Own survey.

FIGURE 7.7. OBSERVED ADOPTION RATE (%) OF SOWING DIRECTION ON DRYLAND PLOTS, BY VILLAGE (*kharif* 1992/3-1993/4).

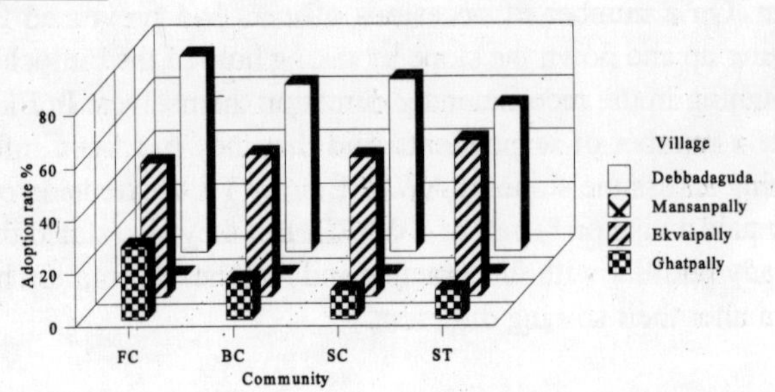

Source: Own survey.

FIGURE 7.8. OBSERVED ADOPTION RATE (%) OF SOWING DIRECTION ON DRYLAND PLOTS, BY COMMUNITY, BY VILLAGE (*kharif* 1992/3-1993/4)

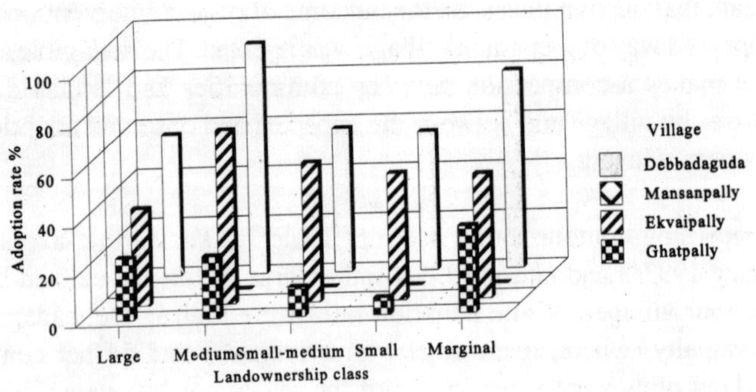

Source: Own survey.

FIGURE 7.9. OBSERVED ADOPTION RATE (%) OF SOWING DIRECTION ON DRYLAND PLOTS, BY LANDOWNERSHIP CLASS, BY VILLAGE (*kharif* 1992/3-1993/4).

Almost all of these plots were sown across every year and contour sowing was limited to only a couple of plots in each of the two *kharif* seasons. The comparison of these observed adoption rates at village level (Figure 7.7) with the hypothesis on expected adoption rates (Figure 4.1), indicated that with the exception of Mansanpally, the projects had a positive effect on the direction of sowing. However, because of the special position of Mansanpally, the hypothesis was neither confirmed nor rejected.

On the basis of the observed average adoption rates by community in the four study villages (Figure 7.8) and the hypothesis on the expected adoption rate (Figure 4.2), the following conclusions can be drawn. In Debbadaguda, the highest average adoption rates were for FC respondents, followed by SC respondents, BC respondents and ST respondents.[5] Since SC respondents performed better than was anticipated, this ranking was not completely in line with what was expected and the hypothesis was neither confirmed nor rejected. Things were very clear in Mansanpally. Since not a single respondent had adopted the recommendation the hypothesis was rejected. In Ekvaipally, on average ST respondents performed best, followed by BC, SC and FC respondents, whose averages did not differ that much. Because of these marginal differences, the hypothesis was neither confirmed nor rejected. The hypothesis for Ghatpally was also neither confirmed nor rejected as the observed ranking (on average FC respondents ranked first, followed by BC respondents, while no difference existed between SC and ST respondents) deviated slightly from the expected ranking.

The observed average adoption rates of each community by village, differed in such a way from what was expected, that the hypothesis was rejected. As was expected, the averages of FC and BC respondents were highest in Debbadaguda. But while the averages of these communities were expected to be similar in the other three villages, they were not. This

TABLE 7.6. COMPARISON BETWEEN OBSERVED ADOPTION RATES AND HYPOTHESES FOR CONTOUR/ACROSS SOWING

Intervention level	Confirmed	Neither confirmed nor rejected	Rejected
Village		Yes	
Community in village		D, E, G	M
Community between villages			Yes
Landownership class in village			D, M, E, G
Landownership class between villages		Yes	

Note: D = Debbadaguda; M = Mansanpally; E = Ekvaipally; G = Ghatpally.
Source: Own survey.

applies especially to Mansanpally. Whilst it was expected that the averages of SC and ST respondents would be highest in Ekvaipally, this assumption was only valid for ST respondents. Again, Mansanpally lagged behind, while Debbadaguda performed better.

As regards the expected (see Figure 4.3) and observed average adoption rates by landowner class in the four study villages (Figure 7.9), the hypotheses regarding the effect of project intervention on the various holding-size classes were rejected for all villages. In Debbadaguda, because large farmers performed worse and marginal farmers performed better than expected.[6] In Mansanpally, it had to be rejected because none of the respondents adopted the recommendation. In Ekvaipally, because small and marginal as well as large farmers performed worse than expected. In Ghatpally, because the average adoption-rate of marginal farmers was much better and that of large farmers worse than expected.

On the basis of the difference between the expected and the observed average adoption rates of each landowner class by village, the hypothesis was rejected. For all classes, the averages were highest in Debbadaguda, while it was expected that these would be highest for large, medium and small-medium farmers only. Thus, the expectation that for small and marginal farmers, the highest averages would be found in Ekvaipally, was incorrect. Also, the averages for most classes were rather low in Ghatpally and negligible in Mansanpally.

Similar to the other recommendations, one might assume a negative relationship between the area of wetland owned and adoption of across sowing. Table 7.7 shows that such is not the case. With the exception of Mansanpally and to a lesser extent Ekvaipally, highest average adoption rates were found amongst those farmers owning more than 1 ha of wetland, while in most villages pure dryland farmers performed worst.

The question arises whether the differences between the villages can be attributed to the activities of the projects or to some other factor?

TABLE 7.7. RATE OF ADOPTION OF ACROSS SOWING, BY TYPE OF FARMER, BY VILLAGE

Type of farmer	Debbadaguda	Mansanpally	Ekvaipally	Ghatpally	Total	N plots
Pure dryland farmers	62	0	41	14	27	319
Farmers with <1 ha wetland	58	0	61	17	35	322
Farmers with ≥ 1 ha wetland	78	0	57	25	50	117
Average	63	0	54	16	34	758

Source: Own survey.

Ghatpally and Mansanpally are, for instance, located in less undulating terrain than Debbadaguda and particularly Ekvaipally. Therefore, a plausible explanation concerning the different adoption rates for contour/across sowing might be the terrain. That the gradient of the plot influenced the sowing direction of a crop is shown by the fact that while 27 per cent of the more or less level plots were sown in the recommended way, the same happened with 55 per cent of the plots located on a steep gradient. Hence, project intervention in Debbadaguda and Ekvaipally was partly supported by these natural circumstances. In fact, respondents in Ekvaipally and Debbadaguda complained more often about the more undulating topography than their counterparts in Ghatpally and Mansanpally. Moreover, while relatively more respondents in the latter two villages argued that the more or less flat terrain made them prefer the indigenous way of sowing, relatively more respondents in the first two villages confirmed that the somewhat larger gradient made them sow across the slope.

Concerning interculturing along the contour or across the slope the following remarks can be made. As plants of most crops were sown at a very short distance from one another in the same furrow, farmers found it almost impossible to interculture against the direction of sowing. The only exceptions were a few farmers who intercultured by ploughing against the sowing direction of castor. As a result, plots sown up and down the slope were almost always intercultured up and down, while plots sown across the slope were almost always intercultured across. It means that adoption rates for interculturing on the contour or across the slope are similar to the rates of contour/across sowing. Consequently, everything that has been written on project success on the direction of sowing applies to the direction of interculturing as well.[7]

That both the Maheswaram project and AWARE had only limited success inducing respondents to change their indigenous ways of land preparation, sowing and interculturing is not a surprise given that they concentrated largely on the promotion of sowing across the major slope. Thus, lack of information was one of the reasons why respondents were not following all the recommendations for contour/across cultivation. But had they been better informed, it is highly doubtful that they would have accepted the advices. After having informed respondents about the contents and benefits of all recommendations myself, they maintained that most indigenous technologies were superior to the alternatives. Thus, as explained by the farmer in the introductory section of this chapter, all agricultural operations along the contour were thought to be too difficult, not compatible with other practices or too complex, not only for the farmer but also for the plough bullocks. Moreover, the indigenous way of criss-cross land preparation was regarded to be superior to preparing the land across the slope year in and year out. Respondents contended that only crosswise ploughing and harrowing would result in a good tilth, in a good mix of soil nutrients and in high yields. They were not convinced that the recommendation was superior to their own technology since they had never heard of any farmer trying it, nor had they seen such a trial themselves.

Concerning the direction of sowing, many respondents preferred to continue their indigenous criss-cross practice since they thought that slope was not much of a problem. They even felt that sowing across the (small or absent) slope every year was not a suitable recommendation since it would be detrimental to yields. In addition, the refusal of farmers in Mansanpally to sow across the slope seemed to be based on village custom or an informal rule which prescribed farmers to sow in the same direction as their neighbours. Although this custom prevailed in Ghatpally and Debbadaguda also, farmers in these villages did not adhere to this rule as strictly as in Mansanpally. But for farmers in this latter village it meant that all farmers sowed up and down in one year and across in the next. A sideeffect of this custom was that those who ploughed and sowed up and down the slope while having graded/contour bunds or *khus* lines in their plot, destroyed these structures and plants.

7.3.3 DEAD FURROWS

A so-called dead furrow was a final element of the *in-situ* soil and water conservation practices in drylands. It had to be ploughed on the contour or across the major slope in a standing crop, not carrying a crop itself. Being a rather recently developed practice, AWARE never introduced it in Ekvaipally, while the Maheswaram project predominantly propagated it in the extended project area. As with the other *in-situ* soil and water conservation practices, the Maheswaram project tried to induce farmers to adopt dead furrows by informing them about the contents and benefits. Financial incentives were not given.

Except for two respondents in Debbadaguda who had been informed by the Maheswaram project, none of the other persons attending the open interviews stated that they had heard of or had seen dead furrows in dryland. None of the respondents had ever ploughed such a dead furrow. Since this finding was contrary to what was expected, the hypothesis concerning the effect of the Maheswaram project on the adoption of dead furrows was rejected. This conclusion shows more or less the demise of the project's extension service, at least for this recommendation. This negligible adoption rate makes a comparison between communities and landholding-size classes by village and between the expected and observed adoption rates not very relevant.

After having informed respondents about the dead furrows and their supposed advantages myself, they were not convinced of the benefits. The most common remark was that, while costs would increase because of a higher labour input, the increase of yields was doubtful. This scepticism concerning yields stemmed from the idea that dead furrows would only increase evaporation.

7.4 CONCLUSION

The effect of both projects on the adoption of recommended soil and water conservation technologies was less than expected. While the recommended indigenous technologies promoted by the projects obtained the highest adoption rates, the three 'real' innovations introduced by the projects gained much lower rates. Of these, post-harvest tillage and dead furrows were not adopted at all. That adoption rates were as low or high as they were, is the result of a combination of factors of which lack of information and unsuitable technology stand out.

While both projects can be criticized for the fact that project staff at local level performed their tasks half-heartedly and were not monitored properly, they are less responsible for the introduction of unsuitable technologies. The introduction by the Maheswaram project of something exotic like *khus* is a good example for showing that a combination of individuals and organizations (agronomists, who were enthusiastic about it; policy makers in New Delhi and Hyderabad; the funding agency (the World Bank) and the project management who agreed on the introduction) were accountable for this introduction and for the low adoption rates scored. Although the introduction was in line with the second objective of the project (being the large scale verification of already developed technological packages), the bad results of the test made many farmers doubt the reliability of the project and its officers.

With regard to AWARE, being largely dependent on agricultural innovations developed at public research stations and on the T&V extension service, this organization can hardly be criticized for the introduction of unsuitable technologies. In fact, it neither introduced *khus* nor dead furrows. This, however, had nothing to do with policy reasons, but was more the result of lack of information on the part of the management and staff. On the other hand, AWARE is accountable for the fact that much of its soil and water conservation programme in Ekvaipally existed on paper only.

The study shows that the area protected by the recommended graded/contour bunds and *khus* in the villages was limited. This in spite of the fact that respondents were aware of soil and water loss in their dryland plots, that many had protected their plots with indigenous border bunds and that many were familiar with these improved soil and water conservation structures. Low adoption rates were the combined result of improper functioning of the projects, the improper design of the technologies and farmers' reluctance. Improper performance of both projects shows from the fact that most respondents became acquainted with graded/contour bunds not by communicating with project officers, but by observing these structures in the field. The improved bunds constructed in Ekvaipally were built by the government before AWARE's intervention. Fortunately, the Maheswaram project fared better with regard to informing farmers about *khus*. That farmers were reluctant to construct graded/contour bunds themselves had everything to do with their relatively high costs and with their incompatibility with existing agricultural practices. The fact that farmers considered this technology unsuited to replace existing structures, casts serious doubts on their sustainability. Unsustainability also applies to the *khus* lines. Although relatively cheap, its incompatibility with the agro-ecological situation in the area resulted in the death of many plants. In addition, because of its incompatibility with agricultural practices and because of lack of interest on the part of the farmers, many lines were wilfully destroyed.

Notwithstanding these low adoption rates, the hypothesis was confirmed. More improved structures were built and *khus* lines were planted in the project villages than in the non-project village of Ghatpally. However, since at most only 10 per cent of the respondents had adopted these recommendations and since AWARE had not built any structures, there was no conclusive evidence. In fact, the comparison of the observed adoption rates of the various communities and landownership classes between the four study villages with what was expected differs to such an extent that the two hypotheses were rejected. None of the other hypotheses were confirmed either.

The study suggests that both the Maheswaram project and AWARE concentrated their endeavours mainly on one of the many practices included *in-situ* soil and water conservation, namely to sow across the major slope. Hardly any information concerning post-harvest tillage, summer tillage for *jowar*, preparatory tillage and interculturing across the slope or along the contour and dead furrows (for the Maheśwaram project) was given to the farmers. Except for summer tillage for *jowar* and sowing and interculturing across the major slope, none of the other recommendations were adopted by respondents. Although non-adoption was largely an effect of negligible project activities, it is highly doubtful whether better project performance would have resulted in better adoption. The characteristics of these recommendations were such that not many respondents were eager to try them out, let alone adopt them immediately.

Summer tillage for *jowar* as well as sowing and interculturing across the major slope were both indigenous technologies. Since this practice had been adopted more in Ghatpally and less in Ekvaipally and Debbadaguda, the effect of both projects was questionable and the hypothesis rejected. All other hypotheses regarding the effects of project intervention on the adoption rates of the various communities and landownership classes in and between the study villages were also rejected. Since summer tillage was indigenous and adopted by many farmers, it is expected that this is a relatively sustainable practice. The study also shows that summer tillage depended largely on the timing of pre-monsoon showers, meaning that most farmers would till their dryland plot only provided the soil was sufficiently moist.

Sowing across the major slope was practised more in Debbadaguda and Ekvaipally than in Ghatpally. Two things need to be kept in mind, however. First, instead of recommending to sow along the contour, both projects more or less adapted their recommendation to the indigenous practice of sowing across the slope. Second, both Debbadaguda and, in particular, Ekvaipally were located in undulating terrain meaning that farmers were more responsive to the recommendation. However, since it was not adopted in Mansanpally, the effect of both projects is doubtful and the hypothesis neither confirmed nor rejected. None of the hypotheses concerning the effects of project intervention on the different communities and landownership classes in and between the study villages were confirmed either. Since the direction of interculturing was largely determined by the direction of sowing, above remarks apply to this recommendation as well. Because both sowing and interculturing were indigenous practices and adopted by many farmers, it is expected that these are sustainable.

A final remark relates to the negative relationship that was assumed between the area of wetland owned and adoption of the recommendations for dryland agriculture. Such a relationship was only found in bunding, less so in summer tillage of *jowar* and not in sowing across the major slope. Hence it seems that, at least for these soil and water conservation measures, whether a recommendation will be adopted depends more on its characteristics than on the wetland area owned by a farmer.

NOTES

1. This and other figures refer solely to dryland plots which were owned, plots in joint ownership and to *poramboke* plots.
2. This and following analyses make allowance for the fact that the sample included only one large farmer from Mansanpally, who stopped cultivating during the research period.
3. Adoption rates for summer tillage for *jowar* of FC farmers in Debbadaguda, Mansanpally, Ekvaipally and Ghatpally were 25 per cent, 86 per cent, 5 per cent and 47 per cent respectively; of BC farmers 21 per cent, 85 per cent, 20 per cent and 44 per cent respectively; of SC farmers 32 per cent, 89 per cent, 16 per cent and 36 per cent respectively; and of ST farmers 50 per cent, 90 per cent, 26 per cent and 56 per cent respectively.
4. Adoption rates for summer tillage for *jowar* of marginal farmers in Debbadaguda, Mansanpally, Ekvaipally and Ghatpally were 40 per cent, 84 per cent, 8 per cent and 39 per cent respectively; of small farmers 30 per cent, 85 per cent, 22 per cent and 41 per cent respectively; of small-medium farmers 41 per cent, 93 per cent, 23 per cent and 49 per cent respectively; of medium farmers 18 per cent, 67 per cent, 18 per cent and 62 per cent respectively; and of large farmers 25 per cent, not applicable, 0 per cent and 0 per cent respectively.
5. Adoption rates for contour/across sowing of FC farmers in Debbadaguda, Mansanpally, Ekvaipally and Ghatpally were 74 per cent, 0 per cent, 51 per cent and 28 per cent respectively; of BC farmers 63 per cent, 0 per cent, 54 per cent and 15 per cent respectively; of SC farmers 65 per cent, 0 per cent, 53 per cent and 11 per cent respectively; and of ST farmers 55 per cent, 0 per cent, 59 per cent and 11 per cent respectively.
6. Adoption rates for contour/across sowing of marginal farmers in Debbadaguda, Mansanpally, Ekvaipally and Ghatpally were 75 per cent, 0 per cent, 50 per cent and 35 per cent respectively; of small farmers 56 per cent, 0 per cent, 51 per cent and 7 per cent respectively; of small-medium farmers 58 per cent, 0 per cent, 56 per cent and 12 per cent respectively; of medium farmers 93 per cent, 0 per cent, 70 per cent and 25 per cent respectively; and of large farmers 57 per cent, not applicable, 39 per cent and 25 per cent respectively.
7. For the additional discussion of intercultivation, being an element of improved crop management as well, I refer to the next chapter.

8. Adoption of Recommended Crop Management Technologies for Dryland Crops

M. Put

Agronomists are convinced that the (*in-situ*) soil and moisture conservation measures described in the preceding chapter are only initial impetus to increase yields of dryland crops. They argue that to get higher and more stable yields, these measures have to be combined with improved crop management technologies. In fact, they claim a kind of synergy: i.e. the benefits of improved crop management technologies for dryland crops are highest if undertaken on 'treated' plots. In this chapter an analysis is made of the rate of adoption of the various recommendations made by both projects. It focuses on *jowar* and castor, two of the most important dryland crops in the study area and crops which were the main focus of both projects. However, to put the adoption of these innovations in a wider perspective, when applicable, the outcome will be compared with the outcome on the adoption of recommendations for irrigated paddy and sometimes for redgram and *bajra*, two other dryland crops. For the description of both indigenous and recommended technologies, I refer to Chapter 6.

As in Chapter 7, for each recommendation separately, the following questions will be answered. Did respondents know about (the exact contents of) the recommendation and what was their main source of information? How many respondents did adopt it? Was it full or partial adoption or did respondents adapt it? Was (partial, adapted) adoption the result of project intervention? What were the reasons for adopting, adapting or rejecting it? This last question deals with how respondents have assessed each recommendation on characteristics like relative advantage, compatibility, complexity, trialability and observability. Observed adoption rates per recommendation will be compared with the hypotheses as for-

* This chapter is included in Put (1998). Courtesy of Thesis/Thala Publishers, Amsterdam.

mulated in Chapter 4 and a summary of the outcome is presented in a separate table. As in Chapter 7, the analysis of adoption rates is made at plot and crop level; where necessary, the information is consolidated for the entire three-year research period, using holding-size and community as intervening variables to differentiate between the respondents in the study villages. In addition to this quantitative plot and crop level data, qualitative information from 88 open interviews is used. For now, the two farmers who were introduced in Chapter 7 are given the floor again.

Mr X.Y. Reddy, the large and influential FC farmer from Debbadaguda, on the topic of improved crop management for dryland agriculture:

The most important crops grown by me on dryland are *jowar* intercropped with redgram, and castor. Usually I intercrop *jowar* by sowing four or five lines of *jowar* and one line of redgram. Because redgram was heavily attacked by a pest last year, I prefer to cultivate *jowar* as a monocrop this *kharif* 1993/4 season. I have never heard of the recommendation to sow *jowar* and redgram in a 2:1 ratio. Anyhow, I do not think that this is a suitable recommendation. It will never give good results since *jowar* will be overgrown by redgram.

I have cultivated different improved *jowar* varieties like CSH-5, CSH-7 and CSH-9 but I was not very satisfied with the returns. Compared to the farm-gate price for the indigenous *jowar* variety, the price for improved *jowar* is low. Similar to *jowar*, I prefer the indigenous redgram variety above a certain ICRISAT variety, of which I do not remember the name. Since not all pods ripen at the same time you have to send labourers into the field a couple of times which makes these improved redgram varieties labour intensive. I know various improved castor varieties like Gauch-1, GCH-4 and *Sowbhagya*. Since the soil in my plots is less suited for these varieties, I cultivate *Aruna*, another improved variety. *Hamsa*, BPT and IR-64 are improved paddy varieties. I no longer grow the local paddy variety, but cultivate *Hamsa* and BPT in *kharif* and *Hamsa* and sometimes IR-64 in *rabi*. I learnt about most of these improved seed varieties some fifteen to twenty years ago. My informants were my father, the Andhra Pradesh Agricultural University, the Village Development Officer (VDO), shopkeepers, seed agencies, fellow Reddy farmers, radio, newspapers and, lately, television.

I know and use many types of chemical fertilizers of which DAP, Growmore, Urea and Potash are the most important ones. I learnt about chemical fertilizers from my father who received information from other Reddies and newspapers. Nowadays, seed and fertilizer companies as also shopkeepers and sometimes television, give information as well. Nobody ever told me about the exact recommendations for *jowar* and castor, neither about the quantities to apply per hectare, nor about the timing or the manner of application. On average, depending on the fertility of the soil and the time of sowing, I apply some 100 kg of Growmore or a mixture of DAP and Urea per hectare in dryland. Wetland crops get more. Dryland crops get chemical fertilizers only at the time of sowing. I do not know about top-dressing for dryland crops. I apply Urea as a top-dressing only in wetland, in particular to paddy.

If the rains are good, my labourers will sow *jowar* immediately after preparatory tillage is completed in the second or third week of May. In case rainfall is insufficient, they will sow during the last week of May or the first week of June. If the monsoon is late, I will not cultivate *jowar*. I prefer to sow castor in the last week of June, but if the rains are late it is possible to postpone the sowing until September.

Interculturing with a *dante* or a *guntaka* (both blade harrows), is not only beneficial to improve the moisture content of the soil and the strength of the plant, it is also the best way to remove weeds in dryland crops. Although one time interculturing is sufficient to remove most weeds, I prefer to interculture two times in *jowar* and up to five times in castor because rain is more easily absorbed in a soft soil. More interculturings in *jowar* are not possible because plants will get damaged. It takes a lot of labour to remove weeds by hand. Because yields of dryland crops are low it is not very remunerative. In contrast, since it is remunerative in paddy and other wetland crops I employ many agricultural labourers to remove the weeds in these crops.

I know about pesticides since some twenty years but have started using it on a rather regular basis since the last six years only. I learnt about pesticides from my father, seed dealers, fellow Reddies, newspapers, television and the VDO. Before using it on a large scale I did some experiments. If needed, I apply pesticides on most crops but concentrate my efforts on wetland crops. Indigenous *jowar* does not need pesticides but improved *jowar* does. I spray castor only in case of a large attack of caterpillars and apply pesticides in paddy if it is infested with flies. Large scale use of pesticides, in the shape of a precautionary measure, is practised only in high value crops like cotton and grapes, not in the other crops.

Mr Y.Z. Jangaiah, the marginal SC farmer from Ghatpally:

In my dryland plot I grow *jowar* intercropped with redgram, and castor. Almost every year I intercrop *jowar* with redgram: four to five rows of *jowar* and one row of redgram. This is beneficial for *jowar* which, without redgram, will give lower yields. In the last couple of years our village was struck with a pest attacking redgram which made me decide to sow in a 5:1 instead of a 4:1 ratio. Because this did not prevent the pest from attacking again, I decided to cultivate *jowar* as a monocrop last year. I have never heard that the government recommends a 2:1 ratio and I have never seen it in practice. I have my doubts about it as the yield of *jowar* will be less because there are less *jowar* plants and because redgram will suffocate *jowar*.

I do not know any name of improved *jowar* seeds, nor have I ever used any. I have heard from other farmers and seen in some neighbouring fields that the yield of these improved varieties is better compared to our local variety, at least if rainfall is good and the fertility of the soil is good too. However, I was told that the taste of the *rotis* (bread) made from flour of these improved seeds is bad, that you get pains in your joints after eating, that animals do not digest the fodder and, finally, that the farm-gate price is also less than the price for the indigenous *jowar* variety. Hence I decided not to sow these improved varieties, not even on a trial basis. I do not know that improved varieties of redgram are available but I am quite aware of improved castor varieties. In fact I cultivate *Aruna*. Its yield is much higher than our local variety and it is as easy to cultivate. Although I do not have wetland I know some improved paddy varieties. Nowadays, neighbouring farmers prefer *Hamsa* and BPT above the local varieties because they give higher returns in both grain and fodder and their taste is fine as well.

I know that since *Aruna* seeds lose part of their higher yielding capacity it is better to buy new *Aruna* seeds every year and not to use seeds from your own harvest. Still, I buy new seeds only after two or three years. I prefer to use seeds from my own harvest, which I select very carefully, instead of buying every year. Seed is expensive and why should I pay so much if I am not sure about the yield? Besides, there is a lot of inferior seed in the market and I do not trust the shopkeepers. If the quantity of my own seeds is insufficient, I borrow or buy seeds from Reddy farmers, like I did the first time I used *Aruna* seeds.

I know about Urea, DAP and Growmore since about fifteen years and I am using it since some ten years. I learnt about these chemical fertilizers while working as a permanent labourer for Reddy farmers. I had to apply it in their fields. It took me about five years to get used to these inorganic fertilizers, to get an idea whether they are beneficial or not. I do not know about any recommendations, neither for the quantity to be used nor for the timing of the operation. All I know is what Reddy and other farmers do.

I will sow *jowar* after the arrival of good rains, mostly in the first week of June. Some one and a half months later, again depending on the rains, I will sow castor. *Karthes* are important, guiding the timing of all sowing operations.

Interculturing in dryland is necessary to remove weeds, to make the soil loose so that rain and fog can penetrate more easily and to make the plants stronger. I use *dante* twice in *jowar* and *guntaka* twice in castor. More interculturings in *jowar* are not possible as one month after sowing the plants have grown rather high and a third interculturing will destroy them. Interculturing in castor is possible up to four or five times, but I prefer to look for other types of employment instead of investing so much time in an activity of which I cannot predict the outcome. I remove weeds by hand only if too many weeds have come up. Since proper preparatory tillage operations and interculturing often prevent this from happening, I almost never weed by hand.

Indigenous methods to avoid a pest attack in your fields are not very popular any longer. Most farmers prefer pesticides nowadays, but only in case of a severe pest attack. I know about the existence of pesticides since some five years when I saw Reddy farmers using it in their fields. I have used it only once, at a time that caterpillars attacked my castor crop. I do not know why, but these pesticides were not very beneficial and most caterpillars did not die. Perhaps the shopkeeper gave me bad pesticides. Because they are very expensive and not very reliable, I do not think that I will use pesticides very often.

8.1 An Overview

Similar to Table 7.2, in a nutshell Table 8.1 demonstrates that the effects of the Maheswaram project and AWARE on the adoption of recommended crop management technologies for dryland crops were anything but spectacular. Calculated for all recommendations, the average adoption rates hardly differed between the project villages and Ghatpally. In fact, while respondents in this latter village lagged behind on certain recommendations, they had the highest average adoption rates on other recommendations.

Some of the highest adoption rates of respondents were found in indigenous technologies like early sowing, intercropping and interculturing. On the other hand, some of these technologies had rather low adoption rates, of which manual weeding is a case in point. One of the most dynamic elements of agricultural change in the study villages relates to the use of improved seeds of specific crops and chemical fertilizers, recommendations already introduced before the commencement of both the projects. The next sections will show that farmers had a distinct preference about which seed variety was best, the indigenous or the improved, a preference which was based on sound reasoning.

8.2 Intercropping

Since agricultural research stations had concluded that the existing cropping pattern on dryland suited the agro-ecological situation in the area, neither the Maheswaram project nor AWARE tried to alter it.[1] Instead, they tried to induce farmers to intercrop *jowar* with redgram (which was an indigenous technology) and, which was considered even better, to adapt it following new insights from agronomic research. Because the recommendations made by both projects were not always very clear and drifted now and then (see Chapter 6), the effects of their intervention have been measured in two ways. Assuming that both projects advised intercropping *jowar* with redgram in a 2:1 or 3:1 ratio, this recommendation is analysed first. The second analysis is made regardless of the ratio and deals with intercropping of *jowar* with redgram as such.

TABLE 8.1. RATE OF ADOPTION OF RECOMMENDED CROP MANAGEMENT TECHNOLOGIES, BY VILLAGE

Recommendation	Debbadaguda	Mansanpally	Ekvaipally	Ghatpally	Total
Intercropping[1]	59	63	67	63	63
Improved *jowar* seeds[2]	2	2	3	3	2
Improved redgram seeds[2]	1	0	0	1	1
Improved *bajra* seeds[2]	0	0	9	0	9
Improved castor seeds[2]	94	67	54	86	77
Seed-rate castor[2]	29	51	46	9	30
Seed-cum-fertilizer drill[2]	0	0	0	0	0
Iron plough[2]	3	3	0	1	2
Timing FYM (*jowar*)[1]	33	22	5	57	33
Timing FYM (castor)[1]	7	0	25	35	24
Quantity FYM (*jowar*)[1]	10	15	38	33	24
Quantity FYM (castor)[1]	14	9	37	25	17
Timing of fertilizers[2]	2	0	1	0	2
Quantity fertilizers (*jowar*)[2]	17	18	15	9	15
Quantity fertilizers (castor)[2]	14	17	13	10	13
Fertilizer placement[2]	0	0	0	0	0
Early sowing (*jowar*)[1]	91	96	83	100	97
Interculturing (*jowar*)[1]	85	64	95	69	77
Interculturing (castor)[1]	65	61	91	15	53
Weeding (*jowar*)[1]	1	1	16	0	4
Weeding (castor)[1]	0	1	4	4	3
Pesticides (*jowar*)[2]	1	1	3	2	1
Pesticides (castor)[2]	5	4	6	10	6
Average	26	26	29	28	28

Note: Since not all recommendations were innovations, I distinguish: 1 = indigenous technology; 2 = technology introduced before the start of the projects; and 3 = technology introduced by the projects.

Source: Own survey.

Similar to the recommendations concerning soil and water conservation, the Maheswaram project spread its information through project officers who visited the villages and through the involvement of village volunteers. In addition, since the Department of Agriculture was responsible for crop production, information on this and other recommendations for crop management had to reach farmers through the T&V system and by taking them to agricultural research stations. AWARE informed farmers predominantly about this and other recommendations during village meetings with T&V extension staff.

That intercropping is an indigenous practice is confirmed by the benchmark study (APAU, nd: 9) and the evaluation study undertaken by the

Maheswaram project (Sivamohan et al., 1994: 51). The latter study even stated that most farmers adopted intercropping not on advice of the Maheswaram project but on advice of neighbours. The open interviews indicated that, except for eight respondents, most respondents had never heard about the recommendation to intercrop in a 2:1 or 3:1 ratio. Hence, it was no surprise to find that during three seasons (*kharif* 1991/2-1993/4) only eight farmers had intercropped *jowar* with redgram in a 3:1 ratio. All were small and small-medium farmers, seven ST farmers living in Mansanpally and one SC farmer living in Ghatpally. This outcome together with the fact that it was a once-only event suggests that although the Maheswaram project performed somewhat better than AWARE, the effect of both projects on this specific recommendation was negligible.

Limited adoption of the 2:1 and 3:1 intercropping ratio was largely a consequence of project failure to inform farmers about the recommendation. Yet, in the eyes of the farmers, the recommendation itself seems to have major shortcomings. After having informed them during the open interviews about the recommendation, almost all respondents said that it was not feasible to follow them for reasons of relative advantage and compatibility. Their main concern was that yields of *jowar*, both for food and fodder, would decrease because: (1) the number of *jowar* plants would decrease; (2) the chance that pests would attack redgram would increase; and (3) the multitude of branches of redgram would get entwined, thereby strangling the *jowar* plants. The few respondents, all FC farmers, who were more positive about the recommendation, felt that it was only feasible provided improved seeds of both *jowar* and redgram were sown. However, the less favourable characteristics of these improved seed varieties (see next subsection) made many disapprove of a 2:1 or 3:1 ratio.

Finding that a lower ratio was unacceptable to farmers, it was interesting to know whether the projects had succeeded in stimulating intercropping of *jowar* with redgram as such. They did not, though. Irrespective of ratio, in the same *kharif* 1991/2-1993/4 period, on average some two-thirds of all *jowar* plots were intercropped with redgram. Intercropping was more prevalent in Ekvaipally (67 per cent of the plots), followed by Mansanpally and Ghatpally (both 63 per cent) and Debbadaguda (59 per cent). Comparing the observed average adoption rates at village level (Figure 8.1) with the hypothesis on expected adoption rates (see Figure 4.1), leads to the only conclusion that project interventions did not generate the expected effect, meaning that the hypothesis was rejected..

Not a single hypothesis on the expected adoption rates of the various communities in the villages was confirmed. In Debbadaguda, highest average adoption rates were for BC respondents, followed by ST respondents, FC respondents and SC respondents (Figure 8.2).[2] Except for FC respondents who performed worse than expected, this ranking was more or less in line with what was anticipated (see Figure 4.2). Hence, the hypothesis for Debbadaguda was neither confirmed nor rejected. Since the various communities in Mansanpally had almost similar average adoption rates and since the observed ranking (highest averages for SC respondents, followed by ST, BC and FC respondents) differed from the expected ranking, the hypothesis was rejected. As expected, the highest average adoption rate in Ekvaipally was for ST respondents. However, since the ranking of the other communities (BC respondents were next in line, followed by FC and SC respondents) was not as expected, the hypothesis was rejected. In Ghatpally, ST respondents had the highest averages, followed by SC, FC and BC respondents, a ranking that was not expected. Hence, the hypothesis was also rejected.

It was expected that the highest adoption rates of FC and BC respondents would be found in Debbadaguda, while the highest adoption rates of SC and ST respondents, albeit at a lower level, would be found in Ekvaipally (see Figure 4.2). However, Figure 8.2 shows that the observed average adoption rates of FC and ST farmers did not differ appreciably between the four study villages. Moreover, highest average adoption rates of FC and ST farmers were found in Ghatpally and not in Debbadaguda (for FC farmers) and Ekvaipally (for ST farmers). For SC farmers, the highest average adoption rates were also found in Ghatpally, not in Ekvaipally. For BC farmers, the highest average adoption rates were found in Debbadaguda, which was in line with the expectation. Nevertheless, since all other outcomes differed considerably from what was expected, the hypothesis was rejected.

As regards the expected (see Figure 4.3) and observed average adoption rates by landowner class in the four study villages (Figure 8.3), the hypotheses were not confirmed for any of the four study villages. Since large farmers in Debbadaguda performed much worse than expected, while the ranking of the other landownership classes was as expected, the hypothesis for this village was neither confirmed nor rejected.[3] Both medium and small farmers in Mansanpally performed better than expected and, as anticipated, had more or less similar adoption rates. However, since marginal farmers scored much lower than expected and had a much

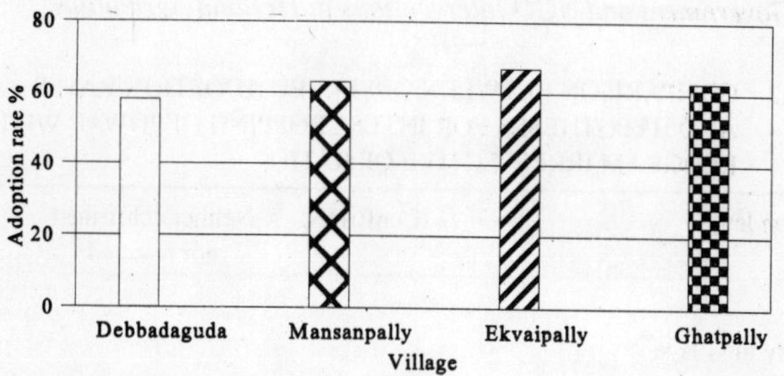

Note: Intercropping relates to all kinds of ratios.
Source: Own survey.

FIGURE 8.1. OBSERVED ADOPTION RATE (%) OF INTERCROPPING OF *JOWAR*, BY VILLAGE (*kharif* 1991/2-1993/4).

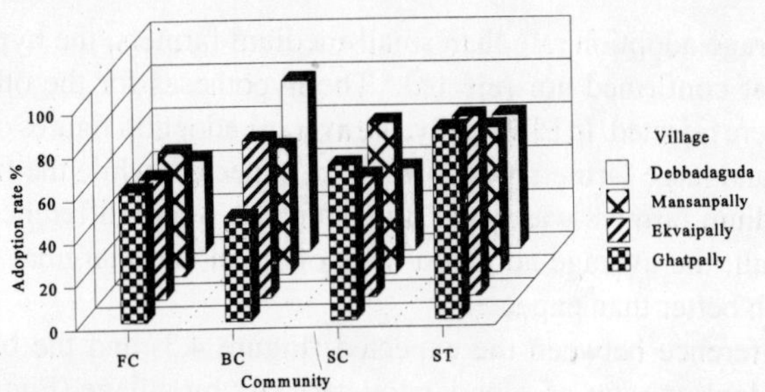

Note: Intercropping relates to all kinds of ratios.
Source: Own survey.

FIGURE 8.2. OBSERVED ADOPTION RATE (%) OF INTERCROPPING OF *JOWAR*, BY COMMUNITY, BY VILLAGE (*kharif* 1991/2-1993/4).

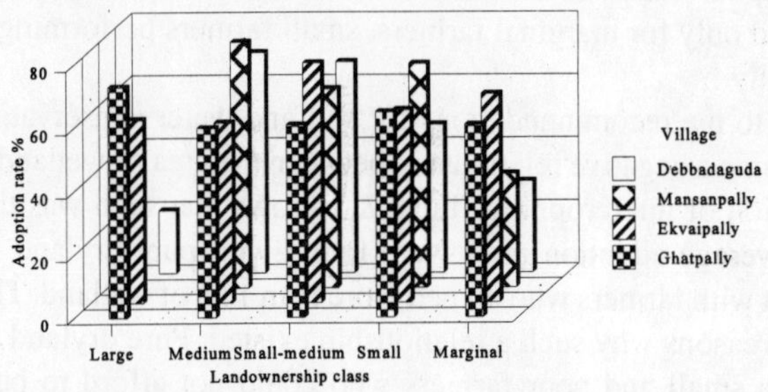

Note: Intercropping relates to all kinds of ratios.
Source: Own survey.

FIGURE 8.3. OBSERVED ADOPTION RATE (%) OF INTERCROPPING OF *JOWAR*, BY LANDOWNERSHIP CLASS, BY VILLAGE (*kharif* 1991/2-1993/4).

TABLE 8.2. COMPARISON BETWEEN OBSERVED ADOPTION RATES AND HYPOTHESES FOR INTERCROPPING OF JOWAR WITH REDGRAM IRRESPECTIVE OF RATIO

Intervention level	Confirmed	Neither confirmed nor rejected	Rejected
Village			Yes
Community in village		D	M, E, G
Community between villages			Yes
Landownership class in village		D, M	E, G
Landownership class between villages		Yes	

Note: D = Debbadaguda; M = Mansanpally; E = Ekvaipally; G = Ghatpally.
Source: Own survey.

lower average adoption rate than small-medium farmers, the hypothesis was neither confirmed nor rejected.[4] The hypotheses for the other two villages were rejected. In Ekvaipally, the average adoption figures of small, marginal and large farmers was lower than expected, while the figure of small-medium farmers was better. In Ghatpally, where differences were rather small, the average adoption rates of small and marginal farmers were much better than expected.

The difference between the expected (Figure 4.3) and the observed average adoption rates of each landowner class by village (Figure 8.3), was such that the hypothesis was rejected. While it was expected that the averages for large, medium and small-medium farmers would be highest in Debbadaguda, these were highest in Ghatpally, Mansanpally and Ekvaipally respectively. In addition, while it was expected that the highest averages for small and marginal farmers would be found in Ekvaipally, this applied only for marginal farmers, small farmers performing best in Mansanpally.

Similar to the recommendations for soil and water conservation, one might assume a negative relationship between the area of wetland owned and adoption of intercropping. Table 8.3 shows that such was the case. Highest average adoption rates were found with pure dryland farmers and lowest with farmers who owned more than 1 ha of wetland. There are two main reasons why such a relationship existed. Pure dryland farmers were often small and poor farmers who could not afford to buy large quantities of such an expensive food crop (redgram being much more expensive than *jowar* or rice), hence preferred to cultivate this crop themselves. Moreover, for lack of sufficient money to buy sufficient quantities of chemical fertilizers (see later) and having insufficient quan-

TABLE 8.3. RATE OF ADOPTION OF INTERCROPPING OF *JOWAR* WITH REDGRAM IRRESPECTIVE OF RATIO, BY TYPE OF FARMER, BY VILLAGE

Type of farmer	Debbadaguda	Mansanpally	Ekvaipally	Ghatpally	Total	N plots
Pure dryland farmers	65	65	71	64	66	240
Farmers with < 1 ha wetland	55	61	71	63	63	235
Farmers with ≥ 1 ha wetland	53	63	51	56	55	73
Average	59	63	67	63	63	548

Source: Own survey.

tities of FYM as well, dryland farmers felt the need to use redgram more to improve soil fertility.

That such an indigenous technology as intercropping was not more prevalent in the villages was not caused by a lack of knowledge (all respondents were aware of this indigenous technology) and was only partially related to a lack of interest among farmers owning wetland besides dryland. In fact, the most important reason for not intercropping was the pot-borer which struck the region regularly, as a consequence of which yields of redgram decreased dramatically. Farmers had two options to limit the possibility of an attack; either they sowed some extra rows of *jowar* and increased the ratio to 6:1 or even higher, or they cultivated *jowar* as a monocrop. Monocropping was somewhat more prevalent in *kharif* 1992/3 when 41 per cent of the plots were cultivated with *jowar* only, an increase of 5 per cent compared to *kharif* 1991/2 and 1993/4.[5] It is indicative of the fact that intercropping might be sustainable as far as the 4:1 and higher ratios are concerned, but that its use depends on natural factors like the presence of pests.

8.3 Improved Seed Varieties

Although it seems easy to study the rate of adoption of improved seed varieties, there are many pitfalls in the way. These arise, for instance, from the fact that not all farmers knew the name of the seed variety they had sown (this applies in particular to improved seeds). Those who did not know the name quite often discriminated between 'local' and 'hybrid'. It was found, however, that the term 'local' did not necessarily mean 'indigenous' and that 'hybrid' was not necessarily equivalent to hybrid or sterile. In addition, imperfect knowledge of farmers (for instance about the size and colour of improved seeds) permitted suppliers to cheat in seeds.[6] I have tried to evade these pitfalls by asking the name of the seeds, making sure whether 'local' meant indigenous and 'hybrid' meant improved, by inquiring about the size and colour and their origin.

The use of improved seeds was advocated long before the start of both the Maheswaram project and AWARE. Walker and Ryan (1990: 51) reported that many improved seeds were introduced in the Amangal area in the early 1970s. Sitapathi Rao et al. (1990: 63) stated that by 1985 the use of improved seeds had become a routine practice in the study area. In addition to the regular T&V extension service, the Lab to Land Programme promoted improved seeds in large areas of Andhra Pradesh. They most certainly did in the area that was selected, not much later, for the Maheswaram project and they most probably also did in the Amangal area where AWARE had already become active. Small and marginal, SC and ST farmers, were entitled to receive improved seeds on a subsidized basis. Moreover, under the government's seed exchange programme, farmers were encouraged to exchange indigenous for improved seed varieties.

Thus, respondents guessed that some twenty years ago, somewhere in the early 1970s, the first improved seeds were introduced in the villages, paddy first, followed by castor and some years later by *jowar*, *bajra* and seeds of other crops. Since the way in which farmers were informed about improved seeds and how they have dealt with that information is such a typical example, it will be discussed at greater length. In most cases it was large farmers, mostly Reddies and a couple of BC farmers, who were informed about the existence and benefits of improved seeds (paddy first and castor later) first by their regular sources of information such as other innovative Reddy and BC farmers outside the village, shopkeepers, newspapers and the radio and, to a much lesser extent, by extension officers.[7] Unfamiliar about the way these seeds had to be cultivated and about their yields in their own plots, most of these farmers tried to get additional information first. Others, convinced that the agro-ecological situation in their village was not suited to these improved seed varieties, paid no heed to the information. Consequently, it took a couple of years before the first farmers became convinced about the advantages of these seeds. Once convinced, most innovators experimented first, sowing part of their plot with the indigenous and part of the plot with the improved seed variety.

The experiments with paddy and castor were promising, yields were much higher than with indigenous seeds while costs did not increase too much, and these innovative farmers slowly gave up using their indigenous seeds. The good news travelled around quickly and most farmers, also those having no irrigated plots, were informed about the improved paddy and castor seeds within a couple of years. Some of these, in particular other high caste farmers who had good personal contacts with these innovative farmers, decided to take up the cultivation of improved paddy and castor. Others, mostly small and marginal, ST and SC farmers, afraid to contact Reddy farmers in person, kept sitting on the fence, watching these higher caste farmers cultivating the improved seeds of these crops. Insufficient knowledge seems to have been one of the more important reasons playing this waiting game. Since the extension service did not work properly and since most farmers believed only their eyes, they wanted to see the improved crop being cultivated in their neighbourhood and so most of them filled their information gap by observing neighbouring farmers. In addition, for most poor farmers, insufficient capital to buy improved seeds was another reason for its slow adoption. Once familiar with the cultivation practices and having the money to buy the seeds, many farmers followed suit and began to cultivate these improved seeds. Most of these farmers, having seen these improved seeds grow in neighbouring plots and being familiar with the cultivation practices and the benefits, did not experiment but cultivated the entire plot with the new seeds.

Information about improved *jowar* seeds reached the same innovative farmers in the study villages some years later than information about paddy and castor seeds. As the news was less spectacular than that of other improved seeds, only a few farmers decided to cultivate improved *jowar* and of those who did, most preferred to experiment first. For reasons to be explained later in this section, these experiments were not promising at all. In fact, the rather negative experiences of the innovative farmers influenced the adoption behaviour of the other farmers to such a large extent that adoption rates remained low.

Although information on improved seeds of *jowar* and castor had reached the area a couple of years before the commencement of the projects and although a considerable number of farmers had already started to cultivate improved seeds of various crops, both projects tried to increase the cultivation of improved seeds of various dryland crops. They used a two-way strategy of informing farmers and distributing improved seeds at subsidized prices or through the seed exchange programme. Since both projects were not able to trace the names of those who had received improved seeds, it was not possible to check what these farmers received and what their experiences were. During the open interviews only eighteen out of the eighty-eight respondents (nine in Debbadaguda, two in Mansanpally and Ekvaipally each, and five in Ghatpally) stated that they had received improved seeds of *jowar* or castor from the projects or the regular extension service in the *mandal* office.

During the open interviews, almost all respondents were able to mention the names of a couple of improved seeds. This applied also to those who did not use a particular seed or did not cultivate a certain crop at that time. Thus all respondents, whether they cultivated wetland or not, knew about *Hamsa*, the most popular improved paddy seed in the study villages. Almost all respondents knew of some improved castor seeds of which the name *Aruna* was mentioned most, followed by *Gujarati*, *Gauch-1* and *Sowbhagya*. Compared to these two crops, the number of respondents knowing improved varieties of *jowar* was less, 61 per cent. Most of them mentioned varieties like CSH-5, CSH-6 and CSH-9 (often referred to as no. 5 or no. 9). However, names of other improved *jowar* varieties, like the SPV varieties propagated by the Maheswaram project, were hardly ever given. Conspicuous by their absence were the names of improved redgram varieties. The few respondents (totalling 8 per cent for the four villages) who claimed to know about improved redgram seeds were unable to mention their names, though. With 35 per cent, the number of respondents who knew of improved seeds of other crops like sunflower and maize (often referred to as Pioneer seeds, the name of the seed company), cotton and tomatoes, was more than with redgram but much less compared to paddy, castor and *jowar*.

During the entire study period, from *kharif* 1991/2 to 1993/4, improved *jowar* was cultivated on only 2 per cent of the dryland plots sown with this crop (18 plots). The fact that a difference hardly existed between the study villages (2 per cent of the *jowar* plots in Debbadaguda and Mansanpally and 3 per cent of these plots in Ekvaipally and Ghatpally were cultivated with improved seeds), together with the low adoption rate, clearly indicates that both projects had hardly any effect at all. On this basis the hypothesis was rejected. These low adoption rates for improved *jowar* seeds makes a comparison between communities and landholding-size classes by village and between the expected and observed adoption rates less relevant.[8] All that can be said is that the difference between the various communities and landownership classes in the four study villages was small. Aggregating for all study villages and on the basis of absolute figures, marginal and small farmers belonging to the FC and BC community cultivated improved *jowar* on a somewhat larger scale than other farmers.

Every study village had a couple of predominantly large FC farmers, who cultivated improved *jowar* more or less regularly. Quite often they did not grow it on dryland but on wetland, arguing that improved *jowar* seeds needed more chemical fertilizers and water than the indigenous varieties which performed better on dryland. In addition to these farmers, a small minority had tested improved *jowar* seeds once or twice on dryland. Because most of these trials were a failure, it made most of them switch back to their indigenous varieties. Nevertheless, it indicated that in every village a couple of farmers annually experimented with these improved seeds, something that happened during the research period as well.[9] Most respondents, however, stated during the open interviews that they had never cultivated improved *jowar* varieties. Horrified as they were by the stories told by those who had made such an endeavour, most respondents had never done a trial with these improved seeds.

In defiance of the recommendations to buy improved seeds every year, 14 of the 18 plots sown with improved *jowar* seeds were sown with 'local' seeds: eleven plots were cultivated with seeds obtained from their own harvest and three plots with seeds received from other farmers living in their village. Only in *kharif* 1991/2, three farmers had bought improved *jowar* seeds in the *mandal* office and one farmer had bought the seeds in a shop.

Similar to *jowar*, the number of farmers who cultivated improved seeds of redgram and *bajra* during the study period was almost negligible. Improved redgram was sown on 3 dryland plots or on less than 1 per cent of the plots cultivated with this crop, 2 plots located in Ghatpally and one plot in Debbadaguda. Improved *bajra* was sown on 9 dryland plots or on 9 per cent of the plots grown with this cereal crop. All plots were located in Ekvaipally. However, most farmers cultivating improved redgram and *bajra* did not know which variety they had sown. The fact that most had not bought these seeds in a shop makes it highly doubtful whether these were improved seeds at all. This leads to the conclusion that neither project had a measurable effect. Hence, the hypothesis concerning these two crops was rejected. Similar to *jowar*, the low adoption rates do not make a comparison between communities and landholding-size classes cultivating improved redgram and *bajra*, and between the expected and observed adoption rates very meaningful.[10]

In contrast to *jowar*, the number of plots cultivated with improved castor constituted a large majority of 77 per cent. Figure 8.4 shows that the use of improved castor seeds was particularly high in Debbadaguda (94 per cent of the plots were sown with improved seeds) and Ghatpally (86 per cent of the plots). It was somewhat lower in Mansanpally (76 per cent of the plots) and lowest in Ekvaipally (54 per cent of the plots). While the use of improved castor seeds did not differ much between the years in Debbadaguda, Mansanpally and Ghatpally, it increased considerably in Ekvaipally. This increase might be spurious, a consequence of a misunderstanding between farmers and interviewers about the term 'local' seed. The fact that Ghatpally scored relatively better than expected while Ekvaipally scored relatively worse, suggests that the effect of both projects was less than expected. A comparison between Figures 4.1 and 8.4 leads to no other conclusion than the rejection of the hypothesis.

Not a single hypothesis regarding the effect of project intervention on the various communities in the study villages was confirmed (compare Figure 4.3 with Figure 8.5). In Debbadaguda, the average adoption rates did not differ very much between the various communities. They were highest for ST respondents, followed by FC and SC respondents (having similar rates) and BC respondents.[11] Since the observed adoption rates were not very different and since the ranking was not anticipated, the hypothesis for Debbadaguda was rejected. In Mansanpally, the observed ranking (highest averages for FC respondents, followed by BC, ST and SC respondents) was more or less similar to the expected ranking. However, since the average adoption rate of ST respondents was lower than expected, the hypothesis was neither confirmed nor rejected. Since the difference in average adoption rates between the four communities in Ekvaipally was relatively small and since their ranking was not as expected (BC respondents ranked first, followed by FC, ST and SC respondents), the hypothesis was rejected. In Ghatpally, the ranking was more or less expected (FC respondents had the highest averages, followed by ST, BC and SC respondents). However, since the difference in the observed average adoption rates was small, the hypothesis was neither confirmed nor rejected.

The expectation that the highest adoption rates of FC and BC respondents would be found in Debbadaguda, proved right (compare Figure 4.2 with Figure 8.5). The expectation that the highest adoption rates of SC and ST respondents would be found in Ekvaipally, proved wrong. Again, the averages were highest in Debbadaguda. Hence, the hypothesis was rejected.

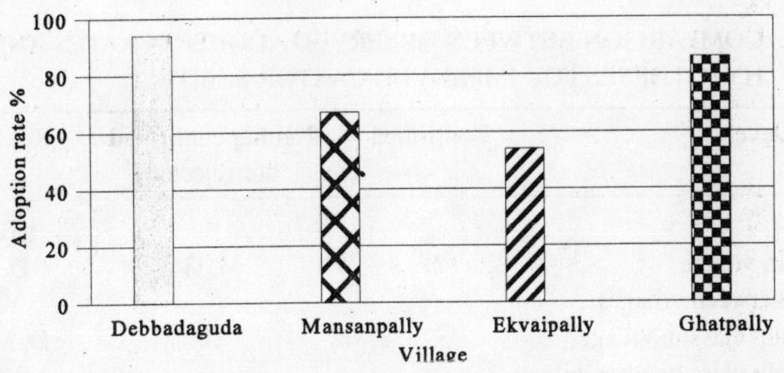

Source: Own survey.

FIGURE 8.4. OBSERVED ADOPTION RATE (%) OF IMPROVED CASTOR SEED, BY VILLAGE (*kharif* 1991/2-1993/4).

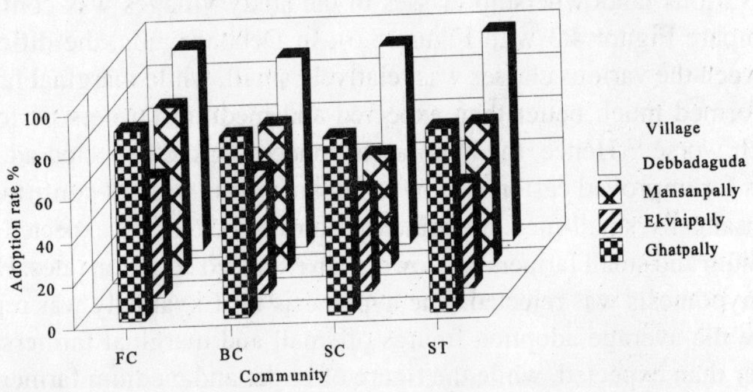

Source: Own survey.

FIGURE 8.5. OBSERVED ADOPTION RATE (%) OF IMPROVED CASTOR SEED, BY COMMUNITY, BY VILLAGE (*kharif* 1991/2-1993/4).

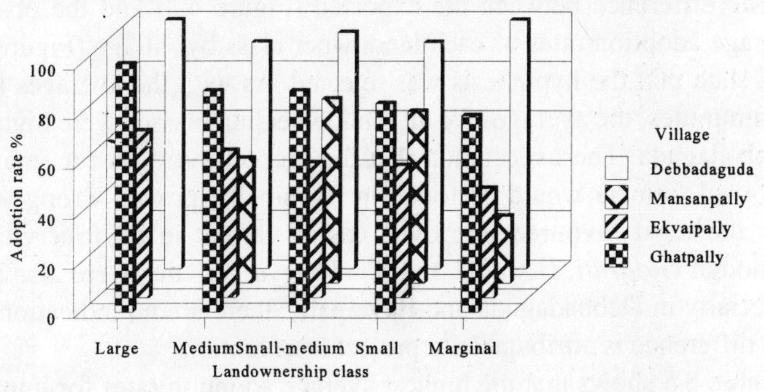

Source: Own survey.

FIGURE 8.6. OBSERVED ADOPTION RATE (%) OF IMPROVED CASTOR SEED, BY LANDOWNERSHIP CLASS, BY VILLAGE (*kharif* 1991/2-1993/4).

TABLE 8.4. COMPARISON BETWEEN OBSERVED ADOPTION RATES AND HYPOTHESES FOR IMPROVED CASTOR SEEDS

Intervention level	Confirmed	Neither confirmed nor rejected	Rejected
Village			Yes
Community in village		M, G	D, E
Community between villages			Yes
Landownership class in village		G	D, M, E
Landownership class between villages			Yes

Note: D = Debbadaguda; M = Mansanpally; E = Ekvaipally; G = Ghatpally.
Source: Own survey.

Not a single hypothesis regarding the effect of project intervention on the various landownership classes in the study villages was confirmed (compare Figure 4.3 with Figure 8.6). In Debbadaguda, the difference between the various classes was relatively small, while marginal farmers performed much better than expected and medium farmers performed much worse.[12] Hence, the hypothesis concerning the expected adoption rates for improved castor seed by landowner class was not confirmed. In Mansanpally, small-medium farmers performed better than expected while medium and small farmers had lower than expected adoption rates. Hence, the hypothesis was rejected. The hypothesis for Ekvaipally was rejected since the average adoption figures of small and marginal farmers were lower than expected, while the figure of larger and medium farmers was better. In Ghatpally, with the highest adoption rates for large farmers and the lowest for marginal farmers, the ranking was more or less expected. However, since the observed average adoption rates for the various classes hardly differed, the hypothesis was neither confirmed nor rejected.

The difference between the expected (Figure 4.3) and the observed average adoption rates of each landowner class by village (Figure 8.6), was such that the hypothesis was rejected. As with the averages for all communities, the averages for all landownership classes were highest in Debbadaguda. The expectation that the highest averages for small and marginal farmers would be found in Ekvaipally, proved wrong. *Aruna* was the most favoured improved castor variety in all four villages. Although *Gujarati*, *Gauch-1* and *Sowbhagya* varieties were also sown, especially in Debbadaguda and Ekvaipally, there are no indications that this difference is attributable to project intervention.

Table 8.5 shows that the highest average adoption rates for improved castor seeds were found with farmers who owned more than 1 ha of wetland. As a commercial dryland crop, the improved variety was lucrative

TABLE 8.5. RATE OF ADOPTION OF IMPROVED CASTOR SEEDS, BY TYPE OF FARMER, BY VILLAGE

Type of farmer	Debbadaguda	Mansanpally	Ekvaipally	Ghatpally	Total	N plots
Pure dryland farmers	91	61	46	79	72	225
Farmers with < 1 ha wetland	96	74	53	93	79	216
Farmers with ≥ 1 ha wetland	95	75	65	96	83	100
Average	94	67	54	86	77	541

Source: Own survey.

for this type of farmers. That pure dryland farmers had the lowest adoption rates was less a consequence of lack of information about these seeds, but had more to do with their limited financial resources which did not allow them to buy improved seeds.

Farmers' aversion to buying improved castor seeds was partly based on the consideration that fluctuations in yield were more related to rainfall than to seeds alone. Only after two to three years, when the decrease in yield had passed a certain threshold level, farmers considered the buying of seeds worthwhile. This aversion to buying every year was also based on farmers' bad experience with seeds bought in shops or received from the *mandal* offices. This experience was also reported by the Maheswaram project: 83.8 per cent of the respondents were dissatisfied with the inferior quality of the castor seeds they had received from the project (Firdausi et al., 1991: 1-11). Sunitha Raju (1990) also reported about dealers who cheated due to which farmers depended on fellow farmers. To avoid illegal practices, in particular, large FC farmers bought seeds only from renowned, dedicated shopkeepers or from the Agricultural University in Hyderabad. Another reason why farmers were reluctant to buy seeds every year was their relatively high price.

What strikes one about Table 8.1 is the large difference between the adoption rates for the improved *jowar*, redgram, *bajra* and castor varieties. In fact, it was quite common to find a farmer who simultaneously cultivated indigenous *jowar* and redgram varieties as well as improved castor and paddy varieties. Improved paddy seeds were a tremendous success in all four study villages, even greater than those of castor. In fact, almost all farmers used improved paddy seeds during *kharif* 1992/3 and 1993/4 and indigenous seeds were hardly used any longer.[13] Walker and Ryan (1990: 51), reporting about the success of *Hamsa* in one of their study villages in

Amangal *mandal*, also showed that the diffusion of this improved paddy variety went much faster than that of *Aruna*. It indicates that, at least for improved seeds, innovations spread faster in wetland than in dryland agriculture.

This comparison shows that farmers' reluctance to grow improved *jowar*, redgram or *bajra* had nothing to do with their resistance to agricultural innovations, with their conservatism or their lack of progressiveness. Non-adoption of these improved seed varieties was partly a consequence of insufficient and unreliable information passed on to farmers, especially about the SPV varieties of *jowar* and the improved redgram and *bajra* varieties. For this, both projects and the T&V system of extension are accountable. In addition, since non-adoption was a response to the inherent shortcoming of these improved seeds, agricultural research is responsible for developing unsuitable varieties.

Most probably, farmers' aversion to improved *jowar* was based on the varieties which were released first. Farmers' experiments with these varieties failed to produce the desired effects. In fact, the results were so bad that a large majority of the farmers in the four study villages became suspicious of all improved *jowar* varieties and more or less decided not to cultivate any improved *jowar* at all. Thus, these 'early generation' seeds produced a barrier against the acceptance of 'later generation' *jowar* varieties. It is quite possible that this barrier may even restrict the diffusion of varieties which are still to come. Thus, although agronomists claimed that some improvements were introduced in 'later generation' cultivars like CSH-9 and CSH-11, farmers' scepticism did not disappear. Although hardly any respondent had ever cultivated these varieties, they claimed that these still did not meet the criteria they thought to be important.

Farmers' resistance to improved *jowar* seeds was not so much based on grain yield performance but on other elements of its relative advantage over indigenous *jowar* seeds and on its compatibility. Although the few farmers who had cultivated improved *jowar* seeds said that the first generation seeds were more afflicted by diseases and insect pests and consequently had relatively lower grain yields than the varieties which were released later, they argued that these yields were much better than yields of the indigenous varieties. In fact, superior grain yield was the one and only reason for those farmers to cultivate improved *jowar*. However, those who produced these *jowar* varieties were confronted with two main, related problems. First of all, as *jowar* was hardly marketed at supra-local level (something mentioned by Von Oppen and Rao (1982: 669-72) as well), they found it difficult to sell their produce. Second, low demand for grain of improved *jowar* varieties, a consequence of its bad taste (see later), resulted in lower farm-gate prices for this crop compared to grain of indigenous varieties. These two marketing problems certainly did not stimulate other farmers to produce improved *jowar*.

For most farmers, higher yields of grain were not a sufficient condition to cultivate improved *jowar*. Being a multi-purpose crop, *jowar* varieties were judged on more criteria than yields of grain only. A large majority of respondents asserted that these improved seeds did not meet, or only partly met, these other criteria. Almost all respondents, whether they had cultivated and tasted the improved *jowar* or not, stated that its palatability was bad and that it could not compete with the taste of the indigenous *jowar* varieties. In addition, they stated that it caused pain in the stomach and, in particular, in the joints. This deep-rooted dislike against the taste of this grain was also noticed by Walker and Ryan (1990: 50) who mentioned that in one of their study villages in Amangal *mandal* improved *jowar* was the only cereal crop for which daily labourers refused to accept daily wage payments in kind.

Another argument respondents had against improved *jowar* was the fact that its fodder was of inferior quantity and quality. Respondents claimed that since the taste of this fodder was not good, animals refused to eat it. The few animals who did, had not digested it properly and had fallen ill. Other respondents mentioned that they had been advised to chop the stalks into smaller pieces, making it more digestible for the animals. Most had refused to do so because this would take too much time. The few farmers who sold excess quantities in urban fodder markets stated that the price fetched by stalks of improved *jowar* was much lower than the price of stalks of the indigenous varieties (see also Von Oppen and Rao, 1982: 669-72).

The positive reaction of most farmers towards improved castor seeds (for paddy see note 13), in particular towards *Aruna*, had everything to do with the fact that it performed (much) better compared to the indigenous varieties. It not only gave higher yields, but it did better on all other properties which were considered important by farmers. Thus, its leaves were digestible for animals and the stalks could be used as fuel. In addition, its drought resistance, its duration and its resistance to pests and diseases were considered to be comparable or better than the indigenous varieties. These positive properties of *Aruna* paved the way for other improved castor varieties as well.[14] The high adoption rates and the fact that other improved castor varieties were also adopted make it highly plausible that the use of these kind of seeds is sustainable.

8.4 Optimum Plant Population

There were two recommendations on optimum plant population: the seed-rate (8-10 kg/ha for *jowar* and redgram each and 10-12 kg/ha for castor) and the advice to use an implement like the seed drill or the seed-cum-fertilizer drill. Although the various Action Plans of the Maheswaram project (Action Plans, 1986/7-1992/3) report about the need to use improved implements and about their distribution, they remain silent about the distribution of these specific seed-cum-fertilizer drills in Debbadaguda and Mansanpally. Moreover, project staff were unable to recollect whether or not these drills had been distributed. AWARE documents only mention the distribution of 'suitable agricultural implements', but do not elaborate on it. AWARE staff, however, claimed that seed-cum-fertilizer drills had not been distributed in Ekvaipally.

During the open interviews almost all farmers were unaware of the advised seed-rates and plant geometry. In contrast, thirty respondents said that they had heard about, or had seen, seed or seed-cum-fertilizer drills. Most of them lived in Debbadaguda (twelve farmers), followed by Ekvaipally (nine farmers), Ghatpally (six farmers) and Mansanpally (three farmers). Almost 50 per cent of these respondents were FC farmers. Most of these thirty respondents had been informed about these implements by farmers living outside the study area or had seen it on television, had read about it in the newspapers. Only three respondents (all from Debbadaguda) had been informed by the Maheswaram project, none had been informed by AWARE and some had seen it while visiting an agricultural research station. All of those who knew about these drills had never considered buying one, feeling that it was not only too expensive but also not suited for dryland cultivation. Since both projects were not very successful in informing farmers about the seed-cum-fertilizer drill and since not a single respondent owned and used this implement, the hypothesis on the effect of the two projects on adoption of these drills was rejected. The negligible adoption rate makes a comparison between communities and landholding-size classes by village and between the expected and observed adoption rates less relevant.

For lack of drills, sowing of most dryland crops was done in the indigenous way, by hand, behind the plough. While sowing, farmers kept a certain crop geometry in mind, meaning that the distance between the furrows for *jowar* and castor was such that it allowed interculturing by the *dante* and the *guntaka* respectively. Since the size of both *dante* (roughly 40 cm) and *guntaka* (roughly 70 cm) agreed more or less with the recommended inter-row distance of 60 cm for *jowar*/redgram and 90 cm for castor (some space had to be left between the implement and the plant), all farmers followed more or less the recommended inter-row spacing for both crops. Since these two implements were indigenous to the area and since the forefathers of the present farmers used the same inter-row spacing as their sons today, it is obvious that neither the Maheswaram project nor AWARE were involved in the introduction and acceptance of this particular recommendation.

The easiest and most reliable way to measure the other element of optimum plant population, the number of plants/ha, is to compare the recommended seed-rate with the actual quantity used. However, all recommended seed-rates related to improved seed varieties and for *jowar* and redgram to plots sown in a 2:1 or 3:1 ratio. Since almost none of the respondents either cultivated improved *jowar*, redgram and *bajra*, or sowed in the prescribed ratio, it is not very meaningful to measure the effect of both projects for these crops. Hence, the next section focuses on castor, for which the projects advised a seed-ratio of 10-12 kg/ha.

Averaged for two seasons, *kharif* 1992/3 and *kharif* 1993/4, and calculated for all plots and all seed varieties, 30 per cent of the castor plots were sown with the recommended seed-rate. Large differences existed between the villages: the highest was in Mansanpally (51 per cent), followed by Ekvaipally (46 per cent), Debbadaguda (29 per cent) and finally Ghatpally (9 per cent). While the observed ranking of Ghatpally was anticipated, the ranking of the other villages was not. Since it was expected that Debbadaguda would rank first, Ekvaipally second and Mansanpally third, the hypothesis was not confirmed.

Most plots received more than the recommended quantity of castor seeds. It was highest in Ghatpally (where 90 per cent of all castor plots were sown with more than the recommended quantity), followed by Debbadaguda (55 per cent), Ekvaipally (53 per cent) and finally Mansanpally (10 per cent). A clear relationship between seed-rate and holding-size or community was not discernible. Respondents seem to prefer a larger than recommended number of plants/ha because they reason that the higher the number of plants, the higher the yield will be. Except for a few plots in Ekvaipally and Debbadaguda, only farmers in Mansanpally (some 40 per cent of the plots) and in Debbadaguda (some 15 per cent) sowed less seeds than advised. Most of these plots belonged to small and small-medium farmers, farmers who could not afford to invest in relatively expensive seeds.

Not a single hypothesis regarding the effect of project intervention on the seed-rate for castor used by the various communities in the study villages was confirmed. In Debbadaguda, the highest average adoption rates were for FC respondents, followed by SC, ST and BC respondents.[15] Since BC respondents performed worse than was expected, this ranking was not completely in line with the hypothesis. Hence, it was neither confirmed nor rejected. In Mansanpally, while FC and SC respondents performed better, BC and ST respondents performed worse than expected. Thus, the hypothesis was rejected. In Ekvaipally, on average BC respondents performed slightly better than FC respondents, followed by SC and ST respondents. Because this observed ranking differed completely from what was expected, the hypothesis was rejected. In Ghatpally, the observed ranking (on average SC and ST respondents ranked first, followed by BC and FC respondents) deviated from the expected ranking as well. Hence, the hypothesis was rejected too.

Based on the comparison of the expected and observed average adoption rates of the communities between the four study villages, the hypothesis was rejected. Contrary to expectations, the averages of FC and BC respondents were not highest in Debbadaguda but in Mansanpally. The idea that the averages of FC and BC respondents were more or less similar in Mansanpally, Ekvaipally and Ghatpally, proved wrong too. Moreover, while it was expected that the averages of SC and ST respondents would be highest in Ekvaipally, they were also highest in Mansanpally.

Not a single hypothesis regarding the effect of project intervention on the seed-rate for castor used by the various landownership classes in the study villages was confirmed either. In Debbadaguda, the ranking of the various classes was almost up to expectations.[16] However, since marginal farmers performed better than expected, the hypothesis was neither confirmed nor rejected. The hypotheses for Mansanpally, Ekvaipally and Ghatpally were all rejected. In Mansanpally, medium, small-medium and marginal farmers performed better than expected while small farmers performed worse. In Ekvaipally, small-medium and medium farmers performed better than anticipated, while small farmers performed worse. In Ghatpally, the average adoption rates of marginal and small farmers was much better and that of large farmers much worse than expected.

Since a large difference existed between the expected and the observed average adoption rates of each landowner class by village, this hypothesis was also rejected. While it was expected that the highest average adoption rates for large, medium and small-medium farmers would be found in

TABLE 8.6. COMPARISON BETWEEN OBSERVED ADOPTION RATES AND HYPOTHESES FOR SEED-RATE OF CASTOR

Intervention level	Confirmed	Neither confirmed nor rejected	Rejected
Village			Yes
Community in village		D	M, E, G
Community between villages			Yes
Landownership class in village		D	M, E, G
Landownership class between villages			Yes

Note: D = Debbadaguda; M = Mansanpally; E = Ekvaipally; G = Ghatpally.
Source: Own survey.

Debbadaguda, these were highest for large farmers only. While it was expected that the highest averages for small and marginal farmers would be found in Ekvaipally, in reality they were highest in Mansanpally and Debbadaguda respectively.

These results suggest that as far as farmers had adopted the recommendations for an optimum plant population, farmers had not been influenced by the projects. First of all, both projects failed to inform and to induce respondents to use improved implements like a seed drill or a seed-cum-fertilizer-drill. Second, the fact that farmers sowed *jowar* and castor at the prescribed distance had less to do with project intervention and more with the size of their indigenous blade harrows. Third, being unaware of the recommended seed-rate for castor, most respondents sowed more than the advised quantities. Asked whether the recommended quantities would produce similar or better yields, those respondents using higher quantities of seeds argued that they felt that lower quantities would produce less plants and hence lower yields. Whether this feeling is correct or not may be a subject of additional research.

8.5 Improved Tillage

Since the outcome of the recommended preparatory tillage operations (off-season tillage) were analysed in the preceding chapter under *in-situ* water conservation practices (7.3.1), and since part of the outcome of the recommended interculturing operations were analysed in that chapter under contour cultivation (7.3.2), other parts will be analysed in this chapter under weeding and interculturing (8.9), this section will focus on a recommended implement that improves preparatory tillage operations in dryland: the iron plough.[17]

Both the Maheswaram project and AWARE followed their usual two-way strategy to popularize the iron plough. Their regular communication channels had to inform the villagers and they also distributed subsidized ploughs. As with other subsidized inputs, both projects were not able to hand over information about the number of iron ploughs given in the project villages and the names of the beneficiaries. From the discussions with project officers and farmers it became clear that while the Maheswaram project had distributed some iron ploughs to farmers in Mansanpally (particularly to ST farmers) it had not done so in Debbadaguda. Although AWARE stated that it had distributed improved implements in Amangal *mandal*, field staff could not confirm that iron ploughs were given to farmers in Ekvaipally.

Although they were sometimes mixed up with the ploughs used by tractors, most respondents in the four villages stated during the open interviews that they knew about iron ploughs. However, only a few of these farmers said that their knowledge was based on information provided by the extension service of both the projects. In fact, their answers suggested that large FC farmers together with the radio, newspapers and even television, were more important sources of information. The fact that only some of the respondents had seen it being used, mostly in faraway places, also indicates that the projects were a less important source of information. This is in line with the finding that the seven respondents who owned such an implement in *kharif* 1993/4, hardly used it. Of these seven respondents, three lived in Debbadaguda, three in Mansanpally and one in Ghatpally. The one and only respondent in Ekvaipally who had possessed an iron plough, had sold it in 1993/4. All except one respondent, who belonged to the BC community in Mansanpally, were medium to large FC farmers. Considering a low adoption rate of 2 per cent and the fact that these ploughs were hardly used leads to the conclusion that neither project had much of an effect. Hence, the hypothesis concerning this implement was rejected. This low adoption rate makes a comparison between communities and landholding-size classes by village and between the expected and observed adoption rates less relevant.

There were a few factors responsible for this low adoption rate. Since many farmers knew about iron ploughs, lack of information does not seem to be an important one. Yet, many respondents only knew that it existed and were unfamiliar with its advantages. In fact, they only mentioned disadvantages, claiming, for instance, that iron ploughs ploughed too deep, making ploughing difficult and fatiguing for farmers and bullocks. In addition, they said that the deep furrows produced by these ploughs were detrimental to the texture of the soil, stimulated evaporation and forced them to place seeds at too deep a level. Whether these claims are correct or not is not known to me.[18] Whatever the reasons, both projects were not able to convince farmers to the contrary. What was certainly true was the complaint that iron ploughs were 2 to 3 times more expensive than indigenous ones and could not be made from local materials by local artisans. Thus, the relative advantage of these improved implements was not sufficient to induce farmers to adopt them.

Other factors, partly caused by weak project performance (for instance insufficient extension), were disinterest and reluctance of farmers. It is apparent from the fact that in due course all iron ploughs distributed to ST farmers in Mansanpally were sold to larger farmers and that too mainly to blacksmiths. Some ST farmers mentioned that they had sold the ploughs because they needed the money. Moreover, they said they were not in need of iron ploughs and had not asked for it. In addition, not being properly informed about its use and possible advantages, they thought that iron ploughs would be less useful than the indigenous ones. Similar to the ST farmers, the few Reddy farmers in Mansanpally who had received such a plough from the project or had bought it from ST farmers were not always convinced about its advantages. In fact, most had sold these ploughs as well.[19]

8.6 Improved Application of Organic Fertilizers

From discussions with project officers of both projects and agronomists I concluded that the recommended dose of Farm Yard Manure (FYM) for dryland was 5,000 kg/ha per year, to be applied some four to six weeks before sowing.[20] This recommendation related to improved seed varieties of all kinds of dryland crops. Recommendations for indigenous seed varieties were virtually non-existent and did not figure in books or leaflets, for instance.

Both the Maheswaram project and AWARE encouraged the use of FYM by informing the villagers about its benefits using their regular communication channels. In addition, realizing that insufficient and low quality FYM was produced, the Maheswaram project also extended financial support to farmers to dig compost pits and informed them about the most suitable way to make compost. However, except for the number of pits dug in Mansanpally (76, which incidentally again indicates that the project did not limit its activities to the area near the *tanda*) the project was not able to supply any details about the number of pits dug in Debbadaguda (most probably a couple of dozen), or about the names of the beneficiaries.

All respondents living in the study villages indicated during the open interviews that they knew about the recommendation to apply some 10 cart-loads of manure per ha every year. However, they mentioned that this knowledge was indigenous to the area, that they had heard it from their forefathers. A few respondents added that staff of both projects had advised them similarly. Whether they possessed a subsidized manure pit or not, most respondents in the project villages explained that they were not properly informed about composting and threw everything in it, as usual. In addition, most of them said that the projects had not informed them about the proper timing of the application of FYM.

From the open interviews with Mansanpally and Debbadaguda respondents, it was learnt that those who had received a subsidized pit valued it greatly, although they considered the indemnification rather poor. However, irrespective of these pits, none of these farmers had altered his/her way of composting for reasons of insufficient inputs (dung, straw and other crop and household residues) lack of time and interest and, above all, lack of knowledge.

On average during *kharif* 1992/3 and *kharif* 1993/4, 33 per cent of the *jowar* plots were manured with FYM four to six weeks before sowing. The recommendation was followed best in Ghatpally (56 per cent of the *jowar* plots received manure in time), followed by Debbadaguda (33 per cent), Mansanpally (22 per cent) and Ekvaipally (only 5 per cent). For castor, the recommendation was followed on 24 per cent of the plots. Again, more plots received manure in the recommended period in Ghatpally (34 per cent), followed by Ekvaipally (24 per cent), Debbadaguda (7 per cent) and Mansanpally (none). Except for Mansanpally, where respondents preferred to manure their castor plots within three weeks before sowing, a rather high percentage of castor plots located in the other three villages had manure applied together with *jowar* plots. Many respondents stated that they preferred to manure the entire plot in one go, regardless of the crop, for reasons of efficiency. These results indicate that a small minority followed the recommendation to manure in time. With Ghatpally, the village without project intervention, having the highest adoption rates, the observed figures were not in line with what was expected. On this basis, the hypotheses on the effect of the projects on timely application of FYM to *jowar* and castor were rejected.

Mostly, FYM was transported to the plot by bullock-cart, sometimes by tractor. Informants claimed that the average bullock-cart carried some 500 kg of FYM. It meant that the recommended quantity for dryland was some 10 cartsloads. Depending on size of the cart and its contents, all bullock-cart-loads were converted to this average load. The same procedure was followed for trailer-loads which were converted into average bullock-cart-loads.

All respondents valued FYM greatly, even more than inorganic fertilizers. However, lacking sufficient quantities, they applied less than what they considered optimal. That farmers were in short supply of FYM is evident from the data. On average 55 per cent of the dryland plots cultivated with *jowar* and 58 per cent of the dryland plots cultivated with castor in the four study villages did not receive any FYM during the three years of the study (*kharif* 1991/2-1993/4).[21] In all these years, the performance of farmers in Debbadaguda and Mansanpally was the worst. In Debbadaguda on average 79 per cent of the *jowar* plots and 73 per cent of the castor plots did not get manure, while the averages for Mansanpally were 62 per cent for *jowar* plots and 68 per cent for castor plots. In contrast, Ekvaipally (where on average 39 per cent of the *jowar* and 41 per cent of the castor plots did not get manure) and Ghatpally (40 per cent for *jowar* plots and 50 per cent for castor plots) performed better in all years, which is no surprise given the fact that animal husbandry was slightly more important in these villages than in the other two.

If dryland plots received manure, farmers applied on average a little more than the recommended 10 cart-loads/ha. Disregarding those *jowar* plots which did not receive manure, the average quantity applied during the study period was 13 cart-loads/ha.[22] If manure was applied, *jowar* plots in Ghatpally and Ekvaipally received the highest average quantity (14 cart-loads/ha in both villages), while Debbadaguda and Mansanpally lagged behind (11 cart-loads/ha in both villages).

As far as castor was concerned and disregarding those plots which did not receive manure during the research period, on average 13 cart-loads/ha were applied, a quantity similar to that applied to *jowar.*[23] Again, on average, manured plots in Ekvaipally and Ghatpally received more (14 and 13 cart-loads/ha respectively), than in Debbadaguda and Mansanpally (11 and 8 cart-loads/ha respectively).

Not surprisingly given the fact that so many dryland plots did not receive manure, the percentage of *jowar* and castor plots that received the recommended quantity of FYM was small. Averaging for *kharif* 1991/2-

1993/4, 24 per cent of all *jowar* plots and 22 per cent of all castor plots received 10 cart-loads/ha or more.[24] Figures 8.7 and 8.10 demonstrate that on average, most of the *jowar* and castor plots which received the recommended dose were located in Ekvaipally (38 per cent for *jowar* and 37 per cent for castor), followed at a short distance by Ghatpally (33 per cent for *jowar* and 25 per cent for castor). Both Mansanpally (15 per cent for *jowar* and 9 per cent for castor) and Debbadaguda (9 per cent for *jowar* and 14 per cent for castor) lagged behind. As these observed figures were at variance with the expectations (see Figure 4.1), the hypotheses on project influence on the recommended dose of FYM to be applied to *jowar* as well as castor were rejected.

Not a single hypothesis on the effect of project intervention on the recommended dose of FYM applied to *jowar* by the various communities in the study villages was confirmed. Figure 8.8 shows that, except for Debbadaguda, on average more SC respondents had adopted the recommendations for *jowar* than farmers of other communities. What is evident from the figure too is that FC farmers almost always had the lowest adoption rates.[25] In fact, none of the FC respondents in Debbadaguda had adopted the recommendation. The average adoption rates of the other communities did not differ very much. Since this was not anticipated (see Figure 4.2), the hypothesis for Debbadaguda was rejected. In Mansanpally and Ekvaipally, the hypothesis was neither confirmed nor rejected. The average adoption rate of SC respondents in Mansanpally was somewhat higher than expected, while the observed ranking of the other communities was similar to what was anticipated. In Ekvaipally, ST respondents performed worse than expected, while the ranking of the other communities was more or less up to expectations. Because FC respondents in Ghatpally performed worse than expected, while SC respondents performed better, the hypothesis for this village was rejected.

While it was expected that the highest average adoption rates for FC and BC respondents would be found in Debbadaguda, this proved wrong. On the other hand, the expectation that the highest adoption rates of SC and ST respondents would be found in Ekvaipally, proved right. Nevertheless, the hypothesis was rejected.

Not surprisingly, since adoption rates of SC respondents were higher than that of other communities, a significant correlation between the number of cart-loads/ha applied to *jowar* and holding-size did not exist. Neither for the total, nor for individual villages.[26] Figure 8.9 illustrates this finding. In most villages, the middle categories of farmers had higher

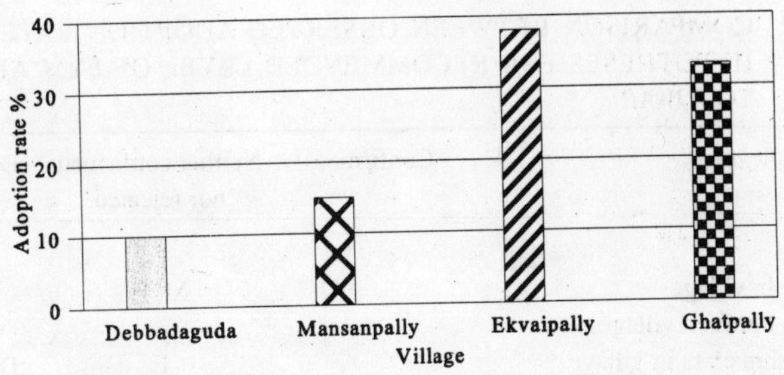

Source: Own survey.

FIGURE 8.7. OBSERVED ADOPTION RATE (%) OF FYM APPLIED TO *JOWAR*, BY VILLAGE (*kharif* 1991/2-1993/4).

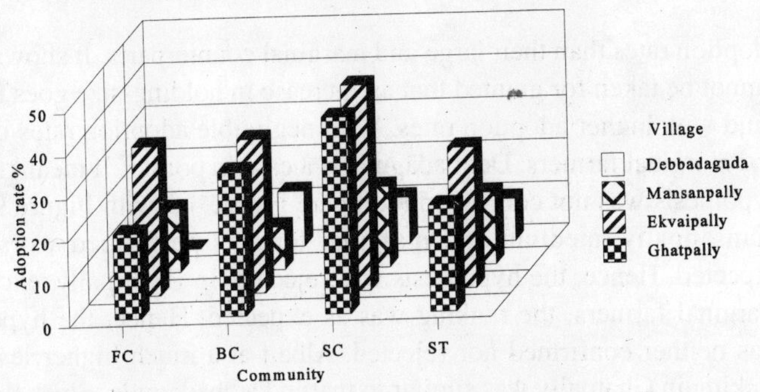

Source: Own survey.

FIGURE 8.8. OBSERVED ADOPTION RATE (%) OF FYM APPLIED TO *JOWAR*, BY COMMUNITY, BY VILLAGE (*kharif* 1991/2-1993/4).

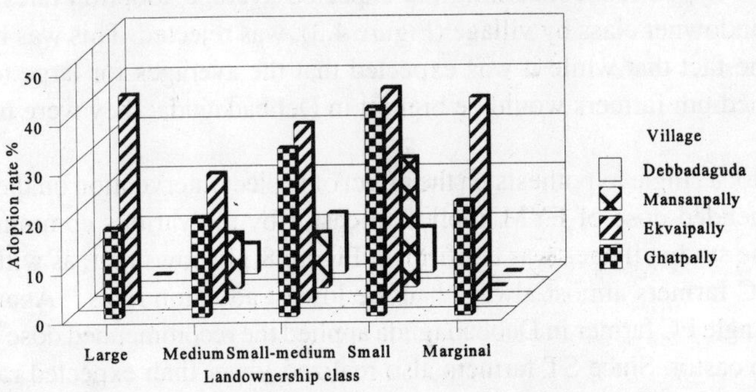

Source: Own survey.

FIGURE 8.9. OBSERVED ADOPTION RATE (%) OF FYM APPLIED TO *JOWAR*, BY LANDOWNERSHIP CLASS, BY VILLAGE (*kharif* 1991/2-1993/4).

TABLE 8.7 COMPARISON BETWEEN OBSERVED ADOPTION RATES AND HYPOTHESES FOR RECOMMENDED LEVEL OF FYM APPLIED TO *JOWAR*

Intervention level	Confirmed	Neither confirmed nor rejected	Rejected
Village			Yes
Community in village		M, E	D, G
Community between villages			Yes
Landownership class in village		E	D, M, G
Landownership class between villages			Yes

Note: D = Debbadaguda; M = Mansanpally; E = Ekvaipally; G = Ghatpally.
Source: Own survey.

adoption rates than their large and marginal counterparts. It shows that it cannot be taken for granted that an increase in holding-size goes hand in hand with higher adoption rates. With negligible adoption rates of large and marginal farmers, Debbadaguda is a case in point.[27] It meant that the hypothesis was not confirmed (compare Figure 4.3 with Figure 9.9). In Mansanpally, medium and marginal farmers performed worse than expected. Hence, the hypothesis was rejected. In Ekvaipally, except for marginal farmers, the ranking was as expected. Hence, the hypothesis was neither confirmed nor rejected. Albeit at a much higher level, the ranking in Ghatpally was similar to that in Debbadaguda. Since this runs counter to expectations, the hypothesis was rejected.

Although the expectation that the highest averages for small and marginal farmers would be found in Ekvaipally, proved right (Figure 8.9), the hypothesis concerning the expected average adoption rates of each landowner class by village (Figure 4.3), was rejected. This was based on the fact that while it was expected that the averages for large to small-medium farmers would be highest in Debbadaguda, they were not.

Not a single hypothesis on the effect of project intervention on the recommended dose of FYM applied to castor by the various communities in the study villages was confirmed. Figure 8.11 shows that, as with *jowar*, FC farmers almost always had the lowest adoption rates.[28] Again, not a single FC farmer in Debbadaguda applied the recommended dose of FYM to castor. Since ST farmers also realized lower than expected rates (see Figure 4.2), the hypothesis was rejected. In Mansanpally, FC farmers had a zero adoption rate too. Since SC respondents in this village performed better than anticipated, the hypothesis was also rejected. In Ekvaipally,

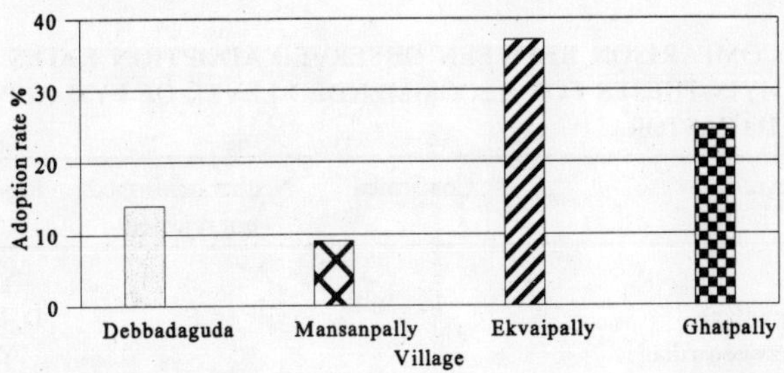

Source: Own survey.

FIGURE 8.10. OBSERVED ADOPTION RATE (%) OF FYM APPLIED TO CASTOR, BY VILLAGE (*kharif* 1991/2-1993/4).

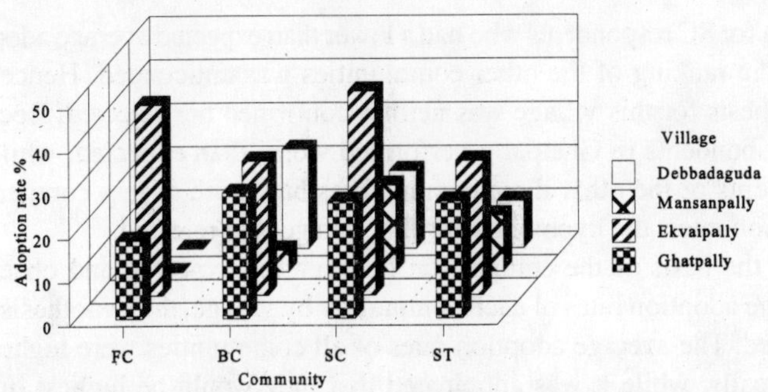

Source: Own survey.

FIGURE 8.11. OBSERVED ADOPTION RATE (%) OF FYM APPLIED TO CASTOR, BY COMMUNITY, BY VILLAGE (*kharif* 1991/2-1993/4).

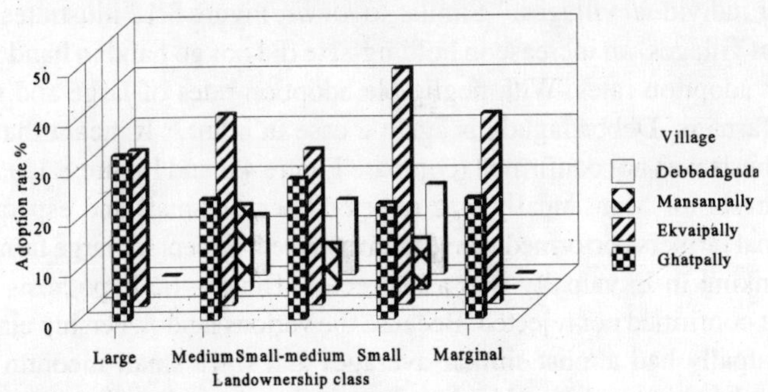

Source: Own survey.

FIGURE 8.12. OBSERVED ADOPTION RATE (%) OF FYM APPLIED TO CASTOR, BY LANDOWNERSHIP CLASS, BY VILLAGE (*kharif* 1991/2-1993/4).

TABLE 8.8. COMPARISON BETWEEN OBSERVED ADOPTION RATES AND HYPOTHESES FOR RECOMMENDED LEVEL OF FYM APPLIED TO CASTOR

Intervention level	Confirmed	Neither confirmed nor rejected	Rejected
Village			Yes
Community in village		E	D, M, G
Community between villages			Yes
Landownership class in village		E	D, M, G
Landownership class between villages			Yes

Note: D = Debbadaguda; M = Mansanpally; E = Ekvaipally; G = Ghatpally.
Source: Own survey.

except for SC respondents who had a lower than expected average adoption rate, the ranking of the other communities was anticipated. Hence, the hypothesis for this village was neither confirmed nor rejected. Because FC respondents in Ghatpally performed worse than expected, while respondents of the other three communities had more or less comparable adoption rates, the hypothesis for this village was rejected.

On the basis of the comparison between the expected and observed average adoption rates of each community by village, the hypothesis was rejected. The average adoption rates of all communities were highest in Ekvaipally, while it was anticipated that they would be highest in this village for SC and ST respondents only.

As with *jowar*, a significant correlation between the number of cart-loads/ha applied to castor and holding-size did not exist; neither for the total, nor for individual villages.[29] Similar to *jowar*, Figure 8.12 illustrates that in most villages, an increase in holding-size did not go hand in hand with higher adoption rates. With negligible adoption rates of large and marginal farmers, Debbadaguda is again a case in point.[30] It meant that the hypothesis was not confirmed (compare Figure 4.3 and Figure 8.12). The hypothesis for Mansanpally was rejected because small and especially marginal farmers performed worse than expected. Except for large farmers, the ranking in Ekvaipally was as expected. Hence, the hypothesis was neither confirmed nor rejected. Because the various landownership classes in Ghatpally had almost similar averages and since small-medium and marginal farmers achieved higher than expected results, the hypothesis was rejected.

Except for large farmers in Ghatpally, all holding-size categories realized the highest average adoption rates in Ekvaipally (Figure 8.12). Since this

TABLE 8.9. RATE OF ADOPTION OF RECOMMENDED QUANTITY OF FYM APPLIED TO *JOWAR* AND CASTOR, BY TYPE OF FARMER, BY VILLAGE

Type of farmer	Debbadaguda		Mansanpally		Ekvaipally		Ghatpally		Total		N plots	
	J	C	J	C	J	C	J	C	J	C	J	C
Pure dryland farmers	5	12	16	11	44	40	27	24	22	21	81	64
Farmers with < 1 ha wetland	15	21	17	6	36	37	43	24	27	23	101	63
Farmers with ≥ 1 ha wetland	5	5	5	8	33	34	25	36	19	21	25	25
Average	10	14	15	9	38	37	33	25	24	22	207	152

Source: Own survey.

ran counter to the expectations (see Figure 4.3), the hypothesis concerning the expected average adoption rates of each landowner class by village, was rejected.

Table 8.9 demonstrates that for almost all villages, farmers owning more than 1 ha of wetland applied less FYM to *jowar* and castor than other farmers. Together with the insignificant correlations and the finding that FC respondents (who owned most wetland) applied less FYM to these dryland crops than other farmers did, it is another indication that farmers preferred to manure wetland crops above dryland crops. In fact, during *kharif* 1992/3-1993/4, paddy received a disproportionate share of FYM. On average, if manure was applied, it received twice the amount applied to *jowar* or castor: 25 cart-loads/ha or somewhat more than recommended. Paddy plots in Debbadaguda received the highest quantity. If given manure, they received on average 30 cart-loads/ha. The averages per hectare for Ghatpally, Ekvaipally and Mansanpally were 24 cart-loads, 23 cart-loads and 20 cart-loads respectively.

In spite of these higher than recommended doses for paddy, FYM was in short supply. Although more paddy plots received manure according to the recommendations than *jowar* or castor plots (averaging 27 per cent for all study villages, against 24 and 22 per cent respectively), 33 per cent received an insufficient quantity, while the remaining 40 per cent did not get manure at all. That also large farmers were not always able to apply sufficient quantities of FYM to paddy is clear from the fact that a significant positive correlation between the number of cart-loads/ha and holding-size existed only at the level of all villages and for Debbadaguda.[31]

The analysis suggests that since farmers' adoption rates in Debbadaguda and Mansanpally were less than anticipated, the effect of the Maheswaram project on the level of FYM applied to these two dryland crops was negligible. On the other hand, the fact that adoption rates in Ekvaipally were better than expected, suggests some effect of AWARE. This might be the result of AWARE's animal husbandry programme, through which villagers received plough bullocks and dairy cattle, and indirectly dung. However, since the two hypotheses of this village, concerning community and landownership class, were not confirmed, the effect of AWARE's intervention is somewhat questionable.

That especially the Maheswaram project was not very successful was a result of its focus on information supply and the digging of compost pits. All project officers could do was advise farmers to dig pits and apply manure in suitable quantities. However, most farmers had pits or heaps and already knew by experience how much FYM was necessary. What farmers needed was a higher supply of FYM, that is more livestock to increase manure production, but neither FYM nor livestock were supplied by the project. Since the Maheswaram project could not influence the supply of FYM and since AWARE had only limited means to do the same, both projects focused on the supply of inorganic fertilizers.

The fact that FYM was valued by all farmers, that it was valued more than chemical fertilizers, that it was indigenous to the area and that it was applied by many respondents to dryland crops, makes it highly probable that it is a sustainable technology in dryland. But, whether it will be used by all farmers for all dryland crops in all years, depends largely on its availability.

8.7 Moderate Use of Inorganic Fertilizers

Respondents guessed that chemical fertilizers were introduced in the study villages around 1970 or before the commencement of both the projects. That both the projects did not start from scratch was also indicated by Sitapathi Rao et al. (1990: 78), APAU (nd: 39-40) and Walker and Ryan (1990: 53) who state that inorganic fertilizers were propagated in the Maheswaram and Amangal project areas since the late 1960s and early 1970s.[32] Their diffusion resembled that of improved seeds. After having heard from various sources like fellow farmers in other villages, shopkeepers, newspapers and the radio of the benefits of these inorganic fertilizers, a couple of mostly large Reddy farmers (who had the resources and could bear the risks) began experimenting. The first trials were predominantly undertaken with wetland crops like paddy. After having seen the benefits, other farmers followed and slowly the use of inorganic fertilizers spread in the villages. Getting to know the benefits in irrigated agriculture, the process of experimentation started in dryland agriculture. Again, mostly large Reddy farmers took the initiative to adopt and other farmers followed. Without exception, respondents were very pleased with inorganic fertilizers, arguing that it increased yields while financial returns were better as well. In spite of this positive attitude, many farmers indicated that they felt trapped in a kind of vicious circle feeling that crops had become more or less addicted to these fertilizers. Whether it was crops or farmers who had become habituated to these fertilizers, farmers believed that using less of these would decrease yields.

Principles enhancing the efficiency and effectiveness of fertilizer use for dryland crops, placement, timing and the recommended dose, will be discussed in the following subsections.

8.7.1 MODERATE USE OF INORGANIC FERTILIZERS: PLACEMENT

Placing inorganic fertilizers in or near the rows of dryland crops, at appropriate depth and at safe distance from the seeds, was considered to be more efficient by the projects than broadcasting. Since *jowar* was sown early in the season in a relatively dry soil, they advised application of a basal dose of chemical fertilizers at a safe distance from the seeds, preferably not in one and the same furrow. For castor, sown in a relatively moist soil, they advised farmers to apply the basal dose in the same furrow. Both projects considered the seed-cum-fertilizer drill to be the appropriate implement for this task.

The extension services of the projects had a negligible effect on farmers' knowledge about the placement of inorganic fertilizers. During the open interviews, only a few respondents claimed to have been informed by the Maheswaram project or AWARE about this recommendation. Notably those who had visited the CRIDA agricultural research station (some as a trainee of the Maheswaram project, but other farmers not because of project intervention), said that they were told to apply the fertilizer close to the seed by hand, using a match-box. As explained in the section on optimum plant population, although quite a few farmers knew about the seed-cum-fertilizer drill, it was not available and not used in the study villages.

Whatever the lack of advice, broadcasting of inorganic fertilizers for *jowar* and castor was not the usual practice in the study villages. For quite a long time, and according to respondents before project intervention, fertilizers were placed with the seeds in the same furrow while sowing behind the indigenous plough. Irrespective of whether they sowed *jowar* or castor, this method of fertilizer application was similar and seeds and fertilizers were placed in the same furrow without much spacing. Thus, while broadcasting did not stop because of project intervention neither did farmers alter their usual practice. This negligible adoption rate leads to the conclusion that neither project had an appreciable effect. Hence, the hypotheses concerning fertilizer placement for *jowar* and castor were rejected. This negligible adoption rate makes a comparison between communities and landholding-size classes by village and between the expected and observed adoption rates less relevant.

Fertilizer application did not conflict with, but even fitted perfectly in indigenous sowing operations. The only adaptation was the extra (female) labour input needed while combining sowing and fertilizing behind the plough. Since this additional labour input often originated from their own household, farmers considered the extra costs to be relatively small. Respondents said that ploughing of additional furrows to apply fertilizers in the recommended manner was not worthwhile since the costs of labour and bullocks would double.

8.7.2 MODERATE USE OF INORGANIC FERTILIZERS: TIMING

Since rainfall in semi-arid areas is unpredictable, the projects advised farmers to split the application of chemical fertilizers, particularly of nitrogen. A basal dressing had to be applied at sowing, while the top-dressing had to be applied when the crop was at knee height, some four weeks after sowing for *jowar* and six weeks for castor. Since the probability that inorganic fertilizers would burn the crop was rather high in a relatively dry soil, both doses should be applied only if the moisture content in the soil was adequate.

As with placement, the extension services of both projects had a limited effect on farmers' knowledge about the timing of inorganic fertilizers for dryland crops. During the open interviews fifteen respondents said they knew that top-dressing was advised for dryland crops. Most of them lived in Debbadaguda (six respondents), followed by Ekvaipally (five respondents), Ghatpally (three respondents) and Mansanpally (one respondent). Only two farmers in Debbadaguda mentioned that they were informed by the Maheswaram project, the main source of information for the other respondents being fellow farmers. This did not mean, however, that other farmers were completely unaware of top-dressing; almost all respondents knew that it was used in paddy cultivation. Like so many agricultural innovations, most respondents had heard about top-dressing for paddy from large Reddy farmers, had applied such a top-dressing while working for them as agricultural labourer or had seen it being used in their paddy plots.

Given the fact that not many respondents were aware of top-dressing for dryland crops it is no surprise that during the three-year research period only one respondent (a small BC farmer living in Debbadaguda) applied a top-dressing to *jowar* and four respondents (three in Debbadaguda (one small and one small-medium BC farmer and one medium FC farmer and one small-medium BC farmer in Ekvaipally) gave a top-dressing to castor. Since it suggests a negligible influence of both projects on top-dressing applied to *jowar* and castor, the hypotheses were rejected. The low adoption rate makes a comparison between communities and landholding-size classes by village and between the expected and observed adoption rates less relevant.

After having informed the respondents about top-dressing for dryland crops, most said that the uncertainty of the monsoon would prevent them from following this advice. It was their firm belief that while rainfall was more assured in the last week of May and the first week of June, the fortnight most suitable for sowing *jowar*, it was less certain some four weeks later, when top-dressing had to be applied. For castor, more or less the same reason was given. However, since top-dressing had to be applied relatively late in the season (at a time that the monsoon had progressed), top-dressing for castor was more acceptable to farmers than for *jowar*. By the way, most farmers preferred to split the limited quantity of chemical fertilizers they could afford to apply, in one dose for *jowar* and one for castor. Another reason why respondents were not in favour of top-dressing was the higher labour input involved.

8.7.3 MODERATE USE OF INORGANIC FERTILIZERS: QUANTITY

The projects advised farmers to apply 40 kg N and 30 kg P per ha to *jowar* and 50 kg N and 30 kg P per ha to castor. These were recommendations for improved seed varieties. However, as with FYM, I presume that they applied these to indigenous seed varieties as well. What stands out from this advice, is that it was given in nutrients, a concept unfamiliar to most farmers. All they knew were the names of a limited number of inorganic (complex) fertilizers like (diammonium phosphate) and Growmore, not their exact nutrient content. Formulas like 28:28:28 or

17:17:17 rang only a bell for a small group of educated, large farmers, mostly Reddies.

To encourage the use of inorganic fertilizers for dryland crops, both the Maheswaram project and AWARE used their regular two-way strategy, informing farmers during their normal extension routine and subsidizing this input. Most of the leaflets and brochures distributed by the Maheswaram project as well as the written material distributed by the T&V system (the service called upon by AWARE) gave the recommended quantities in nutrients only. Extension officers of the Maheswaram project, T&V extension officers and the staff of AWARE reported that they used the equivalent in bags of Growmore, DAP and urea as well. From 1984/5 to 1991/2, the Maheswaram project distributed inorganic fertilizers to small and marginal, ST and SC farmers, sometimes free of costs and sometimes with a 50 per cent subsidy. AWARE distributed subsidized fertilizers to the members of the *sanghams* in the Amangal area for some 10 years after 1976. Subsidies were no longer given by the projects at the time of the fieldwork, though. However, as an element of its empowerment strategy, AWARE stimulated *sanghams* to get access to government programmes. Thus, one might expect that these tried to secure subsidized inorganic fertilizers for their members.

When asked for the recommended doses during the open interviews, almost all respondents kept silent or just hazarded guesses. Invariably, the quantities mentioned were below the recommended levels, not only for *jowar* and castor, but for paddy too. A couple of farmers who had visited the CRIDA research centre in Hyatnagar, said that all they knew was to use one match-box full of inorganic fertilizers per seed, an advice they did not follow because the quantity was considered to be too high. Many of those respondents who were informed by the projects stated that the only advice given was to use chemical fertilizers, not what quantity to apply. Hence, most respondents had learnt what was best for a certain crop by observing and asking other farmers and by trial and error. Many farmers, reasoning along the lines of the law of diminishing economic returns, maintained that the dose they applied was the optimum dose. Use of higher quantities of chemical fertilizers was not advisable, they felt, since it would only increase costs and hardly increase yields.

For dryland crops, DAP and Growmore were used most. Although it was more expensive, farmers preferred Growmore to DAP because they felt that it gave better results. However, since Growmore was not always available (according to some respondents because of hoarding by shopkeepers), DAP was used more.[33] Some respondents, notably large FC farmers, tried to 'upgrade' DAP by mixing it with urea.

In striking contrast to organic fertilizers (where on average 55 per cent of the plots cultivated with *jowar* and 58 per cent of the plots cultivated with castor did not receive manure), averaged for the entire study period and the four villages, only 6 per cent of the *jowar* plots and 5 per cent of the castor plots did not receive inorganic fertilizers.[34] A large difference existed between, on the one hand, Debbadaguda, Ghatpally and Mansanpally (with 5 per cent and less of the cultivated plots that did not receive chemical fertilizers during this period) and, on the other hand, Ekvaipally (with 17 per cent of the *jowar* plots and 13 per cent of the castor plots which were not given fertilizers). Almost all dryland plots without fertilizers belonged to marginal, small and small-medium farmers and only very few of such dryland plots were owned by FC farmers. Except for three farmers, all farmers in the entire sample applied chemical fertilizers to *jowar* or castor at least once during the study period.

Disregarding plots that did not get fertilizers, the average quantity applied during *kharif* 1991/2 to 1993/4 to *jowar* was 56 kg/ha.[35] It decreased over the years, from on average 68 kg/ha in *kharif* 1991/2 to 58 kg/ha in *kharif* 1992/3 and 44 kg/ha in *kharif* 1993/4. This might have been the result of the Indian economic liberalization programme which led to an increase in fertilizer prices after 1991/2. If given fertilizers, with averages of 61 kg and 59 kg/ha respectively, *jowar* plots located in Debbadaguda and Mansanpally received somewhat higher doses than plots located in Ekvaipally (56 kg/ha) and especially Ghatpally (50 kg/ha).

If fertilizers were applied, castor plots received on average 59 kg/ha, an average more or less similar to *jowar*. As with *jowar*, the average quantity decreased over the years, from 69 kg/ha in *kharif* 1991/2 to 57 kg/ha in *kharif* 1992/3 and 50 kg/ha in *kharif* 1993/4. Similar to *jowar*, with an average of 65 kg/ha and 61 kg/ha respectively castor plots in Mansanpally and Debbadaguda received somewhat higher doses than plots situated in Ghatpally (56 kg/ha) and Ekvaipally (55 kg/ha).

The average quantities of inorganic fertilizers applied to *jowar* and castor were in contrast to those of FYM. While respondents in Ghatpally and Ekvaipally applied on average more FYM to their dryland plots than respondents in Mansanpally and Debbadaguda, they applied on average less inorganic fertilizers than these latter respondents. It was the other way around in Debbadaguda and Mansanpally, as if respondents in these two villages wanted to compensate for their relatively lower gift of FYM. As stated earlier, this lower gift was a consequence of the fact that animal husbandry was slightly less important in these two villages compared to Ghatpally and Ekvaipally.

To calculate whether respondents followed the advice of the projects, the recommended dose in nutrients/ha was converted into kg/ha of Growmore, DAP and urea. For *jowar* the advice read 66 kg/ha of DAP as basal, 66 kg/ha of urea as top, making a total of 132 kg/ha or 100 kg/ha of Growmore as basal, 25 kg/ha of urea as top, making a total of 125 kg/ha. For castor it meant 66 kg/ha of DAP as basal, 88 kg/ha of urea as top, making a total of 154 kg/ha or 100 kg/ha of Growmore as basal and 50 kg/ha urea as top, making a total of 150 kg/ha. The percentage of *jowar* and castor plots that received this recommended dose was very small, less than 1 per cent of both *jowar* and castor plots to be precise. Since such a percentage does not allow for a meaningful analysis, the analysis is made on the basis of those plots which received at least 50 per cent of the total (basal and top-dressing combined) recommended dose.

Averaging for *kharif* 1991/2-1993/4, 15 per cent of all *jowar* plots and 13 per cent of all castor plots received at least 50 per cent of the recommended dose.[36] Figures 8.13 and 8.16 show that on average, most of the *jowar* and castor plots that surpassed this threshold were located in Mansanpally (18 per cent for *jowar* and 17 per cent for castor) followed closely by Debbadaguda (17 per cent for *jowar* and 14 per cent for castor) and Ekvaipally (15 per cent for *jowar* and 13 per cent for castor). Ghatpally (9 per cent for *jowar* and 10 per cent for castor) lagged behind. As was mentioned earlier, this outcome is somewhat contradictory to the finding concerning FYM, where Ghatpally topped the ranking, followed by Ekvaipally, and where Mansanpally and Debbadaguda fell behind. This ranking was almost as anticipated (see Figure 4.1). However, because Mansanpally performed so much better than expected, the hypotheses for *jowar* as well as castor were neither confirmed nor rejected.

Figure 8.14 reveals that the relationship between community and the 50 per cent threshold level for fertilizer applied *jowar* plots was not very distinctive, not a single community stood out as best or worst performer.[37] While in Ekvaipally FC respondents performed best, in Debbadaguda they ranked last. In Mansanpally, ST respondents topped the list and in Debbadaguda this place was obtained by BC respondents. In Ghatpally the difference between the communities was only marginal. As mentioned, FC respondents in Debbadaguda lagged behind. However, since the ranking of the other communities was as anticipated (see Figure 4.2), the hypothesis was neither confirmed nor rejected. In Mansanpally, the

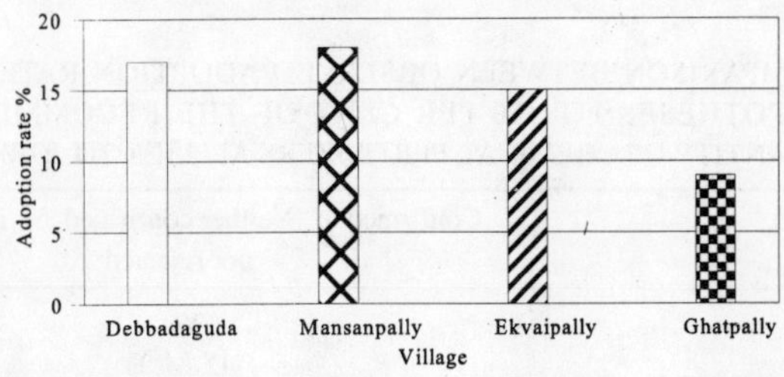

Note: The figure refers to plots which received 50 % or more of the recommended dose.
Source: Own survey.

FIGURE 8.13. OBSERVED ADOPTION RATE (%) OF CHEMICAL FERTILIZERS APPLIED TO *JOWAR,* BY VILLAGE (*kharif* 1991/2-1993/4).

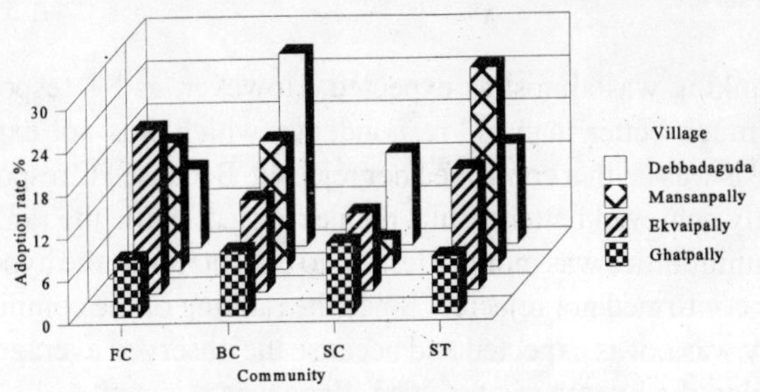

Note: The figure refers to plots which received 50 % or more of the recommended dose.
Source: Own survey.

FIGURE 8.14. OBSERVED ADOPTION RATE (%) OF CHEMICAL FERTILIZERS APPLIED TO *JOWAR,* BY COMMUNITY, BY VILLAGE (*kharif* 1991/2-1993/4).

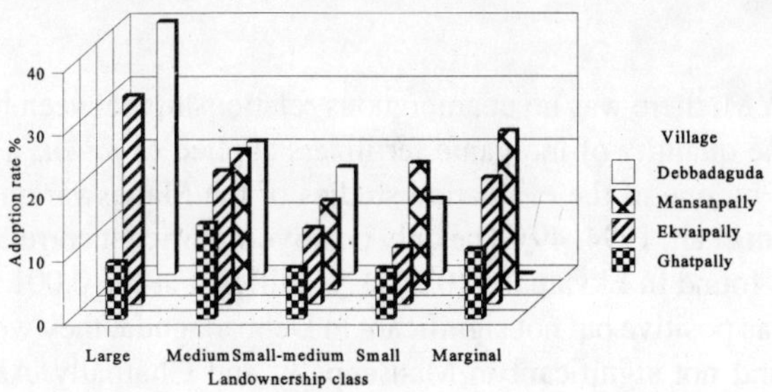

Note: The figure refers to plots which received 50 % or more of the recommended dose.
Source: Own survey.

FIGURE 8.15. OBSERVED ADOPTION RATE (%) OF CHEMICAL FERTILIZERS APPLIED TO *JOWAR,* BY LANDOWNERSHIP CLASS, BY VILLAGE (*kharif* 1991/2-1993/4).

TABLE 8.10. COMPARISON BETWEEN OBSERVED ADOPTION RATES AND HYPOTHESES FOR 50 PER CENT OF THE RECOMMENDED QUANTITY OF CHEMICAL FERTILIZERS APPLIED TO *JOWAR*

Intervention level	Confirmed	Neither confirmed nor rejected	Rejected
Village		Yes	
Community in village		D, M, E	G
Community between villages			Yes
Landownership class in village	D	M	E, G
Landownership class between villages			Yes

Note: D = Debbadaguda; M = Mansanpally; E = Ekvaipally; G = Ghatpally.
Source: Own survey.

observed ranking was almost as expected. However, as ST respondents performed much better than FC respondents, which was not expected, the hypothesis was neither confirmed nor rejected. Because FC respondents in Ekvaipally achieved better results than expected, while the ranking of the other communities was more or less up to expectations, the hypothesis was neither confirmed nor rejected. Since the ranking of the communities in Ghatpally was not as expected and because the observed averages were almost similar, the hypothesis for this village was rejected.

The comparison between each community by village, shows that the expectation that the highest adoption rates of FC and BC respondents would be found in Debbadaguda, proved right for BC respondents only. The assumption that the highest adoption rates of SC and ST respondents would be found in Ekvaipally, proved wrong. Hence, the hypothesis was rejected.

As with FYM, there was no unambiguous relationship between holding-size and the quantity of inorganic fertilizers applied to *jowar*, a finding confirmed by one of the evaluation studies of the Maheswaram project (Sivamohan et al., 1994: 49). The only positive significant correlation for *jowar* was found in Ekvaipally (0.2041, significant at the 0.001 level).[38] While it was positive but not significant in Debbadaguda, they were even negative and not significant in Mansanpally and Ghatpally. Assuming that large farmers had better access to inorganic fertilizers, it did not necessarily mean that they, of all categories of farmers, applied the highest dose to *jowar*. Likewise, Figure 8.15 demonstrates that an increase in holding-size not necessarily went hand in hand with higher adoption rates concerning fertilizer applied to *jowar*. Only in Debbadaguda and

Ekvaipally relatively more large farmers surpassed the 50 per cent threshold level than smaller farmers. For Debbadaguda this positive relationship meant that the hypothesis was confirmed (compare Figure 4.3 with Figure 8.15).[39] In Mansanpally, marginal farmers ranked first which was better than expected. Since the ranking of the other categories of farmers was anticipated, the hypothesis was neither confirmed nor rejected. Since the positive relationship in Ekvaipally was not anticipated, since large farmers ranked first and small farmers had lower than expected adoption rates, the hypothesis was rejected. Because the negative relationship in Ghatpally was not anticipated and since the adoption rates of the various landownership classes differed only slightly, the hypothesis was rejected.

On the basis of the comparison between the expected and observed average adoption rates of each landholding class by village, the hypothesis was rejected. As expected, highest adoption rates of large, medium and small-medium farmers were found in Debbadaguda. However, highest adoption rates of small and marginal farmers were not found in Ekvaipally but in Mansanpally.

Figure 8.17 shows that, similar to *jowar*, the relationship between community and the 50 per cent threshold level for fertilizer used for castor was not very distinctive either. Again, not a single community stood out as best or worst performer.[40] Figures 8.14 and 8.17 have quite some similarities, indicating that adoption rates were hardly influenced by the type of dryland crop. While FC respondents performed best in Ekvaipally, they ranked last in Ghatpally. In Debbadaguda, BC respondents topped the list and in Mansanpally this position was taken by ST respondents. In Ghatpally the difference between the communities was only marginal. In Debbadaguda, the ranking of most communities was as anticipated (see Figure 4.2). However, since FC respondents lagged behind, the hypothesis was neither confirmed nor rejected. In Mansanpally, because SC and ST respondents had higher and FC and BC respondents had lower than expected adoption rates, the hypothesis was rejected. In Ekvaipally, the ranking of most communities was up to expectations. However, because FC respondents achieved better than expected results, the hypothesis was neither confirmed nor rejected. In Ghatpally, the ranking of the communities was not as expected and the observed averages were almost similar. Hence, the hypothesis was rejected.

Similar to *jowar*, the expectation that the highest average adoption rates of FC and BC respondents would be found in Debbadaguda, proved right for BC respondents only. The assumption that the highest adoption

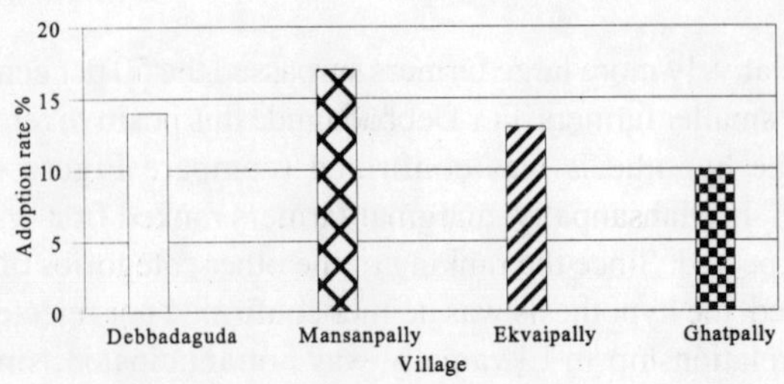

Note: The figure refers to plots which received 50% or more of the recommended dose.
Source: Own survey.

FIGURE 8.16. OBSERVED ADOPTION RATE (%) OF CHEMICAL FERTILIZERS APPLIED TO CASTOR, BY VILLAGE (*kharif* 1991/2-1993/4).

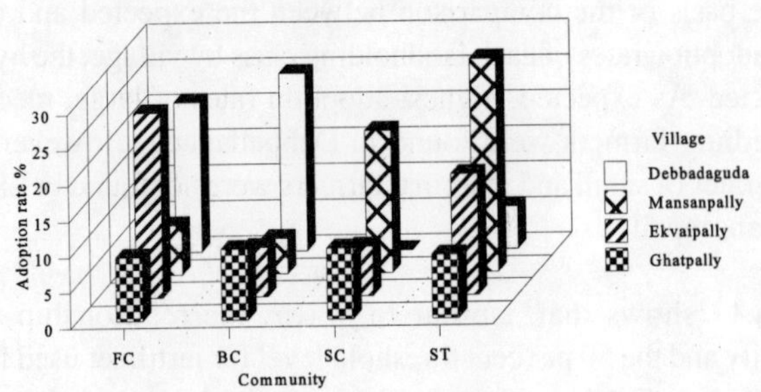

Note: The figure refers to plots which received 50% or more of the recommended dose.
Source: Own survey.

FIGURE 8.17. OBSERVED ADOPTION RATE (%) OF CHEMICAL FERTILIZERS APPLIED TO CASTOR, BY COMMUNITY, BY VILLAGE (*kharif* 1991/2-1993/4).

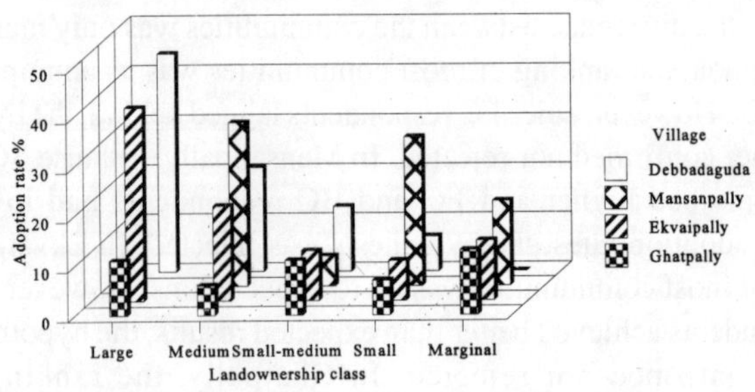

Note: The figure refers to plots which received 50% or more of the recommended dose.
Source: Own survey.

FIGURE 8.18. OBSERVED ADOPTION RATE (%) OF CHEMICAL FERTILIZERS APPLIED TO CASTOR, BY LANDOWNERSHIP CLASS, BY VILLAGE (*kharif* 1991/2-1993/4).

rates of SC and ST respondents would be found in Ekvaipally, proved wrong for both communities. Hence, the hypothesis was rejected.

In contrast to *jowar*, the correlation coefficient for castor for the total four study villages was positive and significant at the 0.001 level: 0.1653. Still, such a positive and significant relationship between holding-size and the quantity of inorganic fertilizers applied to castor only existed in Debbadaguda (0.4656, significant at the 0.001 level) and Ekvaipally (0.2017, significant at the 0.01 level).[41] As with *jowar*, they were negative and not significant in Mansanpally and Ghatpally. Although less apparent than for *jowar*, it is again an indication that even for a commercial crop like castor, large farmers did not necessarily apply larger doses than other farmers. Figure 8.18 indicates that only in Debbadaguda and Ekvaipally relatively more large farmers surpassed the 50 per cent threshold compared to smaller farmers. For Debbadaguda this positive relationship meant that the hypothesis was confirmed (compare Figure 4.3 with Figure 8.18).[42] Although small-medium farmers in Mansanpally performed worse than expected, the difference was small and the hypothesis confirmed. Since the positive relationship in Ekvaipally was not anticipated (large farmers ranked first and small and marginal farmers had lower than expected adoption rates), the hypothesis was rejected. Because the negative relationship in Ghatpally was not anticipated and since the adoption rates of the various landownership classes differed slightly, the hypothesis was rejected.

On the basis of the comparison between the expected and observed average adoption rates of each landholding class by village, the hypothesis was rejected. As expected, highest adoption rates of large and small-medium farmers were found in Debbadaguda. However, the highest adoption rates of medium farmers were not found in this village but in

TABLE 8.11. COMPARISON BETWEEN OBSERVED ADOPTION RATES AND HYPOTHESES FOR 50 PER CENT OF THE RECOMMENDED QUANTITY OF CHEMICAL FERTILIZERS APPLIED TO CASTOR

Intervention level	Confirmed	Neither confirmed nor rejected	Rejected
Village		Yes	
Community in village		D, E	M, G
Community between villages			Yes
Landownership class in village	D, M		E, G
Landownership class between villages			Yes

Note: D = Debbadaguda; M = Mansanpally; E = Ekvaipally; G = Ghatpally.
Source: Own survey.

TABLE 8.12. RATE OF ADOPTION OF 50 PER CENT OF THE RECOMMENDED QUANTITY OF CHEMICAL FERTILIZERS APPLIED TO *JOWAR* AND CASTOR, BY TYPE OF FARMER, BY VILLAGE

Type of farmer	Debbadaguda		Mansanpally		Ekvaipally		Ghatpally		Total		N plots	
	J	C	J	C	J	C	J	C	J	C	J	C
Pure dryland farmers	13	10	21	25	9	11	10	8	14	13	49	40
Farmers with < 1 ha wetland	20	13	14	4	15	8	8	11	14	10	54	26
Farmers with ≥ 1 ha wetland	21	24	21	25	26	26	9	12	20	22	26	26
Average	17	14	18	17	15	13	9	10	15	13	129	92

Source: Own survey

Mansanpally. Contrary to expectations, small and marginal farmers ranked first in Mansanpally, not in Ekvaipally.

Table 8.12 illustrates that especially for Debbadaguda and Ekvaipally and less so for Mansanpally and Ghatpally, farmers owning more than 1 ha of wetland applied more inorganic fertilizers to *jowar* and castor than other farmers. Although these farmers preferred wetland to dryland agriculture, they argued that the financial returns on fertilizers applied to dryland crops were sufficiently high to recover the costs involved. Their financial resources permitted them to invest in this input. Nevertheless, all wetland owning farmers indicated that they favoured the use of fertilizers for paddy. In contrast to *jowar* and castor almost all farmers were aware of, and practised, a split application of chemical fertilizers to paddy, at least once in the nursery and once in the plot at transplanting. On average during *kharif* 1992/3 and 1993/4, some 57 per cent of the farmers applied a top-dressing at weeding. Moreover, the average quantity of inorganic fertilizers applied to paddy during *kharif* 1992/3 and 1993/4 was some four times the quantity applied to the two dryland crops. Only for the plots which had received fertilizers, it was 216 kg/ha, against 56 kg/ha for *jowar* and 59 kg/ha for castor. The averages for paddy were highest in Ekvaipally (307 kg/ha), followed by Debbadaguda (246 kg/ha). Because many respondents in Mansanpally and Ghatpally preferred to give the basal dressing a couple of weeks after transplanting and did not apply a top-dressing at weeding, the average doses applied to paddy were much lower: 129 kg/ha and 118 kg/ha respectively.

Assuming that the recommended dose for paddy (100 kg/ha nitrogen, 60 kg/ha phosphorus and 40 kg/ha potassium) was equivalent to approximately 300 kg of inorganic fertilizers, except for Ekvaipally, above-mentioned averages were lower than the recommended dose. This applies especially to Mansanpally and Ghatpally. For the four villages combined as well as for Debbadaguda and Ekvaipally, a significant positive correlation existed between the dose applied to paddy and holding-size.[43] However, in Mansanpally and Ghatpally these correlations were not significant and negative, indicating that in these villages large farmers did not necessarily apply more fertilizers to paddy than smaller farmers.

The preceding subsections suggest that the effect of both the projects on the quantity of inorganic fertilizers applied to *jowar* and castor was limited. First of all, respondents already knew about these fertilizers before the projects started. Second, most respondents were unaware of recommendations concerning placement, timing and quantities of fertilizers to be used. Third, although the average dose applied to *jowar* and castor by respondents in Ghatpally was lowest in all villages, they did not apply significantly less. Irrespective of the fact that respondents did admit that they were helped by the subsidized fertilizers given by both projects, these were not very successful in increasing the doses of inorganic fertilizers applied to these two dryland crops. Since the projects were not able to determine the price of inorganic fertilizers, many farmers applied what they considered optimal doses, given soil and rainfall characteristics and price. Based on experience, most farmers preferred to apply half a bag per acre or some 60 kg/ha. That they used less after 1991/2, might have been the result of the Indian economic liberalization programme which led to an increase in fertilizer prices. Nevertheless, that farmers kept using these fertilizers, that they felt that their relative advantage was large, makes it plausible that it is a sustainable agricultural recommendation for dryland crops.

8.8 Early Sowing

By advising farmers to sow *jowar* early, before the arrival of the monsoon (in *Rohini karthe*, the last week of May and the first week of June), both the Maheswaram project and AWARE followed an indigenous practice. Nevertheless, discussing this advice with extension officers of the Maheswaram project, a contradiction was found. While most officers knew that farmers were supposed to continue with their preferences, some claimed that they had to advise farmers to sow during or just before the onset of the monsoon, around the third week of June. Being convinced that the hot soil would roast *jowar* seeds sown during *Rohini karthe*, one senior project officer even had toured the project area at the beginning of each *kharif* season, persuading farmers by loudspeaker to stop sowing and to wait for the arrival of the monsoon. He admitted that it had been in vain and that farmers had not listened. Except for this senior and some other officers of the Maheswaram project most Maheswaram and AWARE staff felt that since the advice of early sowing corresponded with farmers' practice, it was not necessary to inform them.

The importance of early sowing of *jowar* is evident from the fact that during *kharif* 1992/3, 72 per cent and during *kharif* 1993/4, even 92 per cent of the *jowar* plots were sown during *Rohini karthe*. The difference between the years was largely a consequence of insufficient pre-monsoon showers in 1992/3. The strong preference of early sowing applies to three out of four villages, Debbadaguda (80 per cent), Mansanpally (96 per cent) and Ghatpally (97 per cent). In Ekvaipally, 50 per cent of the plots were sown during this period. A substantial number of plots (33 per cent) were sown in the following fortnight, while another 9 per cent were sown very early, during the second and third week of May. A reason for relatively late sowing by a considerable number of Ekvaipally farmers might be insufficient rainfall at the proper time. Indeed, in both years, rainfall during this period was less in Ekvaipally than in the other three villages. Remembering that Ekvaipally farmers had the lowest adoption rates for summer tillage as well (see Chapter 7), relatively late sowing might also have been caused by migrant farmers who returned to the village late. Since this similarity of the three villages and the relatively backward position of Ekvaipally was not expected, the hypothesis was rejected.

While the adoption rates of the various communities of early sowing of *jowar* differed considerably in Ekvaipally, this difference was smaller in Debbadaguda and almost non-existent in Mansanpally and Ghatpally.[44] These relatively marginal differences, together with the fact that contrary to expectations, FC respondents in all three villages performed worse than expected, resulted in the rejection of the hypotheses for the latter three villages. Because SC respondents in Ekvaipally achieved better than expected results, while the ranking of the other communities was more or less as expected, the hypothesis was neither confirmed nor rejected.

Comparing each community by village, the highest adoption rates of FC and BC respondents were not found in Debbadaguda but in Ghatpally, followed by Mansanpally. The adoption rates of SC and ST farmers were highest in Mansanpally followed by Ghatpally, not in Ekvaipally. Since this ranking differed from what was expected, the hypothesis was rejected.

Not surprisingly, given the results for the communities, the difference

TABLE 8.13. COMPARISON BETWEEN OBSERVED ADOPTION RATES AND HYPOTHESES FOR EARLY SOWING OF *JOWAR*

Intervention level	Confirmed	Neither confirmed nor rejected	Rejected
Village			Yes
Community in village		E	D, M, G
Community between villages			Yes
Landownership class in village		D, E	M, G
Landownership class between villages			Yes

Note: D = Debbadaguda; M = Mansanpally; E = Ekvaipally; G = Ghatpally.
Source: Own survey.

in adoption rates of the various landownership classes of early sowing of *jowar* was small in Mansanpally and Ghatpally as well; these were larger in Ekvaipally and Debbadaguda.[45] The relatively marginal differences in Mansanpally and Ghatpally, together with the fact that, contrary to expectations, marginal farmers ranked higher than anticipated, resulted in the rejection of the hypotheses for these villages. In Debbadaguda, medium farmers achieved lower than expected adoption rates. However, since the ranking of the other classes was as expected, the hypothesis was neither confirmed nor rejected. Because small farmers in Ekvaipally had lower than expected results, while the ranking of the other landowning classes was more or less as anticipated, the hypothesis was neither confirmed nor rejected.

The comparison of each holding-size class by village shows that the average adoption rates for all landownership classes were found in Ghatpally and Mansanpally. Since it was expected that large, medium and small-medium farmers located in Debbadaguda would have the highest adoption rates and that the same would apply to small and marginal farmers living in Ekvaipally, the hypothesis was rejected.

TABLE 8.14. RATE OF ADOPTION OF EARLY SOWING OF *JOWAR*, BY TYPE OF FARMER, BY VILLAGE

Type of farmer	Debbadaguda	Mansanpally	Ekvaipally	Ghatpally	Total	N plots
Pure dryland farmers	82	99	61	97	88	213
Farmers with < 1 ha wetland	81	96	42	95	80	99
Farmers with ≥ 1 ha wetland	72	85	50	100	76	62
Average	80	96	50	97	83	274

Source: Own survey.

Table 8.14 shows that for all villages except Ghatpally, farmers owning more than 1 ha of wetland had lower adoption rates for early sowing of *jowar* than other farmers. It is an indication that *jowar* production was more important for pure dryland farmers. In spite of their better resource base (more plough bullocks), wetland owning farmers were less interested in sowing *jowar* on the most appropriate time. Many of these farmers also reasoned that if *jowar* was attacked by a pest, they could always apply pesticides. Such an expensive input was out of reach of many of the less affluent pure dryland farmers.

Although early sowing of *jowar* was practised by a large majority of farmers, it goes too far to credit the projects for this success. The fact is that it was indigenous to the area, undertaken by the forefathers of the respondents decades before the projects started. Thus, most respondents were not encouraged by the projects but acted because experience and astrology provided them with a suitable period in which to cultivate *jowar* under the most optimal conditions, given rainfall and pests. Hence, it is expected that early sowing of *jowar* is a sustainable practice. Farmers who did not follow it did not observe it out of ignorance or because the technology did not suit them but because they were prevented from sowing on time. The most important reasons given were insufficient moisture in the soil, lack of bullocks and the inability to hire them at the proper time, preference for wetland crops or because farmers opted for more secure income from wage employment.

8.9 Timely Weeding and Interculturing

By advising farmers to remove weeds in *jowar* through one intra-row manual weeding supplemented with at least two interculturings with the *dante* and to remove weeds in castor through one manual weeding with three interculturings with the *guntaka*, both the Maheswaram project and AWARE largely accepted farmers' indigenous practices. Hence, and similar to early sowing of *jowar*, most extension officers were convinced that they need not inform farmers about already existing practices.

Although indigenous, on average 4 per cent of the *jowar* and 3 per cent of the castor plots were weeded by hand during *kharif* 1992/3 and 1993/4. With 16 per cent of the *jowar* plots and 4 per cent of the castor plots, manual weeding was a somewhat more accepted practice in Ekvaipally only. That all these plots belonged to marginal, small and small-medium farmers of the BC and SC community suggests some influence of AWARE. Manual weeding was hardly undertaken in Mansanpally and Debbadaguda, while Ghatpally scored somewhat better, at least for castor.[46] Since adoption rates were very low in most villages and since those of Debbadaguda and Mansanpally were lower than expected, the hypotheses concerning the effect of both projects on manual weeding of *jowar* as well as on castor were rejected. These very low results do not allow for a comparison between communities and landownership classes by village and make a comparison between the expected and the observed adoption rates not very relevant either.

The fact that a large majority of the respondents did not weed *jowar* and castor by hand suggests that especially the Maheswaram project and to a lesser extent AWARE did not succeed in convincing farmers to adopt this recommendation. Disbelief that pulling weeds by hand was worth the money or the time involved and that the activity had a sufficient positive effect on yields of dryland crops, made most farmers reluctant to invest in this activity. Neither of the projects had been able to change this perception. In contrast, farmers were convinced that manual weeding in a wetland crop like paddy was effective and that the costs involved were less than the incremental returns. In fact, after harvesting, weeding consumed most of the labour input of paddy. On average, some 92 per cent of the paddy plots cultivated in *kharif* 1992/3 and 1993/4 were weeded by hand at least once. In Ghatpally, all paddy plots were weeded, in Debbadaguda 98 per cent of the plots, in Ekvaipally 93 per cent of the plots and in Mansanpally still a substantial 73 per cent.

That interculturing was largely preferred above manual weeding is evident from the fact that all farmers harrowed their dryland crops. In fact, averaged for *kharif* 1992/3 and 1993/4, 77 per cent of the *jowar* plots were even harrowed at least twice and 53 per cent of the castor plots were intercultured at least thrice (Figures 8.19 and 8.22). In all villages, these operations were very much influenced by the monsoon. Insufficient rains at the most appropriate time made many farmers give up one or two interculturings during *kharif* 1992/3. Thus, while on average 55 per cent of the *jowar* plots and 41 per cent of the castor plots were intercultured according to the recommendations in *kharif* 1992/3, this increased to 98 per cent and 65 per cent respectively in *kharif* 1993/4.[47] Average adoption rates for *jowar* were highest in Ekvaipally (95 per cent of the *jowar* plots were intercultured according to the recommendation), followed by Debbadaguda (85 per cent), Ghatpally (69 per cent) and Mansanpally (64 per cent). Since it was expected that Debbadaguda farmers would have the highest average adoption rate and that this rate would be lowest in Ghatpally (see Figure 4.1), the hypothesis was rejected.

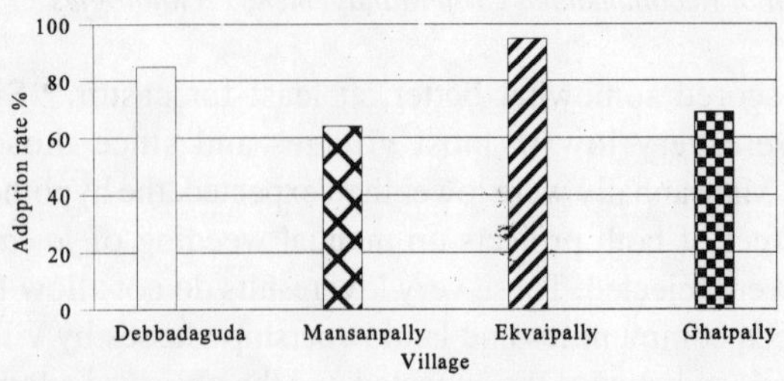

Note: The figure refers to *jowar* plots getting at least two interculturings.

Source: Own survey.

FIGURE 8.19. OBSERVED ADOPTION RATE (%) OF RECOMMENDED INTER-CULTURINGS FOR *JOWAR*, BY VILLAGE (*kharif* 1992/3-1993/4).

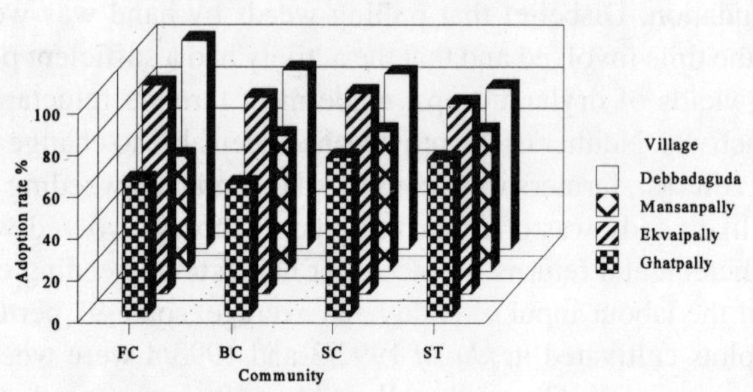

Note: The figure refers to *jowar* plots getting at least two interculturings.

Source: Own survey.

FIGURE 8.20. OBSERVED ADOPTION RATE (%) OF RECOMMENDED INTER-CULTURINGS FOR *JOWAR*, BY COMMUNITY, BY VILLAGE (*kharif* 1992/3-1993/4).

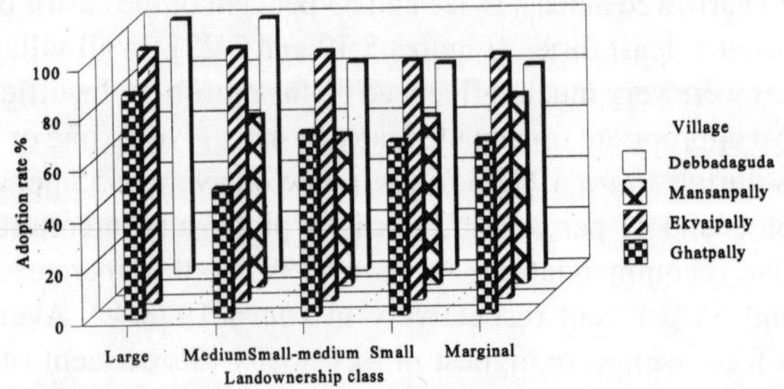

Note: The figure refers to *jowar* plots getting at least two interculturings.

Source: Own survey.

FIGURE 8.21. OBSERVED ADOPTION RATE (%) OF RECOMMENDED INTER-CULTURINGS FOR *JOWAR*, BY LANDOWNERSHIP CLASS, BY VILLAGE (*kharif* 1992/3-1993/4).

For castor, except for Ghatpally, the ranking of the villages was similar to that of *jowar*. Average adoption rates were highest in Ekvaipally (91 per cent of the castor plots were intercultured according to the recommendation), followed by Debbadaguda (65 per cent) and Mansanpally (61 per cent). With an average adoption rate of 15 per cent, Ghatpally lagged very much behind. This low level had nothing to do with the social or economic background of the farmers. Since almost all farmers harrowed castor less than three times, this low adoption rate might be explained by the presence of a certain 'village culture or rule' that made farmers believe that twice was sufficient. Anyway, these low adoption rates suggested considerable influence of the projects on farmers in the other villages. However, since Debbadaguda ranked lower than expected, while the ranking of the other three villages was anticipated, the hypothesis was neither confirmed nor rejected.

Figure 8.20 reveals that concerning the recommended interculturing for *jowar*, the average adoption rates of the various communities in the villages varied marginally. It also indicates that while FC respondents topped the listing in Debbadaguda and Ekvaipally, they ranked last in Mansanpally and Ghatpally.[48] In these latter two villages, ST and SC respondents had slightly higher average adoption rates than farmers of the other two communities. Because the average adoption rates differed marginally, the hypothesis was neither confirmed nor rejected in Debbadaguda, Mansanpally and Ekvaipally. In Debbadaguda, while the ranking of most communities was as anticipated, only ST respondents ranked lower than expected (see Figure 4.2). In Mansanpally, only FC respondents ranked lower than anticipated. In Ekvaipally FC respondents ranked higher than expected. In Ghatpally, because ST and SC respondents ranked higher than was anticipated, the hypothesis was rejected.

The comparison between each community by village, shows that the expectation that the highest adoption rates of FC and BC respondents would be found in Debbadaguda, proved right. Since the assumption that the highest adoption rates of SC and ST respondents would be found in Ekvaipally also proved right, the hypothesis was confirmed.

Similar to Figure 8.20, Figure 8.21 reveals that the adoption rates of the various landownership classes in most study villages did not differ very much.[49] In Debbadaguda for instance, the adoption rates of, on the one hand, large and medium farmers, and on the other hand, small-medium, small and marginal farmers, were more or less similar. Moreover, as far as a ranking was discernible, it was as expected (compare Figure 4.3). Hence, the hypothesis was confirmed. Because of the marginal differences

TABLE 8.15. COMPARISON BETWEEN OBSERVED ADOPTION RATES AND HYPOTHESES FOR INTERCULTURINGS OF *JOWAR*

Intervention level	Confirmed	Neither confirmed nor rejected	Rejected
Village			Yes
Community in village		D, M, E	G
Community between villages	Yes		
Landownership class in village	D	M, E, G	
Landownership class between villages		Yes	

Note: D = Debbadaguda; M = Mansanpally; E = Ekvaipally; G = Ghatpally.
Source: Own survey.

in the other three villages and because the ranking of the classes was almost as expected, the hypotheses for these villages were neither confirmed nor rejected. Marginal farmers in Mansanpally and large farmers in Ekvaipally performed better than anticipated and medium farmers in Ghatpally had lower than expected adoption rates.

The adoption rates of all landownership classes were highest in Ekvaipally. However, it was expected that this would only apply to small and marginal farmers and that the rates of the other categories of farmers would be highest in Debbadaguda (see Figure 4.3). Also because of the relatively small difference between the various categories of farmers by village, the hypothesis was neither confirmed nor rejected.

Figure 8.23 shows that, similar to *jowar*, the average adoption rates on the recommended interculturing for castor did not differ very much between the various communities in the villages.[50] The figure also demonstrates that not a single community stood out as best performer. Because of the marginal difference and because the ranking of the communities in Debbadaguda, Mansanpally and Ekvaipally was almost as expected (see Figure 4.2), the hypotheses for these villages were neither confirmed nor rejected. In Debbadaguda, ST respondents had higher than expected adoption rates. In Mansanpally SC respondents achieved lower than expected results and in Ekvaipally, ST respondents performed worse than was anticipated. In contrast, the hypothesis for Ghatpally was rejected on the basis that both FC and BC respondents had lower than anticipated rankings.

The comparison between each community by village, shows that the expectation that the highest adoption rates of FC and BC respondents

TABLE 8.16. COMPARISON BETWEEN OBSERVED ADOPTION RATES AND HYPOTHESES FOR INTERCULTURINGS OF CASTOR

Intervention level	Confirmed	Neither confirmed nor rejected	Rejected
Village		Yes	
Community in village		D, M, E	G
Community between villages			Yes
Landownership class in village	D	M, E	G
Landownership class between villages			Yes

Note: D = Debbadaguda; M = Mansanpally; E = Ekvaipally; G = Ghatpally.
Source: Own survey.

would be found in Debbadaguda, proved wrong. Like the average adoption rates of SC and ST respondents, they were highest in Ekvaipally. On the basis of this finding, together with the finding that average adoption rates of FC and BC farmers in Ghatpally lagged very much behind, the hypothesis was rejected.

Compared to *jowar*, the difference in the adoption rates for the recommended interculturing for castor of the various landownership classes in Debbadaguda and Mansanpally was larger (Figure 8.24).[51] Because the ranking of the various classes in Debbadaguda was as expected (see Figure 4.3), the hypothesis was confirmed. In Mansanpally, because marginal farmers had lower than expected average adoption rates, the hypothesis was neither confirmed nor rejected. In Ekvaipally, because the difference between the various landowning classes was relatively small and, except for small-medium farmers, because the ranking was anticipated, the hypothesis was neither confirmed nor rejected. In Ghatpally, because the difference between the various classes was small and the ranking not as expected, the hypothesis was rejected.

As with *jowar*, the adoption rates of all landownership classes on this recommendation for castor were highest in Ekvaipally. Since it was expected that this would only apply to small and marginal farmers and that the rates of the other categories of farmers would be highest in Debbadaguda and since the average adoption rates of all categories of farmers in Ghatpally lagged very much behind (see Figure 4.3), the hypothesis was rejected.

Table 8.17 demonstrates that only for the four study villages combined, a positive relationship between average adoption rates and the explanatory variable (ownership of wetland) was found. Yet, this relationship was more obvious for castor than for *jowar*. In fact, the table confirms earlier

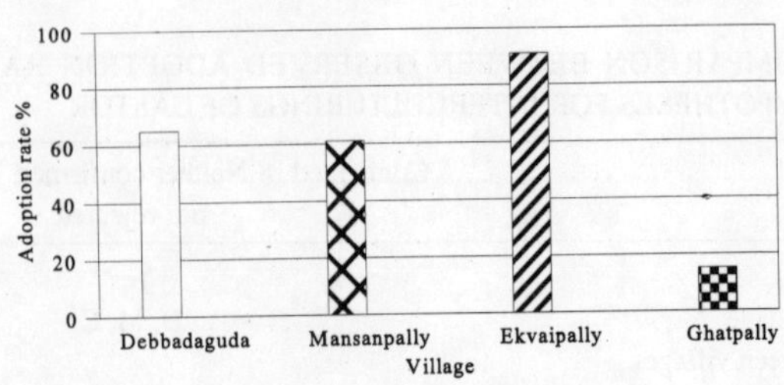

Note: The figure refers to Castor plots getting at least three interculturings.
Source: Own survey.

FIGURE 8.22. OBSERVED ADOPTION RATE (%) OF RECOMMENDED INTERCULTURINGS FOR CASTOR, BY VILLAGE (*kharif* 1991/2-1993/4).

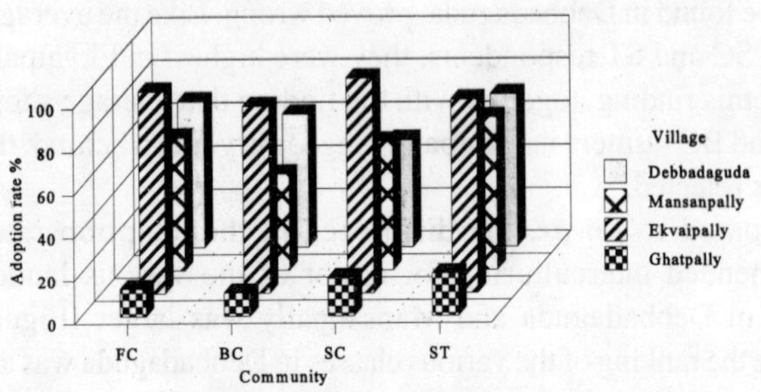

Note: The figure refers to Castor plots getting at least three interculturings.
Source: Own survey.

FIGURE 8.23. OBSERVED ADOPTION RATE (%) OF RECOMMENDED INTERCULTURINGS FOR CASTOR, BY COMMUNITY, BY VILLAGE (*kharif* 1992/3-1993/4).

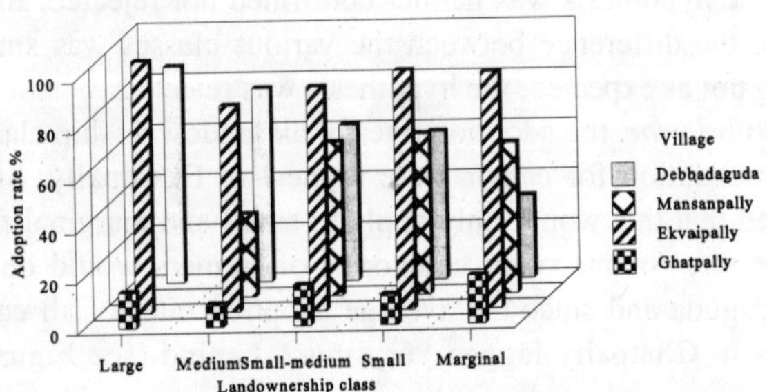

Note: The figure refers to Castor plots getting at least three interculturings.
Source: Own survey.

FIGURE 8.24. OBSERVED ADOPTION RATE (%) OF RECOMMENDED INTERCULTURINGS FOR CASTOR, BY LANDOWNERSHIP CLASS, BY VILLAGE (*kharif* 1992/3-1993/4).

findings, that a negative relationship between adoption levels and ownership of wetland was not always clear. It was quite common in the region and in the villages to interculture twice or thrice depending on the crop, provided time and availability of bullocks permitted and provided the condition of the crop was satisfactory. Thus, if a farmer was convinced that the crop would fail (for instance because it was attacked by a pest or experienced moisture stress), he would leave it. That castor received more interculturings by wetland owning farmers had everything to do with their resource base. Most of these farmers owned sufficient teams of plough bullocks and could utilize permanent labourers to harrow this commercial crop more often than other farmers. That in all villages wetland owning farmers had lower average adoption rates for castor than for *jowar*, had everything to do with the fact that in spite of their more favourable resource base, these farmers still preferred to invest more in wetland than in dryland agriculture.

TABLE 8.17. RATE OF ADOPTION OF RECOMMENDED INTERCULTURING FOR *JOWAR* AND CASTOR, BY TYPE OF FARMER, BY VILLAGE

Type of farmer	Debbadaguda		Mansanpally		Ekvaipally		Ghatpally		Total		N plots	
	J	C	J	C	J	C	J	C	J	C	J	C
Pure dryland farmers	89	57	68	67	90	90	67	13	76	47	184	92
Farmers with < 1 ha wetland	79	71	59	54	97	93	73	16	76	58	190	99
Farmers with ≥ 1 ha wetland	92	67	69	57	100	90	64	17	83	60	68	45
Average	85	65	64	61	95	91	69	15	77	53	442	236

Source: Own survey.

The fact that most farmers harrowed twice in *jowar* and at least thrice in castor, did not mean that both projects laid the foundation of its acceptance. As stated before, interculturing was indigenous to the area and most farmers were aware that *jowar* needed two and castor needed more interculturing. Hence, it is plausible that it is a sustainable practice. Those who failed to follow this practice completely did not do so because of ignorance or because the technology did not suit them, but because they were prevented from doing so (lack of bullocks and the inability to hire them at the proper time), or because they preferred wage employment outside the village, or preferred wetland agriculture or because they felt that the crop would be lost and that harrowing was not worth the investment. As explained earlier, the fact that Ghatpally farmers lagged very much behind with regard to interculturing in castor was most probably a consequence of what might be termed a 'village culture or tradition'. It is an indication that the projects had some success in the other study villages.

Since interculturing is an element of the *in-situ* moisture conservation practices, it ideally should be undertaken along the contour. As was explained earlier (see subsection 7.3.2), while contour cultivation was hardly practised, sowing across the slope was done by one-third of the respondents. This meant that the same majority of respondents did not harrow across the slope either. For a more elaborate discussion on this theme, I refer to the above-mentioned subsection.

8.10 Integrated Pest and Disease Management

The Maheswaram project and AWARE advocated the concept of integrated pest and disease management, recommending both indigenous (like early sowing of *jowar*, removing infested plants and making bonfires) and modern ways (like light-traps and various chemicals) to control pests and diseases. Because both projects put emphasis on the use of chemicals, to be sprayed or dusted, and because some of the other control measures were discussed in earlier subsections, this subsection predominantly deals with the adoption of pesticides.

The first time information about plant protection chemicals reached the study villages must have been in the later half of the 1970s. Similar to so many innovations, most respondents credited larger farmers, mostly belonging to the Reddy caste, for being the disseminators of this information. Most of these large Reddy farmers said that they had heard about pesticides from their usual sources of information: i.e. from other large FC farmers in far away and more neighbouring villages, shopkeepers, newspapers, radio and sometimes television. Some of them also said that they had obtained information from universities and extension officers. It shows that for the first time these had become important sources of information for a small group of farmers too. But not for smaller farmers. Most of them learnt about pesticides from larger farmers. Similar to chemical fertilizers, they had to apply pesticides in Reddy's plots while working for them as agricultural labourer, or had seen others applying it in these plots.

It is unclear whether the government extension service had already started informing farmers about plant protection chemicals and had started supplying these with a subsidy in the Amangal area before AWARE started its activities in 1976. On the other hand, in 1984/5, i.e. before the start of the Maheswaram project, the Lab to Land Programme had started distributing subsidized chemicals particularly to ST and SC farmers in the Maheswaram area (APAU, nd: 39). Subsidized chemicals were also distributed in the Amangal area by AWARE (Arens, 1979: 31). Both the Maheswaram project and AWARE tried to increase the use of plant protection chemicals for dryland crops by using their usual two-way strategy, by informing farmers and by distributing these chemicals at subsidized prices.

During the open interviews, all respondents acknowledged that they knew about pesticides. They said that neither the Maheswaram project nor AWARE were the first to disseminate news about plant protection chemicals and added that these projects were not important as sources of information for pesticides. Against one Debbadaguda farmer who felt that the Maheswaram project was an important source, sixty-eight farmers indicated that they depended on advice from other farmers and another fifteen farmers mentioned shopkeepers, merchants, newspapers, radio and the television as important sources.

A large majority of the respondents had no detailed knowledge about plant protection chemicals. Although almost all were able to mention at least one name (Endosulfan) during the open interviews, they had no idea which particular pesticide to use for which pest or disease, how much to apply, and how and when to apply it. Only larger farmers had more than basic knowledge, although they admitted that they did not always know the recommendations in detail. In fact, these larger farmers quite often complained about the insufficient information given by various channels. Farmers who had met T&V officers and had received advice from shopkeepers often accused them of giving unreliable information.[52] Distrust made many farmers rely on those sources they knew best, fellow farmers.

In spite of project activities, very few plots received plant protection chemicals during the three years of the study. Of the 872 dryland plots cultivated with *jowar*, 11 were treated (1 per cent), 6 in Ekvaipally, 2 in Ghatpally and Mansanpally and 1 in Debbadaguda. Most farmers used Endosulfan, spraying less than one litre/ha. Although more large FC farmers than other farmers treated their *jowar* plots, the difference between the various categories of farmers was small. Pesticides were applied on a somewhat larger scale to castor, i.e. on 43 of the 696 plots (6 per cent).

Most of these were located in Ghatpally (18 plots), followed by Ekvaipally (11 plots), Debbadaguda (9 plots) and Mansanpally (5 plots). Most farmers sprayed less than one litre of Ekalux or Endosulfan per ha. Similar to *jowar*, although relatively speaking more large and medium FC farmers treated their plots with pesticides, the difference between the various categories of farmers was small.

These meagre results were in contrast to expectations and suggested that the effect of both the projects, in particular that of the Maheswaram project, was negligible. Hence, because the number of treated plots was low and because the ranking of the villages differed from the expectation, both hypotheses on the application of pesticides, the one concerning *jowar* and the other concerning castor, were rejected. These very low results do not allow for a comparison between communities and landownership classes by village and make a comparison between the expected and the observed adoption rates not very relevant either.

The open interviews revealed that it was only partly lack of awareness that prevented farmers from using pesticides for dryland crops. Although most respondents knew only a single name and did not know about exact timings and manner of application, most were quite aware of their benefits. They argued that pesticides were less beneficial in *jowar* and that it was better to avoid pests by sowing early. They also said that since it was much more difficult to avoid attacks of pests in castor, pesticides were more beneficial in this commercial crop. Pesticides were also said to be more beneficial in other commercial crops and in paddy. Thus, paddy was treated with pesticides on 25 out of 282 plots (9 per cent) during *kharif* 1992/3 and 1993/4. Almost all plots sown with cotton were treated, often more than once, if only to prevent the outbreak of certain pests. Moreover, grapes, cultivated by only a few large farmers, were also sprayed on numerous occasions, almost always in a precautionary manner. As a matter of fact, specialists from Hyderabad advised farmers about pesticides for grapes during fortnightly visits. In fact, the same farmer who applied pesticides to paddy and cotton did not necessarily treat crops like *jowar* and castor likewise.

In addition, some other reasons explained the limited use of pesticides in dryland agriculture. First, some of the indigenous measures meant to prevent large scale attacks or to cure a pest once it had attacked, were effective. Second, pest infestation was extremely variable, meaning that farmers acted mainly when some threshold level was passed. Once that occurred, they could choose between two alternatives, the indigenous and the modern practices. During the open interviews, most respondents stated that they would use chemical control (sometimes indigenous measures as well) in case of a large-scale attack only. What was meant by a large-scale attack was not only influenced by farmers' perception of the effects of the pest on yields, but was also influenced by their perception of the health of the crop in general, of drought risk and of the money forgone by investing in plant protection chemicals or in labour needed to control the attack. Responding predominantly to large-scale outbreaks of a pest or disease meant that farmers applied pesticides in a late stage of such an attack. Quite often it meant that pesticides were applied too late to be effective. It is no surprise then that many farmers complained that a pesticide had not killed a pest properly. This ineffectiveness of pesticides was the third reason for not using them more often. Many farmers said that this ineffectiveness was caused by the bad quality of Indian chemicals and because shopkeepers sold imitations or mixed the original product with all kinds of other substances. Thus, adulteration of pesticides was the fourth reason. Fifth, since pesticides were extremely costly, many farmers could not afford to buy these.

8.11 Conclusion

This chapter leads to the same conclusion as the one drawn in Chapter 7 on recommended (*in-situ*) soil and water conservation technologies: i.e. the effect of both projects on the adoption of recommended crop management technologies for dryland crops was less than expected. Similar to the recommendations on (*in-situ*) soil and water conservation, it was found that many of the recommendations which were based on indigenous technologies had higher adoption rates than innovations that were introduced before and during the projects. Once again it has to be concluded that low adoption rates were the result of a combination of factors of which lack of information and unsuitable technology seem to be the most important ones.

Many respondents were never informed by the projects or the T&V extension system about the recommendations on crop management. Moreover, many respondents were not able to recount in detail what these recommendations were about. These findings are something the Department of Agriculture of the government of Andhra Pradesh and AWARE, as implementing organizations, have to take seriously. The fact that most respondents were still depending for their information on more experienced Reddy farmers and that many of these Reddy farmers had only fragmented knowledge, is a case in point. That this conclusion applies to both the Maheswaram project and AWARE does not come as a surprise, given the fact that both organizations depended largely on the government sponsored T&V system. The Maheswaram project tried to improve the supply of information by putting on more staff first and using village volunteers later. AWARE tried to do the same by using the village *sanghams* as pressure groups to force T&V extension officers to visit the villages more regularly and also to inform smaller, lower caste farmers; whatever, these attempts, most farmers were not sufficiently informed about the recommendations.

Farmers responded to the climatological predicaments in the region by adapting their cropping pattern. Thus on dryland they cultivated crops which were easy to grow, needed low inputs and yielded reasonable quantities of grain, kernels and fodder. They were prepared to use agricultural innovations and to invest more in dryland crops provided: (1) the short-term benefits (higher yields in grain and fodder and higher monetary returns) were easily discernible; (2) the relationship between the investment and these short-term benefits was clearly visible; (3) the trade-off between the money involved with the investment and the benefit was clearly in favour of the benefit; (4) the innovations were easily put into practice and did not deviate much from the indigenous practices; and (5) they did not use too much additional labour. The recommendations on crop management of the Maheswaram project and AWARE only partly satisfied these conditions. This is not so much a 'mistake' made by the projects, but more a consequence of improper design of the innovations for which agricultural research stations are responsible. Some observers, arguing that many innovations for dryland agriculture are a failure all over semi-arid India, share a view that there has hardly been any substantive breakthrough in new technologies for this type of agriculture.

As a consequence of improper design, farmers in the study villages selected only those recommendations which they considered to be most useful. Moreover, they adapted many innovations to local or individual circumstances. Thus, farmers did not always use the advised quantity or used the recommendation in a (slightly) different way from what was advised. For example, a majority of the respondents in the four villages intercropped *jowar* with redgram, but not in the recommended ratio; they used improved seed varieties for a number of dryland (and wetland) crops, but not for *jowar*; they applied the recommended quantities of FYM to dryland crops, but not in all years; they used inorganic fertilizers for dryland crops, but less than the recommended doses; they sowed *jowar* early; and they intercultured dryland crops, although hardly weeded manually.

Knowing that indigenous technologies like intercropping of *jowar* with redgram, the application of FYM, early sowing of *jowar* and interculturing were not practised in the most optimum way, both projects stimulated their use, although sometimes in an adapted manner. But most respondents were not aware of these adaptations, had not heard of the advice to intercrop in a 2:1 ratio, for instance. Although the Maheswaram project tried to enhance the quality of FYM by subsidizing the construction of compost-pits, most respondents had not been told about the best manner to make compost. Averaging for all four villages, of the 10 recommendations that were based on indigenous technologies, 4 had adoption rates that were above 50 per cent, 4 had rates between 17 and 33 per cent and two recommendations (both concerning weeding) were not acceptable to most farmers. Whatever the activities of both projects towards stimulating these indigenous practices, these average adoption rates of the respondents in the four villages was not as expected. Debbadaguda never ranked first, followed by Ekvaipally, Mansanpally and finally Ghatpally. Hence, all 'village level hypotheses' on these indigenous recommendations were rejected. Of the hypotheses concerning communities and landownership class, 3 out of 60 hypotheses, all concerning interculturing, were confirmed, 23 were neither confirmed nor rejected and 34 were rejected.

Except for the recommended 2:1 ratio in intercropping, not following these recommended indigenous practices had nothing to do with a lack of information on farmers' side. It was more a consequence of lack of resources. Lack of sufficient number of livestock, which in turn had everything to do with insufficient access to fodder (pastures as well as fodder crops), prevented almost all farmers to manure their dryland (and wetland) crops every year. Insufficient quantities of plough bullocks prevented some farmers from harrowing *jowar* and castor according to the recommendation. In addition, a considerable number of respondents preferred a more secure income from daily labour to an insecure higher yield from additional harrowing and especially from manual weeding. In order to understand why farmers opted for certain innovations and not for others, it has to be realized that the average household in the study villages did not rely so much on income from its own dryland farm but that it had multiple (often more reliable) sources of income. Thus, irrespective of the recommendation, be it based on indigenous technologies or be it an innovation, farmers seemed interested in investing in additional labour and capital in dryland agriculture only, provided the returns were more or less assured, these returns were prompt, and provided a clear relationship between incremental labour and yield existed.

Both the projects also tried to induce farmers to use various innovations that had been introduced in the area before they started their activities: i.e. improved seeds of dryland crops, inorganic fertilizers, pesticides and some implements. Since these were not 'real' innovations almost all respondents in the study villages who knew about these recommendations had been informed by their usual sources of information (fellow farmers, friends and sometimes mass communication channels) and hardly at all by project officers or the T&V extension service. Because these usual sources had often supplied only bits of information, knowledge was often superfluous and incomplete. Thus, many respondents were not able to mention the names of improved seed varieties of *jowar*, *bajra* and redgram, or did not know about placement and top-dressing of chemical fertilizers for dryland crops, or about pesticides. Both projects certainly had not succeeded in narrowing the knowledge gap concerning these innovations.

In spite of these project activities, the average adoption rates of the respondents on these 'pre-project innovations' was not as anticipated. Since the observed ranking of the villages was never in line with the expectations, all 'village level hypotheses' were rejected. Of the hypotheses concerning communities and landownership class, only 3 out of 42 hypotheses, all concerning the recommended quantity of fertilizers to be applied, were confirmed, 11 were neither confirmed nor rejected and 28 were rejected.

Besides information, the main reason for the limited success of both projects with respect to these 'pre-project innovations' was founded in the unsuitability of (most of) these innovations. Thus, while improved castor was acceptable to almost all farmers because of its high relative advantage and compatibility, improved *jowar* seeds were not because the multiple objectives (food and fodder) of this crop were not met. Although almost all respondents were convinced of the beneficial effect of inorganic fertilizers on yields of dryland crops, they applied less than the recommended dose because of its high price. In addition, since pesticides were considered to be too expensive as well, most farmers were willing to apply these only in case of an emergency: i.e. once a pest had reached a certain threshold level. Even without having seen a seed-cum-fertilizer drill, the few farmers who had heard about it, were sure that it was too expensive to be used in dryland agriculture. Those farmers who had received an iron plough from the Maheswaram project were convinced that it was less useful compared to the indigenous plough. Opting for immediate profit above an unknown and uncertain profit in the future, they sold it before having tried it out.

The final remark relates to the supposed negative relationship between the area of wetland owned and adoption of the recommendations for dryland agriculture. Such a clear negative relationship was only found in intercropping of *jowar* with redgram and early sowing of *jowar*. A negative, but less obvious relationship, was also found concerning the application of FYM to *jowar* and castor. In contrast, the relationship between the area of wetland owned and adoption of improved castor seeds and the application of chemical fertilizers to *jowar* and castor, was clearly positive. Such a positive, but less obvious relationship, was also found for interculturing of *jowar* and castor. Similar to soil and water conservation recommendations, these findings indicate that adoption depended more on the characteristics of the recommended practice for dryland agriculture than on the wetland area owned by a farmer. More than anything else it clarifies that farmers were in need of simple and more affordable agricultural technologies, multi-purpose in character and suitable to the general agro-ecological and their own socio-economic situation. Projects that do not make an allowance for that, are doomed to fail.

NOTES

1. Both projects tried to diversify the existing cropping pattern, though. The Maheswaram project stimulated the cultivation of vegetables and fruit trees on dryland. AWARE promoted the cultivation of vegetables on both dryland and wetland and mulberry (for sericulture) on wetland. Besides informing farmers about the possibilities of growing these crops, both projects gave financial incentives, for instance, by subsidizing improved vegetable seeds. The Maheswaram project gave financial and technical support to dig pits and to buy saplings of fruit trees, particularly to farmers in Debbadaguda. Because farmers started complaining about low survival rates of the fruit trees, a consequence of insufficient irrigation facilities, the project drilled some bore holes and mounted hand pumps.

   More than AWARE, the Maheswaram project seems to have succeeded in inducing farmers to cultivate these horticultural crops. First of all, while many respondents in Mansanpally and Debbadaguda reported having received seeds or saplings from the Maheswaram project, most respondents in Ekvaipally had never received any information or seeds from AWARE. Secondly, of the plots cultivated with vegetables and fruit trees during 1992/3-1993/4, 55 per cent was located in Debbadaguda, 25 per cent in Mansanpally and only 10 per cent each of the plots were located in Ghatpally and Ekvaipally.

   The most popular horticultural dryland crop was tomato. Although the cultivation of this crop in dryland was taken up in the area before the start of both the projects, the fact that its cultivation was relatively more popular in Debbadaguda and Mansanpally suggests that the Maheswaram project had a favourable effect. In both Debbadaguda and Mansanpally, most of these tomato plots were owned by small and small-medium farmers of the ST and BC communities.

This positive effect of the Maheswaram project was larger for fruit trees: 78 per cent of the plots planted with fruit trees was located in Debbadaguda, 14 per cent in Ekvaipally, 7 per cent in Ghatpally, while in Mansanpally not a single plot owned by a respondent was planted with such trees. In Debbadaguda, 74 per cent of these plots was owned by ST farmers (of which a majority were small and small-medium farmers). This relatively high percentage was the result of the activities of the tribal headman who demanded saplings and handpumps from the Maheswaram project. Another 22 per cent of the plots in Debbadaguda was owned by FC farmers.

2. Adoption rates for intercropping of *jowar* with redgram (irrespective of ratio) of FC farmers in Debbadaguda, Mansanpally, Ekvaipally and Ghatpally were 43 per cent, 57 per cent, 56 per cent and 61 per cent respectively; of BC farmers 79 per cent, 59 per cent, 73 per cent and 48 per cent respectively; of SC farmers 38 per cent, 69 per cent, 56 per cent and 71 per cent respectively; and of ST farmers 62 per cent, 66 per cent, 83 per cent and 88 per cent respectively.
3. Adoption rates for intercropping of *jowar* with redgram (irrespective of ratio) of marginal farmers in Debbadaguda, Mansanpally, Ekvaipally and Ghatpally were 29 per cent, 36 per cent, 66 per cent and 61 per cent respectively; of small farmers 47 per cent, 71 per cent, 61 per cent and 65 per cent respectively; of small-medium farmers 67 per cent, 63 per cent, 76 per cent and 61 per cent respectively; of medium farmers 70 per cent, 78 per cent, 56 per cent and 60 per cent respectively; and of large farmers 20 per cent, not applicable, 58 per cent and 73 per cent respectively.
4. This and following analyses makes allowance for the fact that the sample included only one large farmer from Mansanpally, who stopped cultivating during the research period.
5. For the separate villages, the differences were much larger. *Jowar* was monocropped by 38 per cent, 57 per cent and 29 per cent of the respondents in Debbadaguda in *kharif* 1991/2, 1992/3 and 1993/4 respectively; by respectively 30 per cent, 36 per cent and 44 per cent of the respondents in Mansanpally; by respectively 43 per cent, 35 per cent and 15 per cent of the respondents in Ekvaipally; and by respectively 31 per cent, 35 per cent and 44 per cent of the respondents in Ghatpally.
6. A complicating factor was that irrespective of the fact whether seeds were of an indigenous or an improved variety, many farmers used the term 'local' to characterize seeds which were of local origin, meaning originating from the previous year's harvest. Thus, quite often, improved seeds originating from the previous harvest were considered to be of local origin as well. Farmers

used the term 'hybrid' seeds for all kinds of improved seeds, hybrid or not, which had been bought in a shop or at the *mandal* office. Many farmers just asked shopkeepers for 'hybrid' seeds.

A second complication was that the names of improved seed varieties used by farmers were not always similar to those used by agronomists. While referring to the various GCH varieties of castor, some farmers used the name '*Gujarati*', while others used GCH-2 or GCH-4. Something similar occurred in paddy: not all farmers distinguished *Tellahamsa* from *Hamsa* and hardly any farmer discriminated between the various varieties of BPT, of which the *Mashuri* variety was most popular in the villages. On some occasions, farmers only knew the name of the seed company, stating that they had sown Pioneer or Mayco seeds.

Another complicating factor was that farmers sometimes thought they cultivated an improved seed variety, while in fact they did not. Many farmers were, for instance, not familiar with the fact that genuine improved seeds were certified and sold in bags. Thus, it happened quite often that shopkeepers supplied seeds of a dubious quality, packed in suspicious bags. One of the respondents, having forged an official certification stamp, bought all kinds of seeds from local farmers, packed them in bags marked with that forged stamp and sold the seeds pretending they were certified.

A final complicating factor was the fact that 'local' indigenous and 'local' improved varieties were not necessarily indigenous or improved. The fact that many improved seeds were of an open pollinated variety meant that cross pollination with indigenous varieties took place which resulted in indigenous seeds being less indigenous and improved seeds being less improved. Farmers were very much aware of this problem and knew of the advice to buy improved seeds every year instead of using improved seeds from the last harvest. The fact that a very large majority was reluctant to do so and bought improved seeds only after two or three harvests had everything to do with the distrust they had for the suppliers of seeds.

7. Some of these Reddy innovators said that they had heard about improved paddy seeds from relatives living in Coastal Andhra or saw these seeds while travelling in that area.
8. Adoption rates for improved *jowar* seeds of FC farmers in Debbadaguda, Mansanpally, Ekvaipally and Ghatpally were 5 per cent, 3 per cent, 7 per cent and 2 per cent respectively; of BC farmers 0 per cent, 3 per cent, 0 per cent and 5 per cent respectively; of SC farmers 2 per cent, 0 per cent, 0 per cent and 0 per cent respectively; and of ST farmers 0 per cent, 1 per cent, 6 per cent and 4 per cent respectively.

   Adoption rates for improved *jowar* seeds of marginal farmers in Debbadaguda, Mansanpally, Ekvaipally and Ghatpally were 0 per cent, 0 per cent, 5 per cent and 7 per cent respectively; of small farmers 5 per cent, 1 per cent, 0 per cent and 3 per cent respectively; of small-medium farmers 0 per cent, 2 per cent, 1 per cent and 2 per cent respectively; of medium farmers 0 per cent, 11 per cent, 4 per cent and 0 per cent respectively; and of large farmers 0 per cent, not applicable, 8 per cent and 0 per cent respectively.

9. For instance in *kharif* 1992/3, stimulated by information received from a Hyderabadi friend, one farmer living in a village adjacent to Ghatpally (not one of the respondents) decided to experiment with CSH-11. Having sown part of his dryland plot, some 20 by 20 m, with this seed, he was satisfied with the yield of grain, but displeased with its taste and with the fodder. Next season he cultivated the indigenous *jowar* variety again.
10. The 9 plots cultivated with 'improved' *bajra* in Ekvaipally belonged predominantly to large FC farmers and marginal ST farmers.
11. Adoption rates for improved castor seed of FC farmers in Debbadaguda, Mansanpally, Ekvaipally and Ghatpally were 94 per cent, 78 per cent, 56 per cent and 89 per cent respectively; of BC farmers 89 per cent, 68 per cent, 58 per cent and 85 per cent respectively; of SC farmers 93 per cent, 53 per cent, 47 per cent and 83 per cent respectively; and of ST farmers 98 per cent, 66 per cent, 49 per cent and 86 per cent respectively.
12. Adoption rates for improved castor seed of marginal farmers in Debbadaguda, Mansanpally, Ekvaipally and Ghatpally were 100 per cent, 27 per cent, 44 per cent and 79 per cent respectively; of small farmers 91 per cent, 69 per cent, 53 per cent and 84 per cent respectively; of small-medium farmers 95 per cent, 74 per cent, 54 per cent and 89 per cent respectively; of medium farmers 91 per cent, 50 per cent, 59 per cent and 89 per cent respectively; and of large farmers 100 per cent, not applicable, 67 per cent and 100 per cent respectively.
13. *(Tella) Hamsa* was the most favoured improved variety (on average cultivated on 83 per cent of the plots in the four villages), followed by *BPT/Mashuri* (on 9 per cent of the plots) and varieties like *Krishnaveni*, *Vikas*, *Sureka*, IET-2740 and IR-64, which were cultivated on less than 1 per cent of the plots. The most diversified villages were Ekvaipally and Debbadaguda where five paddy varieties were cultivated during the years of the study, followed by Ghatpally (four) and Mansanpally (three).

    The positive reaction of all farmers towards *(Tella) Hamsa* had everything to do with the fact that it suited their requirements better than the indigenous paddy varieties. *(Tella) Hamsa* combined characteristics like higher yield and good taste of its grain with palatable and good yield of fodder. In addition, it was relatively easy to cultivate. Another advantage was its short duration, meaning that it needed relatively little water, a property considered very important in these dry areas.

    The positive properties of *(Tella) Hamsa* paved the way for other improved paddy varieties too. However, these have also been critically assessed. Although farmers were quite satisfied with the taste and price of *BPT/Mashuri* for instance, which they considered to be superior to *(Tella) Hamsa*, for most farmers it was not a good alternative. First, its yield in grain was relatively low. Second, to get satisfied one needed to eat more while at the same time a feeling of hunger appeared rather quickly. Third, its yield of fodder was relatively low and animals did not like it much. Fourth, it was a medium duration crop, needing relatively more water. These negative properties made it a kind of luxury paddy crop to be cultivated in *kharif* only. It meant that it was a suitable variety for larger farmers, those having sufficient inputs like land and water, but also fodder, at their disposal. However, also for them, *BPT/Mashuri* ranked second after *(Tella) Hamsa*.

Most respondents also took cognizance of the fact that improved paddy seeds lost much of their higher yielding potential after one year and that, therefore, it was best to buy new seeds every year. But even more than with castor, many farmers were not willing to do so and bought new seeds only after two or three years. As with castor, their reluctance was based on the consideration that the decrease in yield was limited and not worth the trouble of buying expensive new seeds every year. Poor experiences with adulterated seeds bought in shops did not stimulate farmers either. This bad experience was responsible for a noteworthy phenomenon in all the four study villages. A couple of large farmers, almost all Reddies, avoided shops and bought improved paddy seeds directly from the Agricultural University. As a result, fresh paddy seed of high quality was introduced into the villages every couple of years. Some of these Reddy farmers distributed these fresh seeds to like-minded farmers. In fact, much more than with other crops, farmers exchanged paddy seeds with one another, sold seed or seedlings for cash or lent seeds to other farmers. Through the *nagu* system, farmers could borrow seeds at a very high interest of 50 to 100 per cent. Despite its exploitative nature, it was an important means to diffuse a new seed variety. Another important means was by paying labourers paddy seeds in kind, assuming that they did not cat or sell all.

14. The reasons given by those who had not adopted improved castor were the following: some believed that since the seeds it produced were much smaller than the seeds of the indigenous varieties, the yield had to be less; others considered the fact that they needed to buy the seeds a disadvantage, especially because they could not afford the price; again others said that the soil in their plot was not suitable for *Aruna*.
15. Adoption rates for the recommended seed-rate for castor of FC farmers in Debbadaguda, Mansanpally, Ekvaipally and Ghatpally were 40 per cent, 63 per cent, 45 per cent and 5 per cent respectively; of BC farmers 20 per cent, 58 per cent, 47 per cent and 7 per cent respectively; of SC farmers 38 per cent, 60 per cent, 40 per cent and 13 per cent respectively; and of ST farmers 29 per cent, 42 per cent, 38 per cent and 13 per cent respectively.
16. Adoption rates for the recommended seed-rate for castor of marginal farmers in Debbadaguda, Mansanpally, Ekvaipally and Ghatpally were 67 per cent, 50 per cent, 44 per cent and 11 per cent respectively; of small farmers 22 per cent, 47 per cent, 32 per cent and 10 per cent respectively; of small-medium farmers 25 per cent, 51 per cent, 54 per cent and 10 per cent respectively; of

medium farmers 30 per cent, 100 per cent, 50 per cent and 0 per cent respectively; and of large farmers 71 per cent, not applicable, 50 per cent and 0 per cent respectively.

17. Although not recommended by the projects, tractors deserve attention. Although numbering few, their presence shows that farmers were not against 'progress' and were willing to invest in innovations. Of the 12 tractors present in the four study villages in *kharif* 1993/4, 8 were owned by respondents, all being medium to large FC farmers, 3 in Debbadaguda, 1 in Mansanpally, 3 in Ekvaipally and 1 in Ghatpally. Even for affluent farmers, tractors were too expensive to keep them idle for part of the year. Hence, to curtail the costs, 3 of the 8 tractors were bought in joint ownership. Moreover, all tractors could be hired for all kinds of (non)agricultural purposes like threshing and to transport manure and other heavy goods. Some tractors were more frequently used in Hyderabad to transport goods than for agricultural purposes in the villages. In fact, tractors were hardly utilized for ploughing since, like with the iron plough, the result was not to the liking of most farmers.
18. I do know, however, that farmers cultivating drylands in a different agro-ecological zone in semi-arid, interior Tamil Nadu, having received iron ploughs on subsidy, parted from their indigenous wooden ploughs.
19. Although these Reddy farmers were not entitled to receive iron ploughs (originally their land was located outside the project area), their good relationship with project officers and the fact that these officers had to meet their targets in the soil and water conservation programme, made these iron ploughs come their way. Information from these Reddy farmers suggested that some bargaining took place. If a Reddy farmer allowed the project to build soil and water conservation structures in his dryland plot, he would get inputs like iron ploughs.
20. The optimum dose for wetland was 10,000 kg/ha each year.
21. Averaged for all villages, in 1991/2 37 per cent of the *jowar* plots and 38 per cent of the castor plots did not receive manure. In 1992/3, these figures were 61 per cent and 68 per cent, and in 1993/4 they were 67 and 70 per cent respectively. The rather large difference between 1991/2 and 1992/3-1993/4 might be caused by a difference in data collection. This difference has no consequence for the ranking of the study villages, though.
22. The difference between the years was small: while farmers applied on average 13 cart-loads/ha in *kharif* 1991/2 and *kharif* 1992/3, they applied on average 14 cart-loads/ha in *kharif* 1993/4.
23. Although somewhat larger than with *jowar*, the difference between the years was small too: in *kharif* 1991/2 it averaged 12 cart-loads/ha, in *kharif* 1992/3 14 cart-loads/ha and in *kharif* 1993/4 13 cart-loads/ha.
24. Averaged for all villages, in 1991/2 33 per cent of the *jowar* plots and 31 per cent of the castor plots received manure according to the recommendation. In 1992/3, these figures were 18 and 16 per cent, and in 1993/4 they were 20 and 17 per cent respectively. The rather large difference between 1991/2 and

1992/3-1993/4 might be caused by a difference in data collection. This difference has no consequence for the ranking of the study villages, though.

25. Adoption rates for the recommended quantities of FYM to be applied to *jowar* of FC farmers in Debbadaguda, Mansanpally, Ekvaipally and Ghatpally were 0 per cent, 15 per cent, 35 per cent and 20 per cent respectively; of BC farmers 13 per cent, 10 per cent, 36 per cent and 34 per cent respectively; of SC farmers 11 per cent, 19 per cent, 48 per cent and 46 per cent respectively; and of ST farmers 10 per cent, 17 per cent, 33 per cent and 25 per cent respectively.
26. For all *jowar* plots in the four villages, irrespective of whether they had received manure or not, the coefficient was negative: – 0.0483. It changes into a positive relationship if account is taken of only the plots which had received manure: 0.0537. Calculated for these plots, the coefficient was positive for Ghatpally and Ekvaipally (0.0690 and 0.0859 respectively), but negative for Mansanpally and Debbadaguda (– 0.0328 and – 0.1684 respectively).
27. Adoption rates for the recommended quantities of FYM to be applied to *jowar* of marginal farmers in Debbadaguda, Mansanpally, Ekvaipally and Ghatpally were 0 per cent, 0 per cent, 41 per cent and 23 per cent respectively; of small farmers 9 per cent, 26 per cent, 43 per cent and 42 per cent respectively; of small-medium farmers 12 per cent, 11 per cent, 36 per cent and 34 per cent respectively; of medium farmers 6 per cent, 11 per cent, 26 per cent and 20 per cent respectively; and of large farmers 0 per cent, not applicable, 42 per cent and 18 per cent respectively.
28. Adoption rates for the recommended quantities of FYM to be applied to castor of FC farmers in Debbadaguda, Mansanpally, Ekvaipally and Ghatpally were 0 per cent, 0 per cent, 44 per cent and 18 per cent respectively; of BC farmers 23 per cent, 5 per cent, 31 per cent and 29 per cent respectively; of SC farmers 18 per cent, 20 per cent, 47 per cent and 27 per cent respectively; and of ST farmers 11 per cent, 13 per cent, 31 per cent and 27 per cent respectively.
29. For all castor plots in all villages, irrespective of whether they had received manure or not, the coefficient was negative: – 0.0336. Calculated for plots which had received manure, it remains negative: – 0.0132. The coefficient concerning these plots was positive for Ghatpally and Ekvaipally (0.0690 and 0.0859 respectively), and negative for Mansanpally and Debbadaguda (– 0.0328 and – 0.1684 respectively).
30. Adoption rates for the recommended quantities of FYM to be applied to castor of marginal farmers in Debbadaguda, Mansanpally, Ekvaipally and Ghatpally were 0 per cent, 0 per cent, 38 per cent and 24 per cent respectively; of small farmers 18 per cent, 10 per cent, 47 per cent and 23 per cent respectively; of small-medium farmers 15 per cent, 9 per cent, 31 per cent and 28 per cent respectively; of medium farmers 12 per cent, 17 per cent, 38 per cent and 24 per cent respectively; and of large farmers 0 per cent, not applicable, 31 per cent and 33 per cent respectively.

31. For all paddy plots, irrespective of whether they had received manure or not, the coefficient was positive and significant at the 0.01 level: 0.2541. Calculated for plots which had received manure, it increases to 0.2877. Calculated for these plots, this coefficient was significant (at the 0.001 level) only for Debbadaguda: 0.4607. For Mansanpally it was 0.0791 and for Ghatpally and Ekvaipally the coefficient even turns negative: – 0.3682 and – 0.0417 respectively.
32. One of the conclusions of the APAU study (nd: 39-40) was that the quantities applied to both *jowar* and castor were less than recommended.
33. On average during the three-year study period, 57 per cent of the *jowar* plots to which fertilizers were applied, received DAP and 33 per cent received Growmore. For castor, the averages were 65 per cent and 24 per cent respectively. DAP was more popular in Mansanpally, Ekvaipally and Ghatpally, while Growmore was used more in Debbadaguda.
34. Averaged for all villages, in 1991/2 12 per cent of the *jowar* plots and 10 per cent of the castor plots did not receive manure. In 1992/3, these figures were 3 and 0 per cent, and in 1993/4 they were 1 and 3 per cent respectively. The rather large difference between 1991/2 and 1992/3-1993/4 might be caused by a difference in data collection. This difference has no consequence for the ranking of the study villages, though.
35. It proved to be impossible to calculate the exact quantities in nutrients NPK. Hence, these averages were calculated by adding the various (complex) fertilizers which were applied.
36. Averaged for all villages, in 1991/2 23 per cent of the *jowar* plots and 18 per cent of the castor plots received at least 50 per cent of the recommended fertilizer dose. In 1992/3, these figures were 15 and 13 per cent, and in 1993/4 they were 7 and 8 per cent respectively. The differences between these years might be caused by a difference in data collection and a result of Indian economic liberalization policy. This difference has only minor consequences for the ranking of the study villages.
37. Adoption rates, based on 50 per cent of the recommended fertilizer dose to be applied to *jowar*, of FC farmers in Debbadaguda, Mansanpally, Ekvaipally and Ghatpally were 11 per cent, 18 per cent, 23 per cent and 8 per cent respectively; of BC farmers 27 per cent, 18 per cent, 13 per cent and 9 per cent respectively; of SC farmers 13 per cent, 4 per cent, 11 per cent and 10 per cent respectively; and of ST farmers 14 per cent, 28 per cent, 17 per cent and 8 per cent respectively.
38. For all *jowar* plots in the four villages, irrespective of whether they had received fertilizers or not, the coefficient was positive and significant at the 0.01 level: 0.0822. It changes into a non-significant positive relationship if account is taken of only the plots which had received fertilizers: 0.0507. Calculated for these plots only, the coefficient was positive and significant for Ekvaipally (0.2041 at the 0.001 level), positive for Debbadaguda (0.1111)

but negative for Mansanpally and Ghatpally (– 0.0476 and – 0.0355 respectively).

39. Adoption rates, based on 50 per cent of the recommended fertilizer dose to be applied to *jowar*, of marginal farmers in Debbadaguda, Mansanpally, Ekvaipally and Ghatpally were 0 per cent, 25 per cent, 20 per cent and 11 per cent respectively; of small farmers 12 per cent, 20 per cent, 9 per cent and 8 per cent respectively; of small-medium farmers 17 per cent, 14 per cent, 12 per cent and 9 per cent respectively; of medium farmers 21 per cent, 22 per cent, 21 per cent and 15 per cent respectively; and of large farmers 40 per cent, not applicable, 33 per cent and 9 per cent respectively.
40. Adoption rates based on 50 per cent of the recommended fertilizer dose to be applied to castor of FC farmers in Debbadaguda, Mansanpally, Ekvaipally and Ghatpally were 21 per cent, 7 per cent, 26 per cent and 9 per cent respectively; of BC farmers 25 per cent, 5 per cent, 7 per cent and 10 per cent respectively; of SC farmers 0 per cent, 20 per cent, 7 per cent and 10 per cent respectively; and of ST farmers 6 per cent, 30 per cent, 17 per cent and 9 per cent respectively.
41. For all castor plots in the four villages, irrespective of whether they had received fertilizers or not, the coefficient was positive and significant at the 0.001 level 0.1787. Calculated for plots which had received fertilizers, it stays positive and significant at that level: 0.1653. The coefficient concerning these plots was positive and significant at the 0.001 level for Debbadaguda (0.4656), positive and significant at the 0.01 level for Ekvaipally (0.2017), but negative and non-significant for Mansanpally and Ghatpally (– 0.0243 and – 0.0548 respectively).
42. Adoption rates, based on 50 per cent of the recommended fertilizer dose to be applied to castor, of marginal farmers in Debbadaguda, Mansanpally, Ekvaipally and Ghatpally were 0 per cent, 17 per cent, 12 per cent and 13 per cent respectively; of small farmers 7 per cent, 30 per cent, 8 per cent and 7 per cent respectively; of small-medium farmers 13 per cent, 6 per cent, 10 per cent and 11 per cent respectively; of medium farmers 21 per cent, 33 per cent, 19 per cent and 6 per cent respectively; and of large farmers 44 per cent, not applicable, 39 per cent and 11 per cent respectively.
43. For all paddy plots, irrespective of whether they received inorganic fertilizers or not, the coefficient was positive and significant at the 0.001 level: 0.3221. Calculated for only the plots which had received fertilizers it was 0.3178 (significant at the 0.001 level). The coefficients for these plots in Debbadaguda was 0.4862 (significant at the 0.001 level), in Ekvaipally it was 0.2825 (significant at the 0.01 level), while they were non-significant and negative in Ghatpally (– 0.1716) and Mansanpally (– 0.1848).
44. Adoption rates for early sowing of *jowar* of FC farmers in Debbadaguda, Mansanpally, Ekvaipally and Ghatpally were 75 per cent, 90 per cent, 60 per cent and 96 per cent respectively; of BC farmers 79 per cent, 98 per cent,

49 per cent and 100 per cent respectively; of SC farmers 78 per cent, 97 per cent, 31 per cent and 96 per cent respectively; and of ST farmers 84 per cent, 98 per cent, 65 per cent and 94 per cent respectively.

45. Adoption rates for early sowing of *jowar* of marginal farmers in Debbadaguda, Mansanpally, Ekvaipally and Ghatpally were 60 per cent, 100 per cent, 77 per cent and 100 per cent respectively; of small farmers 75 per cent, 95 per cent, 45 per cent and 95 per cent respectively; of small-medium farmers 80 per cent, 96 per cent, 39 per cent and 96 per cent respectively; of medium farmers 64 per cent, 100 per cent, 64 per cent and 100 per cent respectively; and of large farmers 100 per cent, not applicable, 67 per cent and 100 per cent respectively.
46. Adoption rates for manual weeding for *jowar* in Debbadaguda, Mansanpally and Ghatpally were 1 per cent, 1 per cent and 0 per cent respectively. For castor they were 0 per cent, 1 per cent and 4 per cent respectively.
47. This increase was visible in all villages. Adoption rates for the recommended number of interculturings in *jowar* in Debbadaguda, Mansanpally, Ekvaipally and Ghatpally were in *kharif* 1992/3, 76 per cent, 25 per cent, 92 per cent and 29 per cent respectively and in *kharif* 1993/4 94 per cent, 100 per cent, 98 per cent and 99 per cent respectively. Adoption rates for the recommended number of interculturings in castor in Debbadaguda, Mansanpally, Ekvaipally and Ghatpally were in *kharif* 1992/3, 46 per cent, 41 per cent, 86 per cent and 9 per cent respectively and in *kharif* 1993/4 84 per cent, 82 per cent, 96 per cent and 20 per cent respectively.
48. Adoption rates for the recommended number of interculturing in *jowar* of FC farmers in Debbadaguda, Mansanpally, Ekvaipally and Ghatpally were 100 per cent, 55 per cent, 100 per cent and 65 per cent respectively; of BC farmers 86 per cent, 65 per cent, 94 per cent and 63 per cent respectively; of SC farmers 84 per cent, 67 per cent, 96 per cent and 76 per cent respectively; and of ST farmers 77 per cent, 67 per cent, 91 per cent and 75 per cent respectively.
49. Adoption rates for the recommended number of interculturing in *jowar* of marginal farmers in Debbadaguda, Mansanpally, Ekvaipally and Ghatpally were 80 per cent, 74 per cent, 96 per cent and 68 per cent respectively; of small farmers 81 per cent, 66 per cent, 94 per cent and 68 per cent respectively; of small-medium farmers 82 per cent, 59 per cent, 98 per cent and 72 per cent respectively; of medium farmers 100 per cent, 67 per cent, 100 per cent and 50 per cent respectively; and of large farmers 100 per cent, not applicable, 100 per cent and 88 per cent respectively.
50. Adoption rates for the recommended number of interculturing in castor of FC farmers in Debbadaguda, Mansanpally, Ekvaipally and Ghatpally were 68 per cent, 62 per cent, 94 per cent and 13 per cent respectively; of BC farmers 62 per cent, 45 per cent, 87 per cent and 11 per cent respectively; of

SC farmers 50 per cent, 60 per cent, 100 per cent and 18 per cent respectively; and of ST farmers 71 per cent, 71 per cent, 90 per cent and 20 per cent respectively.

51. Adoption rates for the recommended number of interculturing in castor of marginal farmers in Debbadaguda, Mansanpally, Ekvaipally and Ghatpally were 33 per cent, 60 per cent, 94 per cent and 19 per cent respectively; of small farmers 59 per cent, 64 per cent, 95 per cent and 12 per cent respectively; of small-medium farmers 62 per cent, 61 per cent, 89 per cent and 16 per cent respectively; of medium farmers 77 per cent, 33 per cent, 82 per cent and 9 per cent respectively; and of large farmers 86 per cent, not applicable, 100 per cent and 14 per cent respectively.
52. Some farmers were, for instance, advised by a shopkeeper to mix several pesticides and to dilute it with cow urine.

9. Summary, Conclusion and Recommendations

M. Put, M.P. van Dijk, A.J. Dietz and G.A. de Bruijne

9.1 Summary and Conclusions

Although agriculture is still an important economic sector in India and although it is still said that the monsoon not only influences agricultural production but also the well-being of the Indian government, it seems that it plays a less important role in the discussion about development now compared with the first decades after Independence. Whereas a major part of the Indian population lives in rural areas and, as farmers or agricultural labourers or as a combination of both, earns a living from agriculture, most of the Five Year Plans allocated more funds to economic sectors other than agriculture. Even the economic liberalization programme of the 1990s, is even basically directed to the industrial/export sector. Out of the total Public Sector Outlay, the share of agriculture, irrigation and rural development as well as the rate of investment in agriculture has decreased in recent years (Hanumantha Rao and Rohini Nayar, 1994: 9; GoI, 1993: 15); and within agriculture much attention is given to export-oriented production. However, as the Department of Economic Affairs, Ministry of Finance, states: 'No strategy of economic reform in India can succeed without sustained and *broad-based* agricultural development' [italic by us] (GoI, 1993: 35).

Most of the government funds allocated to agriculture are earmarked for irrigated or wetland agriculture. An emphasis on a 'radical restructuring of the pattern of public expenditure to favour durable and productive investments' (GoI, 1993: 36) may not lead to less promotion of the more reliable wetland agriculture, hence will not lead to more funds for agriculture undertaken in poorer, dry areas. It seems, however, that a full-scale promotion of wetland agriculture can be counterproductive, can be self-destructive. It seems as if a kind of limit has been reached as far as irrigated agriculture is concerned. In some areas the silting of the soil seems to have negative effects on paddy yields and sometimes even makes agriculture impossible. In many areas, farmers face a declining ground water level as well.

Dryland agriculture, though covering some two-thirds of the net sown area and producing around fifty per cent of the agricultural output, still gets relatively less government funds than irrigated agriculture. In the first Five Year Plans, stimulated by the drive for food self-sufficiency, wetland agriculture flourished to the detriment of dryland agriculture. Only with the implementation of the DPAP in 1970 more funds became available for this type of agriculture. Dryland agriculture also benefited from the watershed approach, launched during the Sixth Plan in 1982, the financing of various watershed projects by the World Bank in 1984, and the NWDPRA, launched during the Seventh Plan in 1987. In the present situation the two major policies of the Indian government concerning dryland agriculture seem to be: the spreading of agricultural innovations, such as soil and water conservation and land management activities; and the active involvement of 'external' donors, be it foreign governments or Indian NGOs.

Notwithstanding the increase in government funds for dryland agriculture, many observers feel that the involvement of efficient NGOs can be an alternative to stimulate and improve this type of agriculture. These observers doubt the effectiveness of the government programmes and projects, as most are very bureaucratic, top-down oriented, often technocratic and without much interest in the 'real life' circumstances of farmers-cum-labourers-cum-non-farm entrepreneurs.

Our analysis of the involvement of NGOs in improving dryland agriculture leads us to conclude that this interest is relatively recent. For many 'old' NGOs, their dryland programme or their ecological programme is only a part of their wider involvement in awareness building, adult education, health and nutrition projects, for instance. Only some of the more recently established NGOs, show a clear focus on 'the environment' as their central issue. Both the 'old' and the 'new' NGOs managed to get financial support from environmentally-minded foreign donors and recently from the central and state governments in India. The government tries to involve more and more NGOs on an experimental basis to use a small part of the relatively huge sums of money available in, for instance, the IRDP.

Reviewing the government approach to dryland development and the approach(es) of NGOs, reveals some differences; not only between the government and the NGO sector, but also within the NGO sector as such.

These show that NGOs differ from one another (not only in ideology, size, strategy, programmes of activities and objectives, but also in the approach followed in relation to dryland agriculture) and cannot be grouped together. The government, for instance, advocates the 'area' or watershed approach, which means that all persons living in that area are possible beneficiaries. Notwithstanding the fact that the government uses the area approach, it also has specific programmes targeted at SCs and STs. Many NGOs, however, advocate a group approach, meaning that they concentrate their activities on certain people, often the most poor and oppressed in society, like small and marginal farmers who mostly happen to belong to the SC and ST community. Many NGOs try to create groups (*sanghams*) which have various functions of which dryland management activities is one. Notwithstanding the fact that many NGOs use the group approach, others prefer an area approach, while again, some others do not concentrate their group formation activities on specific target groups but persuade all kinds of people to join.

Another difference is in regard to the type of agricultural technology promoted. The government as well as NGOs are confronted with the strong attempts of all the farmers to develop irrigation facilities and to get modern inputs like seeds and fertilizers. It took most NGOs quite some time to discover the negative ecological consequences of the one-sided support for irrigation and fertilizers. They found that investments in irrigation only meant that farmers changed their cropping pattern from dryland crops to irrigated crops like paddy, weakening instead of strengthening the dryland focus of their programme. The wells they financed often led to the lowering of ground-water tables, thereby jeopardizing the water conservation elements in their dryland farming programme. Many NGOs, however, afraid to lose contact with their clients, comply with the wish of their target groups to start micro-irrigation projects.

Looking at the dryland technology actually propagated, we find that NGOs do not necessarily propagate a technological package that differs from the one propagated by the government. While some NGOs encourage farmers to construct indigenous soil and water conservation structures on their plots, others introduce improved methods like contour bunds and advocate contour ploughing, while some adapt these improved technologies and introduce solutions like the 'adjusted contour'. The same conclusion can be made for crop management technologies: some NGOs build on the indigenous technological knowledge available in the farming community, while others are very much in favour of the innovations developed at agricultural research stations, while some try to adapt these innovations and propagate a kind of middle-of-the-road strategy.

The objective of our study was to measure the impact of a government and an NGO project in Andhra Pradesh on the adoption of new technologies for dryland agriculture. For that reason two projects were selected: the Maheswaram project as an example of a government project, and the activities carried out by the NGO AWARE in Amangal *mandal*. The Maheswaram project was selected for our study because it was considered to be the opposite of everything a typical NGO stands for. This World Bank financed project started in 1985 and was primarily executed by the Government of Andhra Pradesh. It is an example of a top-down approach, with local participation of the farmers limited to the implementation phase only. Farmers were, for instance, not consulted about the technologies which were to be introduced; neither before nor during the implementation phase of the project, farmers were asked about their perception of the environmental problems and the difficulties related to dryland agriculture in the area or their priorities and wishes. The Maheswaram project was an integrated project, introducing not only various activities to improve soil and water conservation, and crop management, but also active in the realm of horticulture, fodder production and (social) forestry, for instance. Its goal was not so much the improvement of specific target groups, as concentrating on the upgrading of the watershed as such, hence, it did not have a target group but an area approach. To induce farmers to adopt the improved technologies, it used subsidies and various sources of information to spread the message. The World Bank propagated T&V system was used, for instance, as well as the services of so-called Village Volunteers. One of the elements to improve local participation and to stimulate farmers to adopt the recommendations, the Village Watershed Development Committees, failed right from the start. Our study suggests that the extension methods were not very effective. The attitude of many of the field staff towards the farmers, especially towards the smaller farmers of the SC and ST communities, was not always amiable. The top-down thinking of the staff resulted, for instance, in the questioning of elements of the indigenous technologies.

AWARE was selected for our research as the typical example of an NGO because various evaluation reports wrote very favourably about this organization and because AWARE scored high on criteria which were said to explain to a large degree the success of NGOs over government agencies [see e.g. Robinson, 1991; Farrington et al., 1993(a); Farrington et al., 1993(b)]: proximity to the target groups; possessing location-specific knowledge about their social and economic position and the agro-ecological conditions of their environment; being responsive to their customs, needs and priorities; being more efficient in resource use; promoting the participation of the target groups in project design, implementation and monitoring; being able to develop village and supra-village organizations expressing the interests of the target groups; having a better and more highly motivated staff; having greater flexibility and being innovative. AWARE's scale of activities overcomes one of the comparative disadvantages of NGOs: their often limited area of operations. AWARE followed a bottom-up approach. It concentrated its activities on specific target groups (STs and SCs) who were stimulated to form *sanghams*. These *sanghams* participated from the start of the project identification phase. AWARE started in the villages with social motivation campaigns and concentrated later on economic activities, of which the promotion of dryland agriculture was one. The various improved dryland technologies (on soil and water conservation, and crop management) introduced in the villages hardly differed from the ones introduced by the government. In fact, information on these new technologies was not primarily given by AWARE staff. Because the staff had only limited knowledge about these technologies, they relied heavily on the government extension service.

Keeping in mind that the Maheswaram project was located not far from Hyderabad, that AWARE had implemented several projects in the Amangal *mandal* and that the agro-ecological characteristics of the study villages had to be similar, the southern part of the Ranga Reddy and the northern part of the Mahabubnagar districts were selected for our analysis. The outcome on variables like average yearly rainfall, importance of irrigation, cropping pattern, transport links with Hyderabad and socio-economic situation is more or less similar throughout this region. Within this region, four villages were selected: two villages located in the Maheswaram project area (Mansanpally and Debbadaguda), one in AWARE's project area (Ekvaipally) and one village without any direct project intervention in dryland agriculture or soil and water conservation (Ghatpally). In order to evaluate the success of project interventions, the villages were selected in such a way that many intervening variables were kept more or less constant.

The chapter on climate related-risks suggests that any report on the success or failure of agriculture has to start with a description of the monsoon and the pattern of the rainfall in general. Although not the driest districts in Andhra Pradesh, Ranga Reddy and Mahabubnagar are low rainfall areas and clearly very risky for dryland agriculture. We found that this is most obvious in the study area which is the driest and most risky area in the two districts. An important conclusion of the constructed drought risk index is that such risk is considerably higher in the research area than in the district headquarters. In general, there is a strong correlation between better rainfall pattern and improved food production. When the IDPAD research project started collecting the first data, in 1991, farmers were not in bad shape: the year 1990/1 most probably was one of the best years possible. After 1990/1 the situation deteriorated considerably, with 1992/3 and 1993/4 being rather bad years.

The study suggests that social differences between the villages are small. In all four villages, FC farmers are the dominant community although some BC farmers are important as well. Economic differences between the villages are somewhat larger. Economically, Debbadaguda ranks first, followed by Ekvaipally, Ghatpally and Mansanpally. Although certainly not vanished, the study suggests that the caste hierarchy is slowly eroding in all villages. Reddies still dominate the rural scene, owning considerable proportions of all kinds of resources like dryland, wetland, livestock and implements. Their large resource endowment is expressed in economic and political power at village and sometimes at regional level. Compared to most members belonging to castes lowest in the social hierarchy, the communication network of most FCs is spread over a larger area. Slowly, however, various low caste households are gaining more economic and political power, buying land and getting elected to official bodies, and communicating more with the world outside the village. Nevertheless, there is extreme economic inequality and the livelihood security of marginal and small farmers, of the SC and ST communities in particular, is continuously at risk, more so if they do not own wetland.

It is important to emphasize that for almost all farmers, dryland agriculture is only one of many sources of (cash) income. Besides dryland agriculture, other sources of (cash) income are wetland agriculture, animal husbandry, fruit production, collecting forest produce, agricultural labour or labour outside agriculture. The study indicates that yields of dryland and wetland crops differ considerably for the same farmer. This shows that from a farmer's point of view it is very rational to invest more in wetland than in dryland agriculture. In fact, not many farmers are willing to invest much time and capital in dryland agriculture. Some even prefer an income obtained from agricultural labour to the cultivation of their own dryland plot.

During the last decades various innovations have entered the villages. These innovations have not only been introduced by the various government and NGO programmes, but also by the villagers themselves. This suggests that the study villages are not static.

The chapters on the use of improved technologies for dryland agriculture show that, firstly, farmers have and use many sources of information: mass media like radio, television, newspapers and leaflets, and interpersonal channels like relatives, friends and neighbours. Our study suggests that both the Maheswaram project and AWARE have only been of limited importance as sources of information. Not all farmers use the same sources, however. In a majority of cases large farmers (often Reddies) bring information about innovations which have been developed outside the village, into the local community. Since their contacts with the outside world are superior to those of smaller farmers, they have a lead in information gathering. Large farmers select from the available information whatever they think is useful and bring it into practice, often by trial and error. By selecting the information, these large farmers act as a kind of filter. For many fellow farmers, often owning smaller holdings, Reddies are the most important source of information. Smaller farmers who work in the fields of large Reddy farmers, often learn about an innovation by seeing and often applying it in these fields. For them, seeing is believing. They do not want to run the risk of experimenting since this might result in the loss of the limited resource base available to them. Only sometimes smaller farmers get their information from sources other than the local ones: because of improved transportation facilities and labour migration, some smaller farmers get in touch with other sources of information than large farmers from their own village.

Second, the information gathered by farmers is often incomplete or incorrect. Our research suggests that this is often the result of improper information supplied by the source. Only in a limited number of cases it is the ignorance of the farmer which makes him act in an incorrect way.

Third, farmers have often clear-cut ideas why they favour or do not favour a certain innovation. Our study suggests that being informed about an innovation is not a sufficient condition to adopt it. The decision to use a new technology, to adapt it or not to use it at all, depends on various factors, often more than one, and is clearly based on rational criteria. Economic factors, like, for instance, the resources at the disposal of a farming household, the availability of wetland and the importance of agriculture compared to income obtained outside the agricultural sector, to name a few, seem to play important roles in this decision-making process. On the other hand, social and cultural factors like village tradition and ideas about taste of food, play important roles as well. Hence, to the farmers, an innovation is not necessarily an improvement over indigenous technologies.

Finally, based on the time elapsed before farmers adopt an innovation, they have been labelled 'progressive' or 'laggards'. Our study suggests that such labelling is detrimental to the farming community as a whole and certain farmers (the 'laggards') in particular. There certainly exists a time lag. Often, larger farmers have heard about innovations earlier than smaller farmers, and, consequently, they use such an innovation earlier. Smaller farmers make up for most of the arrears quite quickly, however. If they do not, if they decide not to use a certain innovation, it is because of rational decision-making. The availability of resources, the accessibility to various distribution channels or personal preferences, for instance, all influence this decision. These elements play a role in the decision-making process of all farmers, large and small, however. That the outcome of this decision-making process differs sometimes for different groups of farmers, meaning that large farmers use more innovations and on a larger scale compared to smaller farmers, does not mean that smaller farmers are conservative or 'innocent', a very derogatory word used by many Indian urbanites. In many cases, smaller farmers have decided not to use the innovation or to use it in a way which deviates from the recommendations, because it does not suit them properly. Calling such farmers 'laggards' is a misnomer.

The effectiveness of both the projects was measured by concentrating on farmers' adoption rates of the various recommendations made by them. These observed adoption rates were compared with the hypotheses which were formulated in Chapter 4. While it was hypothesized that Debbadaguda (the 'second phase Maheswaram project village') would have the highest adoption rates, followed by Ekvaipally (the 'AWARE village') and Mansanpally (the 'first phase Maheswaram project village'), while Ghatpally (the 'non-intervention village') would rank last, this materialized only once. Five times the observed ranking differed slightly from what was anticipated and the hypotheses were neither confirmed nor rejected. In a large majority of cases, though, the ranking was such that the hypotheses were rejected. Also, the two tables included in Chapters 7

(Table 7.1) and 8 (Table 8.1) indicate that the results of the two projects were far less than expected. Another indication of limited project influence are the marginal differences of the summarized average adoption rates for all recommended (*in-situ*) soil and water conservation and crop management technologies. These were 26 per cent in Ekvaipally, 25 per cent in Debbadaguda and 24 per cent in both Ghatpally and Mansanpally.

Other conclusions at village level drawn in Chapters 7 and 8 were:

- The comparison between the observed average adoption rates for the recommended (*in-situ*) soil and water conservation technologies at village level with the hypotheses suggest some effect of the projects. More improved soil and water conservation structures were built and more *khus* lines were planted in the project villages than in the non-project village of Ghatpally. However, as adoption rates did not exceed 10 per cent and as AWARE had not built any improved soil and water conservation structures, there was no conclusive evidence. Summer tillage of *jowar* and sowing as well as interculturing along the contour or across the major slope had been adopted by a large number of farmers. However, since the ranking of the villages on the first practice was not as expected the hypothesis was rejected. The projects had more effect on the other two recommendations. However, since the ranking differed slightly from what was anticipated, the two hypotheses were neither confirmed nor rejected. Furthermore, it was argued that farmers in Debbadaguda and Ekvaipally were more responsive to these recommendations than farmers in the other two villages because more than the latter two, the former villages were located in slightly more undulating terrain.
- The comparison between the observed average adoption rates for the recommended crop management technologies at village level with the hypotheses suggest that neither project had any significant effect. Not a single hypothesis was confirmed, three were neither confirmed nor rejected (on the doses of chemical fertilizers applied to *jowar* and castor and on the number of interculturings of the latter crop) and most were rejected. Low adoption rates were partly an effect of the limited reach of both projects. By contracting out agricultural extension on improved crop management both the Maheswaram project and AWARE hardly had any control over an important means of communication.
- In all villages, the average adoption rates for the recommended crop management technologies were considerably higher than the averages

for (*in-situ*) soil and water conservation technologies. Lower adoption rates of some of these latter recommendations was the combined result of lack of information, less suitable technologies and of what Rogers (1995: 70, 217) calls their preventive nature. Farmers were less interested in adopting these recommendations as the immediate benefits were not clearly discernible, while future benefits were also uncertain. In fact, the only preventive recommendation having relatively high adoption rates (interculturing) had multiple objectives, combining soil and water conservation with crop management.

- The fact that the average adoption rate of recommended (adapted) indigenous technologies was higher than that of innovations (whether introduced before or by the projects) is another indication that the projects had only limited effect (see Tables 9.1 and 9.2). Contour bunds, *khus*, post-harvest tillage and dead furrows, innovations which were 'new' to the area and which were introduced by the Maheswaram project, had very low to negligible adoption rates.

Besides 'village level' hypotheses also hypotheses on the ranking of the various communities and the various landownership classes were formulated. The observed adoption rates per community and per class were compared with the expected rates both within (e.g. comparing the observed with the expected ranking of marginal with large farmers in one village) and between the villages (e.g. comparing the observed with the expected ranking of large farmers between the four villages). Again, most hypotheses were rejected, showing that the various communities and landownership classes hardly reacted upon project interventions. Tables 9.1 and 9.2 give overviews. They show that with regard to all recommendations, the adoption rates of the various communities and landownership classes differed slightly. In fact, not a single community stood out as best or worst performer having the highest or lowest adoption rates. In slight contrast, the differences between the various classes were some-what more obvious, with somewhat higher adoption rates for large and medium farmers and somewhat lower rates for small and marginal farmers (except in Ghatpally). Keeping in mind that the differences between the communities and classes were rather marginal, the most important conclusions are:

- *Concerning the recommendations for soil and water conservation*: on average, FC and ST farmers had somewhat higher adoption rates than BC and SC farmers and medium and small-medium farmers had somewhat higher rates than small and marginal farmers.

- *Concerning the recommendations for improved crop management*: on average, ST and FC farmers had somewhat higher adoption rates than BC and SC farmers and large farmers had somewhat higher rates than marginal farmers.
- *Concerning the (adapted) indigenous recommendations*: on average, ST farmers had somewhat higher adoption rates than FC and SC farmers and small-medium farmers had somewhat higher rates than marginal farmers.
- *Concerning the 'real' innovations*: on average, FC and ST farmers had somewhat higher adoption rates than BC and SC farmers and large and medium farmers had somewhat higher rates than small-medium, small and marginal farmers.
- *Concerning the comparison of the various communities between the villages for all recommendations*: while it was expected that FC and BC farmers of Debbadaguda would have the highest average adoption rates of all FC and BC farmers in the sample, this applied only to BC farmers. Of the FC farmers, those of Ekvaipally performed best. While it was expected that SC and ST farmers of Ekvaipally would have the highest average adoption rates of all SC and ST farmers in the sample, SC farmers of Mansanpally and ST farmers of Ghatpally performed best. It is not surprising then that most hypotheses on expected adoption rates of the various communities in the villages were rejected.
- *Concerning the comparison of the various landownership classes between the villages for all recommendations*: while it was expected that large, medium and small-medium farmers living in Debbadaguda would have the highest average adoption rates of these classes in the sample, this applied only to small-medium farmers. Large farmers of Ekvaipally and medium farmers of Mansanpally had the highest rates of their class. The expectation that small and marginal farmers living in Ekvaipally would have the highest adoption rates proved correct. Because it was not always correct for each individual recommendation, most hypotheses on expected adoption rates of the various classes in the villages were rejected.

Arguing that farmers who depended entirely on dryland for their agricultural production would be more inclined to adopt the recommended technologies than farmers who also owned wetland, a negative relationship was assumed between the area of wetland owned and adoption. However, such a relationship was only found in bunding, in intercropping of *jowar* with redgram and early sowing of *jowar*. A negative but less obvious

TABLE 9.1. AVERAGE ADOPTION RATES OF RECOMMENDED TECHNOLOGIES, BY COMMUNITY, BY VILLAGE

	Debbadaguda				Mansanpally				Ekvaipally				Ghatpally			
	FC	BC	SC	ST	FC	BC	SC	ST	FC	BC	SC	ST	FC	BC	SC	ST
SWC	22	15	18	20	15	14	15	16	13	13	12	15	13	10	8	11
CM	26	30	23	27	27	26	28	29	31	28	27	27	26	25	29	30
Ind	37	40	35	41	41	41	39	47	44	46	48	47	36	37	42	44
Innov	18	19	15	17	15	13	13	15	17	13	19	13	16	13	14	16
All	25	26	22	25	24	23	25	24	27	25	24	24	23	22	24	26

TABLE 9.2. AVERAGE ADOPTION RATES OF RECOMMENDED TECHNOLOGIES, BY LANDOWNERSHIP CLASS, BY VILLAGE

	Debbadaguda					Mansanpally					Ekvaipally					Ghatpally				
	L	Me	S-m	S	Ma	L	Me	S-m	S	Ma	L	Me	S-m	S	Ma	L	Me	S-m	S	Ma
SWC	16	20	17	15	21	0	15	16	14	14	12	17	14	12	10	4	15	10	8	12
CM	32	29	27	25	18	0	35	26	25	24	36	28	29	28	30	28	26	26	28	26
Ind	37	39	40	36	27	0	46	44	46	40	44	43	46	48	49	36	37	40	39	38
Innov	23	20	16	18	15	0	22	13	14	12	23	16	14	12	13	15	16	14	15	15
All	28	27	25	23	19	0	30	24	22	22	30	26	23	24	25	22	23	22	23	23

Note: SWC = Recommended soil and water conservation technologies; CM = Recommended improved crop management technologies; Ind = Recommended (adapted) indigenous technologies; Innov = Recommended innovations introduced before or by the projects; L = Large farmer; Me = Medium farmer; S-m = Small-medium farmer; S = Small farmer; Ma = Marginal farmer.

Source: Own survey.

relationship was found in summer tillage of *jowar* and the application of Farm Yard Manure (FYM) to *jowar* and castor. The fact that many positive relationships were found indicates that adoption depended more on the characteristics of the recommended practice for dryland agriculture than on the wetland area owned, thus the 'characteristics' of a farmer.

The low adoption rates found by this study were confirmed by the Maheswaram project (AWARE did not have specific data on this subject). One highly placed project officer stated that the project had not succeeded in bringing about a change in farmers' minds and that most farmers just continued with their indigenous agricultural practices. Furthermore, an evaluation study of the project by Sitapathi Rao et al. (1990: 43-4) reported that:

> The aim of large scale verification of the already developed technological package of practices is very much diluted since many of these practices do not appear to have been accepted by the farmers. Farmers lack interest and have apathy towards the works executed.

It added that farmers were not fully convinced of the need to change their sowing and cultivation pattern and that it was clear that contour cultivation had not yet taken strong roots in farmers' minds (Sitapathi Rao et al., 1990: 61-3):

> It was found that the supply of inputs through different crop production schemes was closely linked with the adoption of the *in-situ* moisture conservation programme and served as an incentive for the farmers to adopt contour cultivation. In most cases, farmers did follow contour cultivation mainly to satisfy the official pressure but not out of real conviction. The reasons were not difficult to assess. The traditional cultivation practice by and large was reported to be an easy way of cultivation (56 per cent of the total of 272 interviewed farmers) as compared to the contour cultivation which was serpentine. Another 33 per cent of the respondents felt that there was no need to change the traditional pattern.

It is an intriguing question why the effects of the Maheswaram project and the project implemented by AWARE were so similar and why these effects hardly differed from a non-project situation? In our view, the low adoption rates and the rejection of so many hypotheses were the combined result of the improper design of some recommendations, improper functioning of the projects and farmers' reluctance and inability.

Public agricultural research focused on irrigated agriculture to the detriment of research into dryland agriculture in India's semi-arid tropics. Even the establishment of specific agricultural research stations in the 1970s could not rectify this imbalance. Furthermore, the lay-out of demonstration plots and on-farm verification trials started only at the end of the seventies. Thus in the early 1980s, while AWARE was already active in Amangal *mandal* and the Maheswaram project was in its planning phase, the World Bank (1983: 3) concluded that the development of innovations for dryland agriculture in Alfisols had not progressed much. Consequently, many soil and water conservation practices which were extended by both projects resembled practices that had been developed in the 1930s, while it was known that farmers found them unacceptable. In addition, the more recently developed *in-situ* soil and moisture conservation and improved crop management measures were less beneficial for farmers than research promised them to be.

Moreover, many agronomists who have joined research in dryland agriculture live in ivory towers, are hardly interested in understanding farmers' objectives, constraints and indigenous practices, hardly focus on innovations that suits farmers' needs like low input technologies. Consequently, the preoccupation of agricultural projects and agricultural extension with soil and water conservation engineering, divert their attention from promoting the adoption of low-cost, quick-return, soil and water conservation and crop management technologies already available for application by cultivators themselves (World Bank, 1988: 15). The same document even concluded that given the availability of a far cheaper and more effective vegetative solution to controlling run-off and soil erosion (*khus*), bunding and land-levelling would best be stopped entirely (World Bank, 1988: 27).

Other observers have expressed doubts whether trials, undertaken to assess the relative advantage of the recommendations for dryland agriculture in India's semi-arid tropics, gave reliable estimates. ESCAP (1988: 56), for instance, stated that while most farmers cultivated under 'resource-poor situations', agricultural research stations were located in the more favourable zones of India's semi-arid tropics, having relatively higher rainfall and better soils. Hence, it is not surprising that various researchers doubted the superiority of certain recommendations over indigenous technologies. Concerning (contour) bunds, Huibers (1985: 99-101) argued that these did not have a markedly favourable effect on the productivity of dryland agriculture. Similarly, Jodha (1985: 206) contended that these bunds were initiated more as a measure for famine relief than to increase agricultural production. In 1988, at a time that both the Maheswaram project and the AWARE project were long on their way, the World Bank (1988: 21) stated that none of the recommendations for *in-situ* moisture conservation were likely to increase yields by more than 15 per cent, irrespective of the base level, and that this was not enough to attract much farmer interest. Furthermore, observers like Von Oppen and Rao (1982: 672), Rangaswamy (1982), and Jodha (1985: 207) had doubts about the profitability of improved seeds of several dryland crops, especially of improved *jowar*. Rao (1991), finding only a modest difference in net returns between indigenous and improved dryland practices, frankly advised dryland farmers to stick to his/her indigenous technologies rather than to adopt the innovations, or, even better, turn to wetland agriculture.

These doubts were confirmed by this research. Examples of recommendations that, according to farmers, had an improper design were contour/graded bunds, *khus*, contour cultivation, intercropping of *jowar* with redgram in a 2:1 or 3:1 ratio and improved *jowar* seeds. Most of these ill-suited recommendations were propagated by both the Maheswaram project and by AWARE in its Amangal project area. This study found that ill-suited innovations are likely to have negative effects on the acceptance of newer generations of similar innovations. Improved *jowar* seeds are a case in point.

Besides above-mentioned unsuited characteristics of the recommendations, both projects can be criticized for the top-down approach of their projects. This applies especially to AWARE, an organization that claimed to have a bottom-up approach. Thus, the study expressed doubts about the manner in which its staff and management as well as the members of its target groups used the concepts of 'participation' and 'empowerment'. More specifically, this study doubts whether the needs and priorities of the members of the *sanghams* could really be included in the projects since these had to be in line with AWARE's programmes and with government programmes. Moreover, the decision-making proces of AWARE's beneficiaries was frustrated because AWARE was a large bureaucracy, having a strict organizational hierarchy controlled by a chairman who wanted to have a final say in almost everything. In addition, both organizations can be criticized for the top-down attitude of most of their staff who were used to lecturing and not to listening.

Both projects relied heavily on the public T&V system. Thus, both organizations transferred agricultural recommendations from outside to the farming communities in the villages without checking with farmers whether these suited their needs and without checking whether other solutions would be more appropriate. Besides, because the extension service in both projects did not function properly, both projects had difficulties reaching farmers. Although there were some major differences in the behaviour of government and AWARE staff (while civil servants preferred to contact large FC farmers, AWARE staff mainly contacted smaller farmers) for both the projects it holds true that communication between farmers and T&V staff was limited and that participation of the farmers was mostly limited to adoption of the recommendations. But many respondents were hardly informed by the projects or the T&V extension system about the recommendations, especially on recommendations for improved crop management. As a result, many respondents were not able to recount in detail the exact contents of these recommendations.

That the village councils failed to get established by the Maheswaram project was mainly a result of indifference by the project staff; but also because farmers were not interested to participate. Although AWARE managed to establish *sanghams* in Ekvaipally, their functioning was debatable. This study shows that these *sanghams* were neither internally democratic, nor led by effective, scrupulous and accountable leaders, who acted for the benefit of the members. The study has also found that although created to develop and empower the target groups and to facilitate the access to government organizations, *sanghams* were mainly used by brokers and gatekeepers to promote their own development and that of their followers. For instance, they did not succeed in inducing T&V extension staff to visit the village regularly. The fact that most respondents in the study villages largely depended for their information on more experienced Reddy farmers, is a case in point. The study also showed that, more than a means to empower the target group, *sanghams* were meant to improve the efficiency during project implementation. However, they were very poor at reaching that goal.

Low adoption rates were also the result of farmers' reluctance and inability to use a recommended technology or to use it at the recommended level. Farmers often had clear-cut ideas why they favoured or did not favour a certain recommendation. The decision to adopt a recommendation, to adopt it partially, to adapt it, or not to adopt it at all, often depended on more than one factor and was clearly based on rational criteria. Economic factors, such as the resources at the disposal of a farming household, the availability of wetland and the importance of agriculture in relation to income obtained outside the agricultural sector, influenced this decision-making process. Thus, it was found that most farmers were prepared to use agricultural innovations and to invest more in dryland crops provided (1) the short-term benefits (higher yields in grain and fodder and higher monetary returns) were easily discernible; (2) the relationship between the investment and these short-term benefits was clearly visible; (3) the trade-off between the investment and the incremental return was clearly in favour of the latter; (4) the innovations were easily put into practice and did not deviate much from the indigenous practices; and (5) they did not take much additional labour. On the other hand, social and cultural factors like village tradition and ideas about taste of food, were important as well. Many of the recommendations satisfied these conditions only partly. Thus, many farmers were unwilling to sow across the major slope each year, because they were afraid that it would result in lower yields. Furthermore, while almost all wetland owning farmers removed weeds in paddy by hand, they were reluctant to do so in dryland agriculture, because benefits were not clearly visible and costs were high. On the other hand, while almost all farmers were willing to use chemical fertilizers for dryland crops, most used less than they wanted to, because they could not afford to buy more. In addition, although they wanted to, most farmers bought improved castor seed once in a couple of years, because of their high price and because they did not trust shopkeepers and others selling these goods.

Besides effectiveness the study also analyzed the sustainability of project activities, which is related to both the agricultural recommendations and the local organizations created by AWARE in Ekvaipally.

Farmers were reluctant to construct graded/contour bunds themselves because of the relatively high costs and the incompatibility of these improved structures with existing agricultural practices. The fact that farmers considered this technology unsuited to replace existing structures, casts serious doubts on the sustainability of the structures constructed by the Maheswaram project. The same doubt is expressed by Sitapathi Rao et al. (1990: 23, 39) who reported that the acceptance of improved bunds by farmers living in the Maheswaram watershed appeared to be marginal, which was reflected in the limited maintenance of the works, the many breaches in the graded/contour bunds and in the fact that farmers ploughed the bunds. Unsustainability applies to *khus* too; although relatively cheap, its incompatibility with the agro-ecological situation in the area resulted in the death of many plants. In addition, because of its incompatibility with agricultural practices and the lack of interest on the part of farmers, many lines were wilfully ploughed and destroyed.

Since summer tillage is indigenous and adopted by many farmers, it appears to be a sustainable practice. However, the study showed that its use depended largely on natural factors, in this case the timing of pre-monsoon showers. Most farmers were only willing to till their dryland plot in summer, provided the soil was sufficiently moist.

Sowing across the major slope every year, an indigenous practice in undulating terrain, was accepted more by farmers having plots in undulating than in relatively flat terrain. Being indigenous, having relative advantages in undulating terrain and being compatible with existing agricultural practices, it is expected to be a sustainable practice. Since the direction of interculturing was largely determined by the direction of sowing, it is expected that interculturing across the major slope is also a sustainable practice.

Adoption rates of the recommended 2:1 or 3:1 intercropping ratios of *jowar* with redgram were low, while they were much higher for the 4:1 and higher ratios. Since farmers had major objections against the lower ratios and since the higher ratios were indigenous to the area, it is felt that intercropping is sustainable as far as these higher ratios are concerned. However, the study showed that, similar to summer tillage, the use of intercropping depended on natural factors like the presence of pests.

Low adoption rates of several improved seeds of various dryland crops like *jowar*, redgram and *bajra* against high adoption rates of castor (and of a wetland crop like paddy) illustrate that farmers were very particular about improved seeds. As long as improved seeds satisfied the multiple objectives of farmers, had higher relative advantages on the characteristics deemed important for that crop and were compatible with existing agricultural practices, farmers were willing to adopt improved seeds. Since improved castor seeds, especially the *Aruna* variety, had been adopted by so many farmers, it is highly plausible that the use of this seed is sustainable.

The application of the recommended dose of FYM to dryland crops is a sustainable practice. This finding is based on the fact that FYM was valued by all farmers more than chemical fertilizers, that it was indigenous to the area and that it was applied by many respondents. However, whether it will be used in sufficient quantities by all farmers for all dryland crops in all years, depends largely on its availability.

Although most farmers appreciated FYM more than chemical fertilizers, many applied fertilizers to their dryland crops because its relative advantages were considered to be high and because it was compatible with existing agricultural practices. On the other hand, insufficient soil moisture later in the season made them abstain from top-dressing, while high prices made them use lower than recommended quantities. That farmers kept using chemical fertilizers after price increases, makes it plausible that it is a sustainable agricultural recommendation for dryland crops.

The fact that early sowing of *jowar* was indigenous to the area and that farmers' experience and astrology taught them to sow *jowar* early in the season, makes it plausible that it is a sustainable practice. The same remark applies to interculturing of *jowar* (twice) and castor (thrice or more). Farmers who did not follow these practices did not observe them out of ignorance or because the technology did not suit them, but because various reasons, mainly natural but also related to resource endowments, prevented them from doing so.

Weeding by hand as well as the application of pesticides were hardly undertaken for most dryland crops. The fact that the same farmer who did not weed his/her dryland crops, weeded his/her paddy crop by hand and the fact that the same farmer who did not apply pesticides to dryland crops, did so in high value crops like cotton and grapes, shows that both practices are crop specific. Consequently, sustainability of this practice depends on the crop for which it is used. Given the present situation, it may be high for wetland and high value crops and low for most dryland crops.

The local organizations created by AWARE in Ekvaipally are hardly sustainable. The study showed that although quite a number of people from Ekvaipally became members of these *sanghams*, these organizations were and still are hardly used for the purpose for which they were supposed to have been created. The study also expressed doubts whether the needs and priorities of the members of the *sanghams* were included in the projects carried out by AWARE during its direct involvement in the village and whether the leaders acted for the benefit of the members. It was also stated that most members participated in meetings very irregularly, in later years only when schemes were distributed, and that quite a few members had resigned. During the fieldwork period, the *sanghams* hardly functioned and existed largely on paper. It was learnt that this also applied to many other *sanghams* in Amangal and other areas.

This malfunctioning, which was and is not corrected by AWARE field staff, is largely related to the problems mentioned in Chapter 6: selfish local leadership who hardly tried to ensure equitable sharing of benefits and costs among all the members; and resource-poor members who participated for as long as it yielded them sufficiently (short-term) benefits to justify the effort/costs and the risk of participating. Many members of the target group were reluctant from the beginning about joining the *sanghams*. Belonging to a faction other than the leadership, they were not convinced about the benefits, only about the costs. In fact, more than being vehicles for empowerment, for social and political development of all SCs and STs, *sanghams* were vehicles for economic development of a small group of people.

9.2 Recommendations

9.2.1 RECOMMENDATIONS TO REACH THE TARGET GROUP

All those who are involved in the development of agricultural innovations or policies for agricultural research and extension, should concentrate more on the characteristics of indigenous soil, water and crop management practices. Specific agricultural research programmes should be implemented which study the rationality behind and the positive and negative effects of these indigenous practices. Interventions of the government and NGOs, for instance, are needed to document, evaluate and, after approval, disseminate these practices. Agricultural extension personnel should be trained in such a way that they learn to appreciate these indigenous technologies and should disseminate the valid practices to the farming community.

Too often, agronomists have a single objective in mind while developing an innovation or a new agricultural practice. Farmers, however, have multiple objectives. While most experts consider only yield and higher profits, farmers are also interested in characteristics like yield of fodder, taste, easiness of cultivation, etc. Consequently, farmers select only those recommendations which suit them best.

For many farmers dryland agriculture is not the only and not the most important source of (cash) income. For a project to succeed, it is imperative to keep this observation in mind when implementing projects. Projects which concentrate on agriculture only and which do not select their beneficiaries properly, will find that their success will be limited. Hence, both projects and research should not only concentrate on agriculture but should also follow an integrated approach; include both agricultural and non-agricultural developmental activities. This remark applies even more to (most) women, who have multiple objectives and multiple tasks in the household and have many sources of income as well.

To improve participation of the local population in agricultural projects, farmers should be consulted first on their needs and wishes, on the rationale behind their agricultural practices and on the importance of agriculture.

Too often farmers are treated as mere receivers of project support. Projects should incorporate them more during the implementation phase.

It is important for extension staff and project officers to get a good rapport with the local population and to get informed about local practices and customs. Without a good rapport and without knowledge about local practices and the reasons behind them, project officers will have a hard time and projects have a chance of failure. It shows for instance from the indigenous way farmers sow their crops and the manner in which project agents reacted upon this indigenous practice. The custom of sowing in the same direction as one's neighbour, one year up the slope, the other year àcross the slope, was rather strong in one of the study villages. This custom put a heavy burden on project officers. Since the individual is part of the system, approaching individual farmers and belittling their indigenous activities, will not work. Project officers have to realize two things; firstly, innovations, because they are new, are not always better than the existing indigenous practices. These latter practices may be the best given the existing circumstances and these should not be treated with disdain. Secondly, in order to stimulate adoption of certain innovations, a community approach seems necessary. The majority or even the entire community has to get convinced about the benefits of certain activities.

9.2.2 RECOMMENDATIONS RELATED TO THE ROLE OF THE GOVERNMENT

The government should stimulate agricultural researchers to record indigenous practices. In addition, scientists should conduct on-station as well as on-farm research. Agricultural extension programmes should concentrate more on farmers' experiments with respect to soil, water and crop management practices. Special interest has to be taken in the soil fertility status of plots and the influence of FYM and other indigenous manuring practices. Experiments should be conducted to improve the technical efficiency during collecting, storing and application of FYM.

The T&V extension system, even in a project like the Maheswaram project, does not function properly. Many extension agents do not visit the villages, and if they pay a visit they prefer to talk to larger farmers only. This has everything to do with lack of motivation, lack of transport facilities like bicycles and mopeds and too much administrative work. It is very necessary to select only very dedicated extension staff.

Our study tells us that informing and convincing farmers is not sufficient. A farmer does not want to be told, he wants to be shown, wants to see the innovations in the field. He is also convinced that the best place for such a demonstration at field level is in his own surroundings, in the environment he knows best, preferably the village. Most of our respondents distrust the agricultural research stations. They are discouraged by the particular agro-ecological conditions, the large amounts of money used, the high-tech paraphernalia and the white-collar scientists at work. As a result, they think they cannot match these stations and think that the innovations are not meant for them. It means that farmers' visits to these stations are more or less counterproductive. Ideally, every village should have its own demonstration plot where innovations are introduced.

The fact that most respondents are not eager to construct contour bunds and *khus* by themselves and wait for government or NGO intervention, either in the form of a loan, preferred particularly by the larger and richer farmers, or in the form of wage employment, preferred in particular by the smaller and poorer farmers, means that it is, and most probably will always be, a task for the government or an NGO. A side-effect of this involvement is that most probably, once constructed, farmers will regard these structures to be government or NGO structures, not theirs. Hence, government and NGOs have to bear in mind that it is quite possible that beneficiaries will not maintain these structures, meaning that outside agencies have to pay for maintenance costs as well. However, if the government or the NGO is convinced that soil and water conservation is one of their tasks, that it is part and parcel of a sustainable agricultural policy, they have to bear these costs to the benefit of the farmers and the area at large.

The role of the rations shops should be reconsidered. Selling subsidized rice has changed the food preferences of the consumers, also of farmers. It has led to an increased demand for irrigated paddy to the detriment of dryland food crops like *jowar*, *bajra*, *ragi* and smaller millets, which, although considered to be of inferior status, are more expensive than subsidized rice. The lower demand for these latter crops, both at farmers' level and in the market, does not stimulate farmers to cultivate dryland food crops.

9.2.3 RECOMMENDATIONS RELATED TO THE ROLE OF FUNDING AGENCIES

There exists a rather big gap between what was stated in monitoring and evaluation reports written by or on behalf of the Maheswaram project and AWARE and what we think is the reality in the villages. Hence, we suggest that the existing monitoring and evaluation systems be improved. Ideally, monitoring and evaluation should not be undertaken by project staff (self-evaluation) or by institutes which are in one way or another involved with or have close ties with the executing agencies. This remark applies to financial auditing as well.

Monitoring and evaluation reports are not always up to the standard: terms of reference regarding these reports are not always adhered to; the content of the reports is not always very precise and the concepts used or the way they are defined are sometimes misleading. This suggests that clear arrangements have to be made in which is stipulated how reporting has to be done. If such an agreement is not kept, if (financial and progress) reports are not up to standard, (financial) sanctions have to be imposed.

Projects having a target group approach and which stimulate the participation of the local population (or better: which have participation as one of the major goals) are necessary. In addition, donors should concentrate on institution building and training of NGO staff at all levels. Funding agencies have to adopt a favourable policy towards rural development in the semi-arid tropics of India, in particular to dryland agriculture.

9.2.4 RECOMMENDATIONS RELATED TO THE ROLE OF INDIAN NGOs

In our opinion, the internal democracy of a substantial number of NGOs needs to be improved. Our study suggests that in a number of cases the leadership of NGOs regard the organization as 'theirs', and any criticism is regarded as interference in internal affairs. The results of our study suggests, that checks on the management, on the way it functions, on the way it decides about the allocation of funds, is necessary.

To get a good rapport with the beneficiaries, in order not to lose its credibility, NGOs should not only stick to the law and the various rules and regulations imposed by the Indian government, but they should stick to their own principles as well. Ideally, NGOs should set an example to others. If during an evaluation it is found, that the NGO does not stick to its own rules and regulations, its own principles, that it, for instance, does not pay minimum wages, that it discriminates against women, that it employs children, that it does not keep the accounts in the stipulated way, funding agencies should question the integrity of such NGOs.

NGOs should improve the selection procedures (only very motivated personnel) and the training of its personnel (not only general subjects but also technical training). An element which needs special attention is the attitude of many NGO employees. Too often these employees have a rather negative impression of and a rather negative attitude towards their beneficiaries, which puts people off. This applies especially in the Indian context, where most staff are of a higher social status than their clients which leads to a patronizing attitude.

Bibliography

ADATS (Agricultural Development and Training Society) (1989). 'Completion Report of the Dry Land Development Project, March 1987 to December 1989' (typescript). ADATS, Bagepalli.

ADATS (1995). 'Coolie Sangha Building in North Kolar District 1995-98'. Bagepalli, Karnataka, India (application for co-financing, European Union).

Alliband, T. (1983). *Catalysts of Development: Voluntary Agencies in India.* Kumarian Press, West Hartford.

Anubhav (1988). Experiences in Community Health: AWARE. Ford Foundation, New Delhi.

APAU (nd). *Pilot Project for Watershed Development in Rainfed Areas: Socio-Economic Survey of Sub-watershed No. 1, K-10/III-I-A, Maheswaram Watershed, Ranga Reddy District.* Andhra Pradesh Agricultural University, Hyderabad.

APAU (1991). *Package of Practices.* Andhra Pradesh Agricultural University, Hyderabad.

Arens, J. (1979). 'Preliminary Report of an Evaluation Study of AWARE in Andhra Pradesh in India'. Unpublished report, Novib, The Hague.

A Social and Economic Atlas of India (1987).

AWARE (1988?). *The Perspective Plan of AWARE 1988-1990.* AWARE, Hyderabad.

AWARE (1990a). *Progress Report of the Year 1989/90.* AWARE, Hyderabad.

AWARE (1990b). *Vision for the Nineties.* AWARE, Hyderabad.

AWARE (1991). *For the People: Human Resources Development Centre.* AWARE, Hyderabad.

AWARE (1992). *Sustainable Agriculture: A Case Study of Excess Use of Chemicals.* AWARE, Hyderabad.

AWARE (1994a). *Progress Report of the Implementation of the Perspective Plan of AWARE for the Period 1988-1993.* AWARE, Hyderabad.

AWARE (1994b). *A Report on Ekvaipalli Village (on the Socio-economic Development of the Community).* AWARE, Hyderabad.

Bailey, H.P. (1979). 'Semi-Arid Climates: Their Definition and Distribution'. In: Hall, A.E., G.H. Cannell and H.W. Lawton, *Agriculture in Semi-Arid Environments*, pp. 73-97. Springer, Ecological Studies 34, Berlin.

Bakker, E.J. (1992). 'Rainfall and Risk in India's Agriculture: An Ex-ante Evaluation of Rainfall Insurance'. Ph.D. Thesis, Wolters-Noordhoff, Groningen.

Bandhyopadhyay, J. and V. Shiva (1989). *Political Economy of Ecology Movements*. IFDA Dossier, 71, May-June 1989: 37-59.

Beets, N., J. Neggers and F. Wils (1988). *Big and Still Beautiful. Enquiry in the Efficiency and Effectiveness of Three Big NGOs (BINGOs) in South Asia.* Programme Evaluation No. 32, DGIS/Novib, The Hague.

Bessy Natuurkommando: De Loonslaven (Bessy Environmental Command: Bonded Labour) (1989). Standaard Uitgeverij, Antwerpen.

Bijoy, C.R., K. Arun and Raju (1991). 'Some Thoughts on "Remote Money-lenders" (Funding Agencies), "NGO Lords, Ladies and Middlemen" (Directors and Consultants), "Joint Families" (NGOs) and "Voluntarism and Bondage" (NGO Workers) (typescript).

Binswanger, H.P., D. Jha, T. Balaramaiah and D. Sillers (1982). *The Impact of Attitudes towards Risk on Agricultural Decisions in Rural India.* ICRISAT, Economics Programme Progress Report, No. 42, Patancheru.

CAPART (1989). *Directory of Voluntary Agencies.* Council for Advancement of People's Action and Rural Technology, New Delhi.

Centre for Monitoring Indian Economy (1993). *Profiles of Districts*. CMIE, Bombay.

Chambers, R., A. Pacey and L. Thrupp (eds) (1989). *Farmer First: Farmer Innovation and Agricultural Research.* Intermediate Technology Publications, London.

Chambers, R., N.C. Saxena and T. Shah (1989). *To the Hands of the Poor. Water and Trees.* Oxford and IBH Publ. Co., New Delhi.

Charyulu, U.V.N. (1992). *Directory of Voluntary Organizations Promoting Rural Development: Volume I (Southern States of India).* National Institute of Rural Development, Hyderabad.

Committee for the World Atlas of Agriculture (eds) (1973). *World Atlas of Agriculture, Vol. 2: Asia and Oceania.* Instituto Geografico de Agostini, Novarra.

Dandekar, V.M. (1986). 'Agriculture, Employment and Poverty'. In: *Economic and Political Weekly*, Vol. XXI, Nos. 38-9, pp. A90-A100.

De Bruijne, G.A. and H. Schenk (1994). 'Development NGOs In India: Between Liberal Policies and Communalist Movements'. In: Harts-Broekhuizen, A. and O. Verkoren (eds). *No Easy Way Out: Essays on Third World Development in Honour of Jan Hinderink.* Netherlands Geographical Studies, Utrecht.

Deccan Development Society (1991 and 1992). *Report 1990-91 and Report 1991-92*. NP: DDS, Hyderabad.

Dietz, A.J. (1987). *Pastoralists in Dire Straits. Survival Strategies and External Interventions in a Semi-Arid Region at the Kenya/Uganda Border: Western Pokot, 1900-1986.* Netherlands Geographical Studies, Utrecht/Amsterdam.

Dietz, T. (1996). 'Entitlements to Natural Resources: Contours of Political Environmental Geography'. Inaugural Lecture, Department of Human Geography, University of Amsterdam. International Books, Utrecht.

Dietz, Ton J., Marcel Put and S. Subbiah (1998). 'Drought Risk Assessment for Dryland Agriculture in Semi-arid Telangana, Andhra Pradesh, India'. In: Hendrik J. Bruins and Harvey Lithwick, *The Arid Frontier. Interactive Management of Environment and Development*. Dordrecht/Boston/London: Kluwer Academic Publishers, pp. 143-61.

DRDA Anantapur (1992). *A Report on the Implementation of Community-Based Convergent Services in Anantapur District*. DRDA's Office, Anantapur.

Drèze, J. (1997). 'Palanpur, 1957-93: Occupational Change, Landownership and Social Inequality'. In: J. Breman, P. Kloos and A. Saith (eds). *The Village in Asia Revisited.* Oxford University Press, New Delhi.

Drèze, J. and A. Sen (1995). *India: Economic Development and Social Opportunity.* Oxford University Press, New Delhi.

Dvorak, K.A. (1988). *Indigenous Soil Classification in Semi-arid Tropical India.* Resource Management Program, Economic Group, Progress Report No. 84, ICRISAT, Patancheru.

ESCAP (1988). *Economics of Dryland and Rainfed Farming: Report of an Expert Consultation and Country Studies from Bangladesh, India, Nepal, Phillipines, Sri Lanka and Thailand.* United Nations Economic and Social Commission for Asia and the Pacific, Bangkok.

Eswaran, V.B., C. Udaya Shankar, J. Mansingh and B. Liongs (1994). *Report of the External Evaluation (1994) of the Eco-Employment Programme of Accion Fraterna, Part I and II.* NP.

FAO (1980). *Report on the Second FAO/UNFPA Expert Consultation on Land Resources for Populations of the Future*. FAO, Rome.

Farrington, J. and A.J. Bebbington (1994). 'From Research to Innovation: Getting the Most from Interaction with NGOs'. In: I. Scoones and J. Thompson, *Beyond Farmer First: Rural People's Knowledge, Agricultural Research and Extension Practice.* Intermediate Technology Publications, London.

Farrington, J. and A. Bebbington with K. Wellard and D.J. Lewis (eds) (1993a). *Reluctant Partners? Non-Governmental Organizations, the State and Sustainable Agricultural Development.* Routledge, London.

Farrington, J. and D.J. Lewis with S. Satish and A.M. Miclat-Teves (eds) (1993b). *Non-Governmental Organizations and the State in Asia: Rethinking Roles in Sustainable Agricultural Development.* Routledge, London.

Feder, G. and G.T. O'Mara (1981). 'Farm Size and the Adoption of Green Revolution Technology'. In: *Economic Development and Cultural Change*, Vol. 30, No. 1, pp. 59-76.

Fernandez, A.P. (1987). 'NGOs in South Asia: People's Participation and Partnership'. In: *World Development*, Vol. 15, Supplement, pp. 39-49.

——— (1993). 'The Interventions of a Voluntary Agency in the Process and Growth of People's Institutions for Sustained and Equitable Management of Micro-watersheds' (typescript). Myrada, Bangalore.

Fernandez, A.P. and J. Mascarenhas (1993). 'Mysore Relief and Development Agency (MYRADA): Participatory Rural Appraisal and Participatory Learning Methods'. In: J. Farrington, D.J. Lewis, S. Satish and A.M. Miclat-Teves (eds) (1993b). *Non-Governmental Organizations and the State in Asia: Rethinking Roles in Sustainable Agricultural Development.* Routledge, London.

Firdausi, A.A., M.V.K. Sivamohan, T.V. Suryanarayana Rao and R. Sudhakar Rao (1991). *Pilot Project for Watershed Development in Rainfed Area, Maheswaram: Socio-economic Survey in Extended Area.* Administrative Staff College of India, Hyderabad.

Foeken, D. (1994). 'Climatic Seasonality in Kenya with Special Reference to Coast Province'. *Eastern & Southern Africa Geographical Journal*, Vol. 5, No. 1: 28-43.

GoAP (nd). *Pilot Project for Watershed Development in Rainfed Areas, Maheswaram Project: Soil and Landuse Survey Report.* Department of Agriculture, Hyderabad.

——— (various years). *Reports for the World Bank Supervision Mission: Pilot Project forWatershed Development in Rainfed Areas, Andhra Pradesh Component.* Department of Agriculture, Hyderabad.

——— (various years). *Action Plans 1986/87-1993/94: Pilot Project for Watershed Development in Rainfed Areas, Maheswaram Project.* Department of Agriculture, Hyderabad.

——— (1981). *Project for Development of Rainfed Areas in Andhra Pradesh.* Department of Agriculture, Hyderabad.

——— (1985c). *Master Plan: Pilot Project for Watershed Development in Rainfed Areas, Maheswaram.* Department of Agriculture, Hyderabad.

——— (199?). *An Overview of Maheswaram Watershed Project: A Pilot Project for Management of Rainfed Areas.* Department of Agriculture, Hyderabad.

——— (1992b). 'Statement Showing the Land Use Particulars of Villages under Watershed 4K-10/III-1'. Handwritten document, not published. Department of Agriculture, Hyderabad.

——— (1993). *Statistical Abstract 1990 and 1991.* Directorate of Economics and Statistics, Hyderabad.

——— (1995). 'Brief Note on Important Items' (typescript). Department of Agriculture, Hyderabad.

GoI (various years). *Five Year Plans.* Planning Commission, New Delhi.

——— (1988). *Mid Term Review of World Bank Aided Pilot Project for Watershed Development in Rainfed Areas.* Ministry of Agriculture, New Delhi.

——— (1992). *Statistical Outline of India 1992/93.* New Delhi.

——— (1992a). *Eight Five Year Plan, 1992-1997, Vol. I.* Planning Commission, New Delhi.

——— (1992b). *Eight Five Year Plan, 1992-1997, Vol. II.* Planning Commission, New Delhi.

——— (1993). *Economic Reforms : Two Years After and the Task Ahead.* Ministry of Finance, Department of Economic Affairs, New Delhi.

——— (1995). *Economic Survey 1994-5.* Ministry of Finance, New Delhi.

GRAM (1995). *Strengthening Alliances of People's Organizations for Empowerment.* Novib project proposal IDA-011-95-001, The Hague.

Groteman, S., T. Haan, M. van der Weide and H. Wiersma (1993). *Beeld'ver'vorming, een Kwestie van Leven of Dood?* (Image 'Trans' Formation, a Matter of Life and Death?). Department of Communication Studies, Free University, Amsterdam.

Hanumantha Rao, C.H. and Rohini Nayyar (1994). 'New Initiatives for Rural Poverty Alleviation'. Paper Seminar Structural Adjustment and Poverty Alleviation in India, The Hague 1994.

Hema Malini, B. (1979). 'Studies in Eco-climatology of Andhra Pradesh'. Unpublished Ph.D. Thesis, Andhra University, Waltair.

——— (1981). 'Patterns of Aridity, Spread and Severity of Drought in Andhra Pradesh'. *Indian Geographical Journal*, Vol. 56, 1, June: 30-6.

Huibers, F.P. (1985). 'Rainfed Agriculture in a Semi-arid Climate: Aspects of Land- and Watermanagement for Red Soils in India'. Ph.D. Thesis, Wageningen Agricultural University, Wageningen.

ICCO/DGIS (1991). *Programme Evaluation ICCO/DGIS. India. Matching Poverty Alleviation with Sustainable Land Use.* The Hague.

India Today (15 Sept. 1992). 'P.K.S. Madhavan: The Missionary Zeal of an Autocratic Reformer'. *India Today Journal*, New Delhi.

IOV (1994). *India: Evaluation of the Netherlands Development Programme with India, 1980-1992.* DGIS/IOV, The Hague.

ISEC (1991). 'PIDOW Self Evaluation' (typescript). Proceedings of Concluding Workshop. ISEC, Bangalore.

Jha, D. and R. Sarin (1984). *Fertilizer Use in Semi-Arid Tropical India.* ICRISAT, Research Bulletin No. 9, Patancheru.

Jodha, N.S. (1979). *Some Dimensions of Traditional Farming in Semi-arid Tropical India.* ICRISAT, Patancheru.

——— (1985). 'Dry Farming Research: Issues and Approaches'. In: ISAE, *Technology Options and Economic Policy for Dryland Agriculture: Potential and Challenge.* Selected Papers form the Proceedings of a Workshop held on 22-4 August 1983 at ICRISAT Centre, Patancheru, Andhra Pradesh. Concept Publishing Company, New Delhi.

Kannabiram, V., S.T. Somasekhara Reddy and H. Schenk (1990). 'Gram Abyudaya Mandal, A Voluntary Organization in Andhra Pradesh' (typescript). Report of the evaluation of Gram's development programme 1986-9. Hyderabad/ Heiloo.

Krishna Reddy, P.R. (1990). *Some Aspects of Agroclimatology, Landuse and Cropping Pattern of the Cuddapah District, Andhra Pradesh, Using Remote Sensing Data.* Sri Krishnadevaraya University, Anantapur.

Krishna Rao, G.V. (1991). 'Impact of Dryland Farming Technology on Production, Income and Employment in Ranga Reddy District of Andhra Pradesh'. Unpublished Ph.D. Thesis, APAU, Hyderabad.

Kumar, Y.V. (1991). 'Agricultural Potential and Development in the Drought Prone Area of Anantapur District, Andhra Pradesh'. Unpublished Ph.D. Thesis, Sri Krishnadevaraya University, Anantapur.

Kumar, V., J. Neggers and F. Wils (1986). *AWARE, Report of a Joint Novib-AWARE Evaluation Study.* AWARE, Hyderabad.

Kushwant Singh (1992). Two articles on AWARE in the *Hindustan Times* of 24 and 31 August 1992.

Laban, P. (1994). *Final Report of a Field Visit to the Dry Land Development Programme of ADATS/DDS in Kolar District, Karnataka, India.* IAC, Wageningen.

Long, N. and M. Villareal (1994). 'The Interweaving of Knowledge and Power in Development Interfaces'. In: I. Scoones and J. Thompson, *Beyond Farmer First: Rural People's Knowledge, Agricultural Research and Extension Practice.* Intermediate Technology Publications, London.

Mahabubnagar District (1985 etc.). *Handbook of Statistics 1983-84, 1985-86, 1986-87, 1987-88, 1988-89, 1989-90, 1990-91, 1991-92.* Office of the Chief Planning Officer, Mahabubnagar.

Malgavkar, P.D. (1996). 'Inter-state Human Development: Supporting Areas'. Paper Prepared for the Seminar on 'Liberalization and the Poor: Impact, Opportunities and Barriers', organized by ICSSR and CESS, 7-9 February 1996, Hyderabad. CESS, Hyderabad.

Mathur, J.S., U.V.N. Charyulu and V. Krishnamurthy Reddy (1984). *Participation of Non-Governmental Organisations (NGOs) in Integrated Rural Development: Andhra Pradesh, India.* Centre on Integrated Development for Asia and the Pacific (CIRDAP), Study Series No. 64, Comilla.

National Commission on Agriculture (1976). *Rainfall and Cropping Patterns: Vol. I, Andhra Pradesh.* GoI, New Delhi.

Nothdruft, F. and V.S.B. Reddy (1991). *Evaluation Report Rural Development Project MOTHE.* AWARE, Hyderabad.

Novib (1987). *AWARE 1987 Interim Programme.* Novib, The Hague.

Panikkar, K.N. (ed.) (1991). *Communalism in India: History, Politics and Culture.* Manohar, New Delhi.

Parthasarathy G. and C. Durgaparameswari (1996). 'Rural Poverty and Economic Reforms; Opportunities and Constraints (A Case Study on Andhra Pradesh)'. Paper prepared for the seminar on 'Liberalization and the Poor: Impact, Opportunities and Barriers', organized by ICSSR and CESS, Hyderabad, 7-9 February 1996.

Penchaliah, T. (1989). 'An Analysis of Rainfall, Droughts and Water Resources of the Cuddapah District'. Unpublished M.Phil Thesis, Sri Krishnadevaraya University, Anantapur.

Prasad, S.D.J.M. (1992). 'Role of Voluntary Organizations in Rural Development'. Unpublished M.Sc. Thesis, University of Hyderabad, Department of Economics, Hyderabad.

Premkumar, P.D. and B. Humbert-Droz (1994). *Farmers are Engineers. Indigenous Soil and Water Conservation Practices in a Participatory Watershed Development Programme.* MYRADA/Swiss Development Corporation, Kamalapur/Bangalore.

Put, M. and M.P. van Dijk (1989). *Semi-arid Indian Agriculture. Land, Water and Crop Management in Southern India.* Bulletin 316, Royal Tropical Institute, Amsterdam.

Put, M. (1998). *Innocent Farmers? A Comparative Evaluation Into a Government and an NGO Project Located in Semi-arid Andhra Pradesh (India), Meant to Induce Farmers to Adopt Innovations for Dryland Agriculture.* Thesis Publishers, Amsterdam.

Radhakrishna, R. and S. Reddy (1990). *Incidence of Poverty in Andhra Pradesh.* CESS, Hyderabad.

Ramakrishna Reddy, V. (1987). *Economic History of Hyderabad State (Warangal Suba: 1911-1950).* Gian Publishing House, New Delhi.

Ranga Reddy District (1985 etc.). *Handbook of Statistics 1982-83 & 1983-84, 1985-86, 1986-87 & 1987-88, 1988-89, 1989-90, 1990-91 & 1991-92.* Office of the Chief Planning Officer, Ranga Reddy District.

Rangaswamy, P. (1984). *Dry Farming Technology in India: A Study of its Profitability in Selected Areas.* Agricole Publishing Academy, New Delhi.

Rao, G.R.S and P.K.S. Madhavan (1988). 'Voluntary Action in Andhra Pradesh: An Appraisal' (typescript). AWARE, Hyderabad.

Rao, V.M. (1991). 'Growth in the Context of Underdevelopment: Case of Dry-land Agriculture'. In: *Economic and Political Weekly*, Vol. XXVI, No. 13, pp. A2-A14.

Reitsma, H., T. Dietz and L. De Haan (eds) (1992). *Coping with Semi-aridity: How the Rural Poor Survive in Dry-season Environments.* Netherlands Geographical Studies, No. 146, Royal Dutch Geographical Society/Department of Human Geography, Faculty of Environmental Sciences, University of Amsterdam, Amsterdam.

Robinson, M.A. (1991). *Evaluating the Impact of NGOs in Rural Poverty Alleviation: India Country Study.* Working paper 49, Overseas Development Institute, London.

Rogers, E.M. (1995). *Diffusion of Innovations.* Fourth Revised Edition. Free Press, New York.

Rupa Kumar (1980). 'Agroclimatic Conditions in Andhra Pradesh and the Effects of Weather Anomalies in Major Crops'. Unpublished Ph.D. Thesis, Andhra University, Vishakhapatnam.

Rural Development Trust (1992). 'Experiences in Health and Community Development'. In: *Anubhav.* Voluntary Health Association of India, New Delhi.

——— (1993a). 'Training on Soil Conservation and Rain Water Harvesting Techniques, 19-24 April 1993' (typescript). RDT, Anantapur.

——— (1993b). 'Functioning of Watershed/ Women/ Labour Groups: A Case Study of Ayyagarlapalle' (typescript). RDT, Anantapur.

Sandee, H. (1995). *Innovation Adoption in Rural Industry: Technological Change in Roof Tile Clusters in Central Java, Indonesia.* Ph.D. Thesis, Free University, Amsterdam.

SEARCH (1990). *Drought: Roles and Perspectives for NGOs, A Workshop Report.* SEARCH Publications, Bangalore.

SEARCH (1993). *Training on Sheep and Goat Rearing* (typescript). SEARCH, Bangalore.

Shah, G. (1991). 'Non-Government Organizations in India' (typescript). Centre for Social Studies, Surat.

Shah, G. and G. Menon (1991). *India, Association of Rural Education and Development Service.* Veldonderzoek Impactstudie Medefinancieringsprogramma, Surat/Amsterdam.

Shah, P. (1994). 'Local Institutions and Para-professionals in Watershed Management'. In: *ILEIA Newsletter*, Vol. 10, No. 2: 17-19.

Shanti, K. (1994). *Empowering Women to Attain Self Sufficiency. The Role of NGOs.* IDPAD Occasional Papers and Reprints, 1994-5, New Delhi/The Hague.

Singh, N.K. (1979). 'Annual Potential Water Deficit in Andhra Pradesh'. In: N.B.K. Reddy (ed.), *Proceedings of All India Symposium on Drought Prone Areas in India. Rayalaseema Geographical Society.* S.V. University, Tirupati.

Sitapathi Rao, C., M.V.K. Sivamohan, S. Chandy and R. Sudhakar Rao (1990). *Pilot Project for Watershed Development in Rainfed Areas, Maheswaram: Special Studies on Soil and Moisture Conservation Activities, Crop Production Programme and Pasture and Fodder Development Programme.* Administrative Staff College of India. Hyderabad.

Sivamohan, M.V.K., M. Rajagopalan and R. Sudhakar Rao (1994). *Pilot Project for Watershed Development in Rainfed Area, Maheswaram: Impact Evaluation Study.* Administrative Staff College of India, Hyderabad.

Sivamohan, M.V.K., C.A. Scott and M.F. Walter (1993). 'Vertiver Grass for Soil and Water Conservation: Prospects and Problems'. In: D. Pimentel (ed.), *World Soil Erosion and Conservation.* Cambridge University Press, Cambridge.

Society for Participatory Research in Asia (1991). *Voluntary Development Organizations in India. A Study of History, Roles and Future Challenges.* SPRA, New Delhi.

Sohoni, N. (1994). 'Action for Welfare and Awakening in Rural Environment'. In: S. Pachauri (ed.), *Reaching India's Poor: Non-governmental Approaches to Community Health.*

Sohoni, N. and M. Singh (1990). *People in Action.* Vikas Publishing House, New Delhi.

Stuurgroep Impactstudie Medefinanciering (1991). *Betekenis van het Medefinancierings-programma: Een Verkenning* (The Importance of the Co-Financing Programme: An Exploration). Gemeenschappelijk Overleg Medefinanciering, Oestgeest.

Subrahmanyam, V.P. and C.V. Sastry (1969). 'A Study of Aridity and Droughts in Vizag'. *Annals of Arid Zone*, Vol. 8, No. 1: 18-20.

Subramanaian, M.J. and V. Vidyanath (1984). 'Moisture Adequacy and Cropping Pattern in Andhra Pradesh, India'. *Indian Geographical Journal*, Vol. 59, No. 2, December.

Sundaram, I.S. (1986). *Voluntary Agencies and Rural Development.* B.R. Publishing Corporation, New Delhi.

Sunitha Raju (1990). 'Fertilizer Use and Development of Institutional Infrastructure: A Preliminary Analysis'. In: *Economic and Political Weekly*, Vol. XXV, 30 June, A79-A85.

The Week (31 Dec. 1989). 'P.K.S. Madhavan: Man of the Year'. The Week Journal, New Delhi.

Tiffin, M., M. Mortimore and F. Cichuki (1994). *More People, Less Erosion: Environmental Recovery in Kenya.* John Wiley and Sons, Chichester.

Trouw (21 Oct. 1995). 'Een Vriend Krijgt een Mes in de Rug (A Friend is Stabbed in the Back)'. Trouw Newspaper, Amsterdam.

——— (11 Oct. 1997). 'ICCO Geeft haar Besmette Indiase Partner weer een Kans (ICCO Gives its Tainted Indian Partner a New Chance)'. Trouw Newspaper, Amsterdam.

UNESCO (1979). *Map of the World Distribution of Arid Regions: Explanatory Note and Map.* UNESCO/MAB, Technical Note, No. 7, Paris.

Usha Rao, N.J., C. Chakrapani and S. Vijaya Kumar (1992). *Rural Development Schemes and their Delivery System: A Case Study of two Mandals in Mahabubnagar District.* C.D. Deshmukh Impart Centre, Council for Social Development, Hyderabad.

Van Bouwel (1993). *Ik Noemde Mijn Man 'God': Emancipatie op het Indiase Platteland* (I Called My Husband 'God': Emancipation in India's Countryside). NCOS, Brussels, and Novib, The Hague.

Venkataraman, I. (1981). 'Agrometeorological Studies for the Optional and Planned Use of Surface Irrigation and Rainfall Resources'. Unpublished Ph.D. Thesis, Andhra University, Waltair.

Venkateswara Rao, M. (1984). 'Some Studies on Climate and Crops in Rayalaseema, Andhra Pradesh, India'. Unpublished Ph.D. Thesis, Andhra University, Waltair.

Von Oppen and Rao (1982). 'Sorghum Marketing in India'. In: ICRISAT, *Proceedings of the International Symposium on Sorghum.* ICRISAT, Patancheru.

Vrienden van Young India Project (1994). *Rondzendbrief.* Landelijke India Werkgroep, Utrecht.

Walker, T.S. and J.G. Ryan (1990). *Village and Household Economies in India's Semi-Arid Tropics.* Johns Hopkins University Press, Baltimore.

Walsh, R.P.D. (1981). 'The Nature of Climatic Seasonality'. In: R. Chambers et al., *Seasonal Dimensions to Rural Poverty.* Francis Pinter, London.

WCADL (Weaker Communities Action for Development and Liberation) (1992). *A Voluntary Experience in Rural Transformation, 1976-1992*. WCADL, Hyderabad.

Wils, F., V.B.N.S. Madduri, K.C. Alexander and N. Sohoni (1994). 'AWARE and its Work with Tribals and Harijans in Andhra Pradesh: An Impact Study of AWARE's Programme among Harijan, Girijan and Mixed Communities, Now Independent from AWARE' (typescript). Institute of Social Studies, The Hague.

World Bank (1983). *India: A Pilot Project for Watershed Development in Rainfed Areas.* Staff Appraisal Report, World Bank, Washington.

——— (1988). *India: Review of Rainfed Agriculture and Watershed Development: Volume I, Main Text.* Staff Appraisal Report, World Bank, Washington.

——— (1994). *The World Bank Atlas 1995*. World Bank, Washington.

WWFN (1992). *Environmental NGOs in India: A Directory 1992.* World Wide Fund for Nature-India, New Delhi.

Youth for Action (1992). 'Papers presented at a workshop on "Management of the Natural Resources by People", 19-20 January 1992 at Hyderabad'. YfA, Hyderabad.

——— (1993). *Annual Report 1992-1993*. YfA, Hyderabad.

——— (1993). *Alternative Strategies for Development*. YfA, Hyderabad.

Contributors

Marcel Put is a geographer, and has been working as a planner in Bhutan. In the 1990s while at the Faculty of Environmental Sciences, University of Amsterdam, he specialized in the fields of sustainable agricultural and rural development, soil and water conservation and crop management in drylands.

Meine Pieter van Dijk is a professor of urban management in emerging countries at the Economic Faculty of the Erasmus University in Rotterdom (EUR), and also as a senior economist at the Institute of Housing and Urban Development Studies in Rotterdam (IHS) the Netherlands. He has published several books and articles. He has co-edited a book on clusters and networks (London: F. Cass, 1997).

D. Adi Lakshmi is a sociologist, working with AWARE, an NGO in Hyderabad.

G.A. de Bruijne is professor in geography of developing countries at the University of Amsterdam. From 1991 to 1998 he has been the Dutch co-chairman of IDPAD.

A.J. Dietz is professor in human, geography with special emphasis on environmental and global issues. He is director of AGIDS, the Amsterdam Research Institute on Global Issues and Development Studies.

S. Subbiah is professor at the Department of Geography, University of Madras, Chennai. His studies include agriculture in India, land use dynamics in rural and urban fringes and GIS. He is also the Editor of the *Indian Geographical Journal.*

Index

Action for Community Service Society 67
ADATS 72
'adjusted contour' 379
Administrative Staff College 63
AIRDS 112
Anantapur District Farmers Forum 65
Andhra Mahila Sabha 70
Andhra Pradesh (Telangana Area) Public Societies Registration Act of 1350 Fasli 65
Andhra Pradesh Agricultural University (APAU) 140, 143, 148, 250, 251, 334
ARADS 112
area approach 42
Area Development Service Society (ADSS) 167, 169
Arens, J. 175
Arthik Samata Mandal 66
Aware (Action for Welfare and Awakening in Rural Environment) 45-50 *passim*, 63, 66-71 *passim*, 106, 112, 113, 125, 126, 127, 192, 193, 216, 229, 231, 237, 238, 241, 254-62, 272-89 *passim*, 293, 294, 295, 296, 303, 304, 305, 318, 320, 321, 324, 325, 334, 335, 338, 347, 350, 351, 358, 359, 361, 362, 380, 381, 383, 385, 389, 390, 391, 392, 393, 395, 398; activities and approach of 157-65; history of 150-4; ideology and objectives of 154-7; internal organization of 165-71; and its means 171-4; and its planning and decision-making 174-5

Bailey, H.P. 88, 90, 91
bajra (pearl millet) 53, 56, 77, 103, 105, 106, 109, 141, 214, 217, 219, 236, 238, 269, 299, 310, 313, 318, 364, 394, 398
bengal gram 107
BGMS 112
Bijoy, C.R. 170
blackgram 53, 107
'bottom-up' approach 42, 45, 381
Brot für die Welt 114

CAPART (Council for the Advancement of People's Action and Rural Technology) 69, 113
castor 53, 56, 77, 107, 109, 137, 141, 213, 215, 216, 217, 219, 220, 232, 233, 234, 236, 238, 242, 243, 246, 247, 248, 269, 270, 271, 293, 299, 300, 301, 310, 311, 312, 314, 317, 320, 321, 323, 326, 327, 330-40 *passim*, 345, 346, 347, 355, 357, 363, 389, 393, 394, 395
Central Planning Commission 30
Centre for Environmental Concerns 67
Centre for Monitoring the Indian Economy 53
Charyulu 67, 69
Chenchus 188, 190
chillies 107, 263
Chittoor Rural Development Society 66
Christian Concern Mission 112
CIDA 172
Cluster Development Service Society (CDSS) 168, 169
'communalism' 39
Community Aid Abroad (CAAO) 150, 171
Community Development Programme 32

community-based organizations (CBOs) 40
Comprehensive Rural Operations Service Society (CROSS) 66, 70
Consumer Education Centre 67
contour cultivation 136, 137, 235, 288-94
contour/graded bunds 60, 62, 74, 115, 129, 135, 145, 162, 231, 232, 272-82, 296, 386, 390, 391, 393, 398; *see also* soil and water conservation structures
cotton 53, 56, 107, 395
CRIDA 117, 134, 135, 140, 141, 237, 250, 335, 338
crop management practices, and the choice of suitable seed varieties 237-9; and organic fertilizers 241-2; and pest and disease control 246-8; and plant population 239-40, 320-3; and tillage implements 240; and the time of sowing 242-5
crop rotation 235-7, 246
CRYD 112
Cuddapah District Rural Welfare Development Society 66
Cuddapah District Yuvajana Harijana Christian Girijana Sarvodaya Sangham 66
cumulative deviation technique 79

DDS 71, 72, 73, 74, 169
dead furrows 136, 162, 234, 235, 294-5, 297, 386; *see also in-situ* soil and moisture conservation technologies
Debbadaguda 47, 49, 75, 79, 130, 141, 175, 176, 177, 179, 182n2, 185, 187, 190, 192, 194, 195, 196, 198, 201, 203, 205, 206, 209-21 *passim*, 229, 243, 244, 247, 248-54 *passim*, 261, 262, 264, 265, 270-81 *passim*, 284, 286, 288, 289, 291-300 *passim*, 306, 308, 313, 314, 316, 320-42 *passim*, 345, 347-60 *passim*, 382, 384, 385, 387
Desert Area Development Project Agencies 36-7
Development Action for Rural Environment 67
District Rural Development Agency (DRDA) 69, 70, 129
District Watershed Development Service Society (DDSS) 167
Drèze, Jean 191, 193
Drought Prone Area Programme (DPAP) 35, 36, 61, 62, 182n4, 378
dryland agriculture 27, 28, 36; and the adoption of new technologies 44; and government activities 58-63; paucity of funds in 378; policies of the Indian government concerning 378

Ekvaipally 47, 49, 75, 79, 99, 106, 171, 174, 175, 176, 177, 182n6, 185, 188, 190, 192, 193, 196, 198, 201-21 *passim*, 229, 231, 236, 237, 243, 244, 247, 251, 254-62, 263, 272-81 *passim*, 284, 286, 288, 289, 291, 292, 294, 296, 297, 305, 306, 308, 313, 314, 316, 320-32 *passim*, 336, 340-60 *passim*, 382, 384, 387, 392, 393, 395
Environmental Society of Tirupati 67
ESCAP 390
European Community 172

FAO 88
Farm Yard Manure (FYM) 136, 138, 229, 233, 241, 242, 266n1, 309, 325-42 *passim*, 363, 365, 389, 394, 397
Farmers Cooperative Society 211
Föeken, D. 83, 84
Ford Foundation 171
Foreign Contribution (Regulation) Act 64
FSCS 69

Gandhi, Indira 67
German Friedrich Ebert Stiftung 117
Ghatpally 47, 49, 75, 79, 176, 177, 178,

179, 185, 190, 191, 192, 194, 203, 205, 206, 209, 210, 213, 217, 218, 221, 244, 247, 251, 263, 272, 273, 274, 275, 280, 284, 286-93 *passim*, 297, 301, 303, 305, 308, 313, 314, 316, 321, 326, 327, 332, 336, 339, 340, 346-60 *passim*, 382, 384, 385, 387
Girijana Prajala Sangham 66
Girijana Seema Welfare Association 66
Golconda Grameen Bank 211
Good Samaritan Evangelical Social Welfare Association 66
GRAM 71, 72, 73
greengram 53, 107
Green Revolution 27, 34, 35, 37, 53, 59, 161
ground nuts 53, 56, 77, 107, 109
group approach 42

Help the Aged 171
High Yielding Varieties Programme (HYVP) 32, 33, 61, 129, 182n4
Hindu fundamentalism 39
horsegram 53, 107, 214, 239
Hüibers, F.P. 390
Hyderabad District Cooperative Central Bank 78
Hyderabad Zilla Mahila Mandalula Samakhya 67

ICCO 151, 171
ICRISAT 117, 135, 140, 188, 300
in-situ moisture conservation practices 36, 37, 129, 133-47 *passim*, 162, 179, 180, 232-5, 245, 249, 264, 282-95, 296, 361, 385, 386; *see also* dead furrows
in-situ soil conservation technologies 37, 129, 133, 134, 135-7, 139, 140, 141, 143, 145, 147, 162, 179, 180, 232-5, 249, 264, 282-95, 296, 361, 385, 386, 390; *see also* dead furrows
incremental approach 43
Indian Council of Agricultural Research 117
Indian National Watershed Development Policy Committee 127
inorganic fertilizers, placement of 335-6; quantity of 337-47; timing of 337-47
Integrated Dryland Agricultural Development Project 35, 61
Integrated Pest and Disease Management 358-61
Integrated Rural Development Programme (IRDP) 35, 39, 62, 67, 164, 172, 200, 211, 257, 378
Integrated State Watershed Development Programme (ISWDP) 62, 63
Intensive Agricultural Area Programme (IAAP) 33
Intensive Agricultural District Programme (IADP) 32, 33
inter-bund treatment 36
intercropping 137, 235-7, 303-9, 362, 363, 366nn2 and 3, 387, 391, 394
interculturing 136, 138, 234, 239, 245-6, 264, 297, 303, 320, 350-8, 362, 363, 385, 386, 394, 395
International Association of Extension for Rural Development 67
Islamic social reform movements 41

Jawahar Rozgar Yojna (JRY) 164
Jodha, N.S. 390, 391
jowar (sorghum) 53, 56, 77, 99, 103, 105, 106, 107, 136, 137, 138, 141, 213-20 *passim*, 232-47 *passim*, 251, 264, 269, 270, 271, 282, 283, 284, 287, 297, 298-305 *passim*, 310-23 *passim*, 326-40 *passim*, 343, 345, 346, 347, 349, 350, 351, 353-65 *passim*, 385, 387, 389, 391, 394, 395, 398
Jowar Research Station, establishment of 59

Kumar, Y.V. 80

Lab to Land Programme 310, 359

Lambadas 187, 190, 219, 248, 252, 254
land improvement approach 42
'liberalization policy' 32, 339, 347, 377
Long, N. 262

Madhavan, P.K.S. 64
Maheswaram project 36, 46, 47, 48, 50, 79, 106, 125, 126, 152, 162, 174-81 *passim*, 188, 206, 216, 229, 231, 234, 235, 237, 238, 240, 241, 242, 244, 248-54, 261, 262, 271-89 *passim*, 293, 294, 295, 296, 303, 304, 305, 310, 312, 317, 320, 321, 324, 325, 334, 335, 336, 338, 342, 347, 350, 351, 358-64 *passim*, 380, 381, 383, 386, 389, 390, 391, 392, 393, 397, 398, activities and approaches of the 132-5; history of the 127-30; ideology and objectives of the 130-2; and the improved management of dryland crops 137-42; and *in-situ* soil and water conservation 135-7; and its internal organization 142-6; means 146-8; planning and decision-making 148-50
maize 53, 56, 141, 219
Malgavkar, P.D. 52
Mansanpally 47, 49, 75, 129, 135, 141, 176, 177, 178, 179, 182n2, 185, 187, 190, 192, 194-206 *passim*, 211-18 *passim*, 221, 229, 243, 244, 247, 248-54 *passim*, 262, 264, 272-80 *passim*, 284-94 *passim*, 305, 306, 314, 316, 320-30 *passim*, 334, 336, 339, 343-60 *passim*, 382, 384, 385, 387, 389
Mass Education and Organization Society 66
Ministry of Environment and Forestry 69
Ministry of Health and Family Welfare 69
Multiple Cropping Programme 33
MYRADA 70, 71, 72, 74

National Commission on Agriculture 75
National Institute of Rural Development 65
National Oil Seed Development Programme 129, 182n4
National Rural Employment Programme (NREP) 35, 62
National Wasteland Development Board 69
National Watershed Development Council (NWDC) 142
National Watershed Development Policy Committee (NWDPC) 142
National Watershed Development Programme (NWDP) 36
National Watershed Development Programme for Rainfed Agriculture (NWDPRA) 62, 182n4, 378
Natural Resources Development Cooperative Society 65
Naxalites 260
'Non-Governmental Development Organizations' (NGDOs) 40
Norad 117
Novib 46, 151, 152, 153, 171, 172

organic fertilizers *see* Farm Yard Manure
OXFAM 117

paddy *see* rice
panchayati raj system 28, 33, 118
Panikkar, K.N. 40
Participatory Resource Appraisal 117
Pinakini Satyagraha Asramamu Trust 68
poverty alleviation programmes (PAPs) 35, 39, 41
Prasad, S.D.J.M. 42, 175, 184n22

ragi (finger millet) 53, 56, 106, 398
Raju, Sunitha 238, 317
Rangaswamy, P. 391
Rao, G.R.S. 64
Rao, Sitapathi 148, 149, 310, 334, 389, 393
Rao, V.M. 391
Rao, Venkateswara 79
RASS 69

Rayalaseema 51, 52, 54, 65, 67, 69, 70, 79, 80
Rayalaseema Seva Samithi 70
RDO 69
Reddies 115, 150, 191, 192, 193, 220, 252, 254, 262, 263, 275, 303, 310, 311, 325, 334, 335, 337, 338, 358, 361, 382, 383
Reddy, Krishna 80
Reddy, Ramakrishna 54
redgram 53, 99, 107, 137, 213, 216, 218, 233, 236, 239, 247, 269, 300, 301, 303, 305, 317, 318, 321, 346, 362, 363, 364, 379, 387, 391, 394
rice 53, 54, 77, 214, 218, 269, 310, 333, 335, 393
Robinson, M.A. 41, 191
Rockfeller Foundation 117
Rogers, E.M. 279, 386
Rohini karthe 243, 245, 348
Rural Development Trust (RDT) 63, 65, 70, 71, 72, 73, 74
Rural Landless Employment Guarantee Programme (RLEGP) 62
Ryan, J.G. 80, 91, 188, 191, 209, 211, 310, 317, 319, 334

Sandee, H. 263
Sanghameshwara Grameena Bank 211
Sastry, C.V. 79
'*seasonality assessment*' 84
'*seasonality index*' 84
seasonality ration 84
Sen, Amartya 193
sesamum 57, 107
Shri Karmika Aivyamata Sangha 66
SIDA 117
Singh, N.K. 79
Singh, Kushwant 153
Sivamohan, M.V.K. 148, 274
Small Farmers Development Agency (SFDA) 35, 61
small millets 53, 56
social 'safety nets' 43
'socialism' 30
Societies Registration Act of 1860 65
Society for Nature Conservation 67
Soil and Water Conservation Programme 61
soil and water conservation structures 133-5; *see also* contour bunds
Sri Satya Sai Central Trust 70
State Watershed Development Cell (SWDC) 142
State Watershed Development Policy Committee (SWDPC) 142
structural reforms 32
Study Action Group for Participative Integrated Development 67
Subrahmanyam, V.P. 79
Subramanian, M.J. 79
sugarcane 56

Telangana 51, 52, 53, 54, 55, 60, 63, 65, 75, 78, 80, 109, 110, 112, 125, 151
Tiffin, M. 263
tillage, improvement in 323-5; off-season 282-8
tobacco 56, 107
top-down approach 46, 148, 149, 380
Transfer-Of-Technology (TOT) approach 139, 163

UNESCO 81
UNICEF 113

VASORD 112
Venkatraman, I. 79
Vidyanath, V. 79
Village Reconstruction Organization 65
Villages in Partnership (ViP) 71, 72, 112, 116-17
Villareal, M. 262
Von Oppen 391

Walker, T.S. 80, 91, 188, 191, 209, 211, 310, 317, 319, 334
Walsh, R.P.D. 83, 84
WAMERT 65
Warangal Consumer Council 67
Wasteland Development programmes 42

Watershed Development Schemes 36; objectives of 37
Watershed Development Team (WDT) 142, 143
Weaker Community's Action for Development and Liberation (WCADL) 66, 70, 71, 72, 73, 112, 113, 114, 115, 116
wheat 53, 215
Wils, F. 175
Women's Association for Development Action 67
World Bank 28, 30, 36, 37, 46, 47, 106, 125, 126, 127, 135, 141, 142, 145, 149, 179, 252, 378, 380, 390
World Missionary Evangelism 70
World Wide Fund for Nature (WWFN) 65, 67

Young India Project (YIP) 66, 70, 72
Youth for Action (YFA) 71, 72, 74, 112, 117-18